Cartography

Cartography
Visualization of Geospatial Data

Fourth Edition

Menno-Jan Kraak and Ferjan Ormeling

CRC Press
Taylor & Francis Group
Boca Raton London New York

CRC Press is an imprint of the
Taylor & Francis Group, an **informa** business

Fourth edition published 2021
by CRC Press
6000 Broken Sound Parkway NW, Suite 300, Boca Raton, FL 33487-2742

and by CRC Press
2 Park Square, Milton Park, Abingdon, Oxon, OX14 4RN

© 2021 Taylor & Francis Group, LLC

Third edition published by CRC Press 2009

CRC Press is an imprint of Taylor & Francis Group, LLC

Library of Congress Cataloging-in-Publication Data
Names: Kraak, M. J., author. | Ormeling, Ferjan, 1942- author.
Title: Cartography : visualization of geospatial data / Menno-Jan Kraak, Ferjan Ormeling.
Description: Fourth edition | Boca Raton ; London : CRC Press, 2020. |
Includes bibliographical references and index. | Summary: "This fourth
edition serves as an excellent introduction to general cartographic
principles and as an examination of the best ways to optimize the
visualization and use of spatio-temporal data. It incorporates all the
changes and new developments in the world of maps such as open street
maps and GPS-based crowdsourcing, the use of new web mapping technology,
and adds new case studies and examples. Printed in full color, this
fully-revised edition provides students with the knowledge and skills
needed to read and understand maps and mapping changes, and offers
professional cartographers an updated reference with the latest
developments in cartography"—Provided by publisher.
Identifiers: LCCN 2020015986 (print) | LCCN 2020015987 (ebook) |
ISBN 9781138613959 (hardback) | ISBN 9780429464195 (ebook)
Subjects: LCSH: Cartography.
Classification: LCC GA105.3 .K73 2020 (print) | LCC GA105.3 (ebook) |
DDC 526—dc23
LC record available at https://lccn.loc.gov/2020015986
LC ebook record available at https://lccn.loc.gov/2020015987

ISBN: 978-1-138-61395-9 (hbk)
ISBN: 978-0-429-46419-5 (ebk)

Typeset in ITC Benguiat
by codeMantra

Contents

Preface ix
Acknowledgements xiii
Authors.. xv

Chapter 1   Geographical Information Science and Maps.. 1

   1.1   The Map as an Interface 1
   1.2   Geospatial Data 4
   1.3   Geographical Information Systems 7
   1.4   Geospatial Analysis Operations 12
   1.5   The Spatial Infrastructure and Maps 18
   1.6   Cartographic Education... 20
   Further Reading... 21

Chapter 2   Data Acquisition 23

   2.1   The Need to Know Acquisition Methods 23
        2.1.1   Terrestrial Surveys 24
        2.1.2   Photogrammetrical Surveys 24
        2.1.3   Lidar or Laser Altimetry 24
        2.1.4   Satellite Data 24
        2.1.5   GPS Data 25
        2.1.6   Digitizing or Scanning Analogue Maps 25
        2.1.7   Using Existing Boundary Files 25
        2.1.8   Socio-economic Statistical Files 25
        2.1.9   (Geo)physical Data Files 25
        2.1.10 Environmental Data Files 25
        2.1.11 Volunteered Geographical Information; Citizen Science 25
   2.2   Vector File Characteristics... 26
   2.3   Raster File Characteristics 27
   2.4   Deriving Data from Existing Maps 29
        2.4.1   Finding the Proper Map: Documentation 29
        2.4.2   Preparation 29
        2.4.3   Digitizing 30
   2.5   Working with Digital Data 32
        2.5.1   Modelling the World 32
        2.5.2   Vector Approach 34
        2.5.3   Raster Approach 34
        2.5.4   Hybrid Use of the Database 37
   2.6   Control and Accuracy... 38
   Further Reading... 42

Chapter 3   Map Characteristics 43

   3.1   Maps Are Unique... 43
   3.2   Definitions of Cartography... 45
   3.3   The Cartographic Communication Process 49
   3.4   Map Functions and Map Types.. 56
   Further Reading... 60

Contents

Chapter 4 GIS Applications: Which Map to Use? 61

4.1 Maps and the Nature of GIS Applications 61
4.2 Cadastre and Utilities: Use of Large-Scale Maps 61
    4.2.1 Cadastral Maps in Use 61
    4.2.2 Utility Maps at Work 62
4.3 Geospatial Analysis in Geography: Use of Small-Scale Maps 65
    4.3.1 Socio-economic Maps 65
    4.3.2 Environmental Maps 66
4.4 Geospatial, Thematic and Temporal Comparisons 67
    4.4.1 Comparing Geospatial Data's Geometry Component 68
    4.4.2 Comparing the Attribute Components of Geospatial Data 70
    4.4.3 Comparing the Temporal Components of Geospatial Data 71
Further Reading 73

Chapter 5 Map Design and Production 75

5.1 Introduction 75
5.2 Symbols to Portray Data Related to Points, Lines, Areas and Volumes 76
5.3 Graphic Variables 77
    5.3.1 Visual Hierarchy 80
    5.3.2 The Use of Colour 81
5.4 Typography: Conceptual and Design Aspects of Text on the Map 84
5.5 Requirements for the Cartographic Component of GIS Packages 86
    5.5.1 Data Manipulation 87
    5.5.2 Output 88
    5.5.3 Graphical User Interface 88
5.6 Map Design and Production 88
5.7 Web Map Design 94
5.8 Web Maps and Multimedia 96
    5.8.1 Sound 98
    5.8.2 Text 98
    5.8.3 Images 98
    5.8.4 Video/Animations 98
Further Reading 99

Chapter 6 Topography 101

6.1 Georeferencing 101
6.2 Map Projections 104
6.3 Geometric Transformations 109
6.4 Generalization 112
    6.4.1 Background and Concepts 112
    6.4.2 Cartographic Generalization 116
    6.4.3 Generalization Processes and Tools 118
6.5 Relief 123
    6.5.1 Introduction 123
    6.5.2 Digital Terrain Models 124
    6.5.3 Terrain Visualization 126
6.6 Topographic Data: Mapping and Charting Organizations 131
    6.6.1 Introduction 131
    6.6.2 OpenStreetMap (OSM) 138
    6.6.3 EuroBoundaryMap 138
    6.6.4 USGS National Nap Concept 138
    6.6.5 Ordnance Survey 138

6.7   Geographical Names 142
Further Reading 145

Chapter 7   Statistical Mapping 147

7.1   Statistical Surveys 147
7.2   Data Analysis 149
        7.2.1   Data Adjustment 155
7.3   Data Classification 155
        7.3.1   Graphic Approach 158
                7.3.1.1   Break Points 158
                7.3.1.2   Frequency Diagram 158
                7.3.1.3   Cumulative Frequency Diagram 158
        7.3.2   Mathematical Approach 159
                7.3.2.1   Equal Steps 159
                7.3.2.2   Quantiles 160
                7.3.2.3   Arithmetic Series 160
                7.3.2.4   Geometric Series 160
                7.3.2.5   Harmonic Series 160
                7.3.2.6   Nested Means 160
7.4   Cartographical Data Analysis 163
7.5   Mapping Methods 166
        7.5.1   Chorochromatic Maps or Mosaic Maps 168
        7.5.2   Choropleth Maps or Choropleths 170
        7.5.3   Isoline Maps 173
        7.5.4   Nominal Point Data 175
        7.5.5   Absolute Proportional Method 176
        7.5.6   Diagram Maps 178
        7.5.7   Dot Maps 179
        7.5.8   Flow Line Maps 181
        7.5.9   Statistical Surfaces 182
        7.5.10   Cartograms 182
        7.5.11   Chorèmes 183
Further Reading 185

Chapter 8   Mapping Time 187

8.1   Introduction 187
8.2   Mapping Change 192
8.3   Animation 193
        8.3.1   Temporal Animations 194
        8.3.2   Non-Temporal Animations 194
8.4   Dynamic Variables 196
Further Reading 197

Chapter 9   Maps at Work: Presenting and Using Geospatial Data in Maps and Atlases 199

9.1   Introduction 199
9.2   Paper Atlases 200
9.3   Electronic Atlases 202
        9.3.1   Electronic Atlas Types 202
        9.3.2   Electronic Atlas Functionality 204
9.4   Map Machines 207
9.5   Story Map 208
9.6   Atlases at Work: Map-Use Functions 208

9.6.1 Explaining Patterns 208
9.6.2 Comparison and Analysis 210
9.6.3 Analysis and Decision-Making 211
9.6.4 Conditions for Proper Use of the Maps 212
9.7 Working with (Web-Based) Electronic Atlases 213
Further Reading 214

Chapter 10 Maps at Work: Analysis and Geovisualization 215

10.1 Geovisual Analytics 220
Further Reading 223

Chapter 11 Cartography at Work: Maps as Decision Tools 225

11.1 Again: Why Maps? 225
11.2 Management and Documentation of Spatial Information 225
11.2.1 Retrieving Geodata 226
11.3 Outdated Data: At Work with the Digital Chart of the World 227
11.3.1 Case 1: The Netherlands' Railroads 227
11.3.2 Case 2: East African Highlands 228
11.4 Accessibility: Cartography, GIS and Spatial Information Policy 228
11.5 Copyright and Liability 230
11.5.1 Copyright 231
11.5.2 Exceptions to the Copyright Law 232
11.5.3 Doubtful Copyright Protection of Geographical Information 233
11.5.4 Freedom of Information Act 234
11.5.5 Copyright and the Internet 234
11.5.6 Creative Commons Licences 234
11.5.7 Right of Possession 234
11.5.8 Public Lending Right 234
11.5.9 International Differences 235
11.5.10 Liability 235
11.6 Map Use and Usability 236
11.7 Maps and GIScience Revisited 237
Further Reading 238

References 239
Index 243

Preface

PREMISES AND OBJECTIVES

This book has been written to assist in cartographic education and intends, as a first objective, to provide an overview of the role that maps will play both today and in the near future in the world of geospatial data handling. It shows the background against which the provision and visualization of geospatial information takes place. It provides awareness of the Web both as a spatial data source and as a means for distributing the results of visualizing this spatial information. To realize that first objective, the nature of geospatial data is described as well as the characteristics of maps and the ways in which they can be put to use. A development stimulated by the Web was the increased use of spatial data infrastructures, for sharing national and global geodata with the professional and general public. The development of the Internet has boosted the possibilities for interaction and for querying the databases behind the maps presented there. The number of databases available via the Web has increased dramatically and so has the ability to interact with them (query, process, etc.) online. Maps have acquired an important interface function in this new cyberspace geo-information distribution environment. If mapmaking with GIS (geographical information system) mainly involved geo-professionals, the World Wide Web potentially allows everyone to have access to this new medium to create maps.

But not everyone is aware of the intricacies of map design and of the characteristics of the various map types and their limitations. That is where our second objective comes up: teaching map design. What types of geovisualization are appropriate, and how do we translate the numbers collected through censuses or the data measured by sensors into images that allow us to draw sensible conclusions? The answers are shared out over topographic, statistical (or thematic) and temporal maps. For all three categories of maps, we intend to provide sufficient relevant knowledge of cartography and geovisualization concepts and techniques to those accessing the World Wide Web for the production and use of effective visualizations of geospatial information.

Showing the manner in which maps function, either independently or combined in atlases, and can be analysed and interpreted, either in stand-alone or in geo-information environments, is the third main objective of this book. Since the position of the Web has strengthened and stabilized itself, it also stimulated a more integrative approach to problem-solving with geo-information (GIScience (geographical information science)). Since the World Wide Web is highly interactive, and since it allows one to integrate data files, and to link distributed databases, this makes maps suitable instruments for exploring these databases.

BOOK HISTORY AND ACKNOWLEDGEMENTS

Most of the design processes advocated in this book are still based on the inspiration provided by Jacques Bertin's book *Semiology of Graphics* (2011). He wrote the original French version in 1967 in order to improve the printed maps he was confronted with in the media. The media have changed, and the Web is our new medium now, but the basic cartographic design rules still apply in the new interactive visualization of geospatial data for the Internet.

The first, 1996, edition of this book published by Longman was developed from a book (*Cartography: Design, Production and Use of Maps*) for cartography students in the Netherlands, published by Delft University Press in 1993. Since then, a Polish edition has been published (1998) as well as a second edition of the Dutch version (1999). Each of these editions influenced the subsequent editions in other languages. The second English edition was published by Prentice Hall in 2003, and a Russian translation was published in 2005, followed by an Indonesian translation in 2007; an inexpensive English edition for the South-East Asian market is already in its second print. In 2010, Pearson published a third edition, and a year later, The Guilford Press published a larger-sized edition for the United States. A Chinese edition was published in 2014.

The illustrations in these books were initially produced by practical cartographers from Delft and Utrecht universities; the illustrations of the third edition have been based on them but have been reprocessed and updated by Wim Feringa of University of Twente, Enschede, the Netherlands. His was the major job now to convert all illustrations to colour mode, as the publishers allowed us to make this fourth edition a full-colour one, so that the full visual impact of maps could be unleashed.

Though the map examples include many references to situations outside Europe, most of them stem from inside that continent, or even the Netherlands. There are some favourite spots where we return again and again, as many maps refer to Maastricht (the Dutch city where in the year 1992 the treaties were signed that led to the creation of the European Union). Over 20 maps refer to the English Lake District, and another favourite spot is Mount Kilimanjaro. In a sense, these illustrations reflect the professional practice and the related movement patterns of the authors.

STRUCTURE

The basic structure of this book has been left unaltered: it has three distinctive parts. The first five chapters offer the context and basics of maps. The second three each deal with the components of geospatial data: location, attribute and time. The last three chapters deal with 'maps at work' and demonstrate how maps and atlases can assist in problem-solving and decision-making. These three parts are structured as follows: in the first part, in Chapter 1, we discuss the place of maps and mapping in the geo-information environment (GIS (geographical information system), GIScience (geographical information science) and the geospatial data infrastructure, of which the Web constitutes an ever more important part). We proceed to show how data are collected (Chapter 2) and present the concepts that are valid in mapping (Chapter 3) and GIScience (Chapter 4). Chapter 5 deals with the necessary analysis of geospatial data prior to their visualization and also offers some basics of map production.

In the second part, Chapter 6 is focused on location. It not only deals with the characteristics of the base map (reference system, projection, relief portrayal, generalization and geographical names) but also deals with the organizational aspects of topographic map production. Chapter 7 shows the visualization options of the attribute data that are to be rendered on these base maps (thematic map types). Chapter 8 discusses the temporal component of geospatial data.

In the third part, the subject matter becomes more advanced as we deal with the intricacies of map use, analysis and interpretation: Chapter 9 describes how to work with maps and atlases, Chapter 10 shows how to work with maps in a highly interactive geovisualization environment, and finally, Chapter 11 deals with maps for decision-making in a wider context.

UPDATING AND ACCESS

In this age and time where new software packages, new institutional set-ups and technical advances impact us almost continuously, coupled to the more ephemeral aspects of the Web, to keep up to date is a challenge. Although the structure of this book remained the same (the only new sections in this edition are on cartographic education, map machines and story maps), its existing sections were also overhauled. The antithesis between commercial packages that try to monopolize geospatial data handling and public initiatives, such as OpenStreetMap, OpenNauticalChart, Open Geospatial Consortium and the World Wide Web Consortium that try to effectuate the reverse, is dealt with. The launch in 2020 of new GPS platforms and the advent of higher resolution lidar or laser scanning are examples of new techniques described. Usability is a new concern we deal with, as well as volunteered geographical information. Volunteers meet in mapathons, in missing map projects to help out in emergency situations. And the goal we ultimately map for, a sustainable future, where big data are analysed to realize our sustainable development goals is also made explicit. Both authors have been engaged in relevant projects undertaken by the United Nations.

As we have rewritten the text in order to accommodate new generations of web browsers, we also use the prime function of the Web to keep this book up to date. Apart from the Web, every chapter has its section on books for further reading, while all the references to printed literature are grouped together at the back.

From a society that was used to having free access to printed maps, we have evolved to a society used to having free access to geospatial data and maps on the Internet. Everyone can process and visualize the geospatial data available there and put the resulting maps on the Web in turn – there is no quality standard against which the material is checked first, before incorporation is permitted, which is acceptable because the very impact of the Web stems from the fact that it is a free medium. But geo-professionals – and cartographers belonging to this group – have the responsibility of convincing as many as possible to keep the tenets of good and responsible design while visualizing the geospatial data, in order to support the process of spatial decision-making; this refers to a large part of all the cybernetic processes. The decisions based on visualized geospatial data remain only as good as the data and the visualizations themselves.

Menno-Jan Kraak
Ferjan Ormeling
December 2019

Acknowledgements

We are grateful to the following for permission to reproduce copyright material: Figure 1.6, Rijkswaterstaat NL (CC); Figure 1.10, Hootsmans & Van de Wel, 1993; Figure 1.12, Kadaster Geoinformatie NL (CC); Figure 1.14, Kadaster Geoinformatie NL (CC); Figure 1.15, Open Topo NL (CC); Figure 1.18, Google LLC; Figure 1.19, Google LLC; Figure 2.4, OpenNauticalChart.org (CC); Figure 2.5, Kadaster Geoinformatie NL (CC); Figure 2.14-2.16, Hootsmans & Van de Wel, 1992; Figure 3.6, ANWB Den Haag; Figure 3.13, Bundesamt für Seeschifffahrt und Hydrographie Hamburg/Rostock; Figure 3.14, Kramers, 2020; Figure 3.15, Google LLC; Figure 3.16, Nationale Atlas NL (CC); Figure 3.17, Rijkswaterstaat NL (CC); Figure 3.18, Georg Westermann Verlag; Figure 4.1, Kadaster Geoinformatie NL (CC); Figure 4.8, Koussoulakou 1990; Figure 4.11, The Copyright Licensing Agency Ltd HMSO; Figure 5.23, Patterson; Figure 5.34, Microsoft; Figure 6.15, Noordhoff; Figure 6.17, Open Street Map (CC); Figure 6.24, Kadaster Geoinformatie NL (CC); Figure 6.25, Open Topo NL (CC); Figure 6.31a, EDK; Figure 6.31b, Noordhoff; Figure 6.32, National Geographic; Figure 6.37, Kadaster Geoinformatie NL (CC); Figure 6.39, Ordnance Survey UK; Figure 6.40, Kadaster Geoinformatie NL (CC); Figure 6.41, French Mapping Agency IGN; Figure 6.42, National Land Survey Open data SF (CC); Figure 6.43, USGS; Figure 6.44, Swiss Topo; Figure 6.45, LGL Baden Wurtemberg; Figure 6.46, ESRI; Figure 6.47, Open Street Map (CC); Figure 6.49, Royal Mail (UK); Figure 6.50, Bolder; Figure 6.51, Kadaster Geoinformatie NL (CC); Figure 7.4, Kadaster Geoinformatie NL (CC); Figure 7.16, Freitag 1992; Figure 7.26, Erkamov & Cholodnyj, 1929; Figure 7.36, Dorling, 1995; Figure 8.4, Iceland Air; Figure 9.1, Noordhoff Figure 9.4, Swiss Topo; Figure 9.5-9-6, Microsoft; Figure 9.7, Centennia Historical Atlas; Figure 9.9, Georg Westermann Verlag; Figure 9.10, ANWB Den Haag; Figure 9.13, Centennia Historical Atlas; Figure 10.3, Google LLC; Figure 11.2, ONC US Government; Figure 11.2, ONC US Government; Figure 11.5a, Alphen aan den Rijn municipality; and Figure 11.5b, Falkplan BV.

In some instances, we have been unable to trace the owners of copyright material, and we would appreciate any information that would enable us to do so.

Authors

Menno-Jan Kraak is a professor of geovisual analytics and cartography at the University of Twente/ITC. Currently, he is the head of ITC's Geo-Information Processing Department.

He was the president of the International Cartographic Association (ICA) for the period 2015–2019, and currently immediate past president, and chair of the UN-GGIM Geospatial Societies (2018–2020). He wrote more than 200 publications, among them the book *Mapping Time*. He is a member of the editorial board of several international journals in the field of cartography and GIScience, and the chair of the Foundation Scientific Atlas of the Netherlands.

Ferjan Ormeling worked as an atlas editor before being nominated professor of cartography at Utrecht University. Currently, he is employed at the University of Amsterdam as a member of the Explokart research group. He chaired the Education Commission of the International Cartographic Association (ICA) for 11 years and was the vice-chair of the UN Group of Experts on Geographical Names 2007–2017. In 2017, he was elected president of the UN conference on the Standardization of Geographical Names in New York.

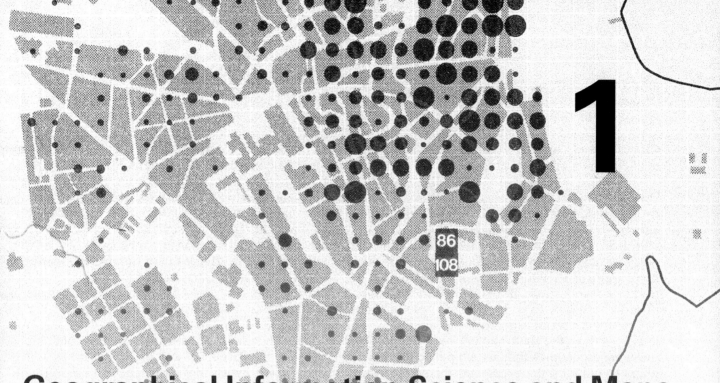

Geographical Information Science and Maps

1.1 THE MAP AS AN INTERFACE

Maps are used to visualize geospatial data, i.e. data that refer to the location or the attributes of objects or phenomena located on Earth (the terms 'spatial data' and 'geographical data' will be used interchangeably). Maps help their users to better understand geospatial relationships. From maps, information on distances, directions and area sizes can be retrieved; patterns revealed; and relations understood and quantified. Since the 1980s, developments in digital geospatial data handling have gained momentum. Consequently, the environment in which maps are used has changed considerably for most users. With the computer came on-screen maps. Through these maps, the database from which they are generated can be queried, and some basic analytical functionality can now be accessed through menus or legends. In the 1980s, these software packages that allowed for queries and analyses of geospatial data became known as 'geographical information systems' (GISs). As their functionality matured, their application spread to all disciplines working with geospatial data. GIS introduced the integration of geospatial data from different kinds of sources. Its functionality offers the ability to manipulate, analyse and visualize the combined data. Its users can link their application-based models to the data contained in the systems,

and try to find answers to questions such as 'Which is the most suitable location to start a new branch of a supermarket chain?' or 'What effect will this plan, or possibly its alternative, have on the surrounding area?'

Maps are no longer only the final products they used to be. The paper map functioned, and functions, as a medium for storage and presentation of geospatial data. The introduction of on-screen maps and their corresponding databases resulted in a split between these functions. To cartographers, it brought the availability of database technology and computer graphics techniques that resulted in new and alternative presentation options such as three-dimensional and animated maps. Geospatial analysis often begins with maps; maps support judging intermediate analysis results and presenting final results. In other words, maps play a major role in the process of geospatial analysis.

The rise of Internet brought the next revolution in mapping. Access to interactive maps is no longer limited to professionals. Products such as Google Maps/Google Earth even allow people to add their own data to the maps and share it with others in a mouse click. The IT-related developments have resulted in a convergence of the different disciplines working with geo-information. GIS is integrated in the workflow of geo-related problem-solving. The disciplines studying related methods and techniques have converged under the header of geographical information science (GIScience). Scientists in this field do research on GIS (e.g. study principles on which GIS is based) and with GIS (e.g. study how GIS can be used in scientific applications (Longley et al., 2015).

The above development also led to spatial or geographical data infrastructure (SDI or GDI). Next to a technical setting, a GDI comprises a set of agreements and arrangements to access, integrate and use geo-information. These new infrastructures for accessing geospatial data are being developed all over the world in order to allow access to the geospatial data files created and maintained in order to monitor the population, resources and environment spatial aspects of our modern societies. Access to the data needed requires complex querying procedures that are simplified when using maps to pinpoint the areas and themes for which data are needed (Figure 1.1).

In a GIScience environment, visualization is applied in four different situations. Firstly, visualization can be used to explore, for instance, in order to play with unknown data. In several applications, like those dealing with remote sensing data, there are abundant (temporal) data available. Questions such as 'What is the nature of the data set?' or 'Which of those data sets reveals patterns related to the current problem studied?' have to be answered before the data can actually be used in a geospatial analysis operation. Secondly, visualization is applied in analysis, for instance, in order to manipulate known data. In a planning environment, the nature of two separate data sets can be fully understood (e.g. the groundwater level and the possible location of a new road), but their relationship cannot. A geospatial analysis operation, like overlay, can combine several data sets regarding the same area to determine their possible geospatial relationship. The result of the overlay operation could, when necessary, be used to adapt the plans. Thirdly, maps are used to synthesize the results of the analysis. Fourthly, visualization is applied to present or communicate the new geospatial knowledge. The results of geospatial analysis operations can be displayed in well-designed maps easily understood by a wide audience. The cartographic discipline offers design rules to do so. As the fourth objective of visualization, we have already mentioned the easier access to the data files behind the maps.

Considering these four different fields of visualization in GIScience (exploration, analysis, synthesis and presentation), it can be noticed that the tools for presentation are the most highly developed. While producing maps to communicate

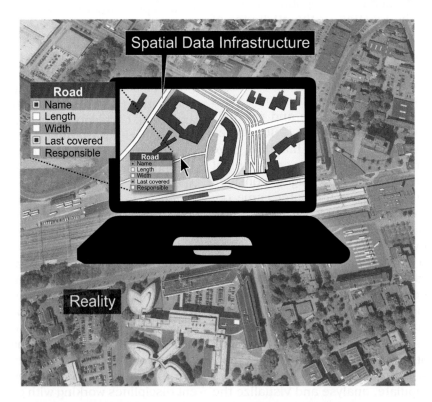

FIGURE 1.1 The interface role of maps in the spatial data infrastructure. Here, the GIS answers a query regarding a clicked object

geospatial information, cartographic rules and guidelines (together called 'cartographic grammar', based on the nature of the data and the communication objectives) are available to make the maps effective. However, as these rules are not part of the mapping software, it allows users to produce their own maps even when they are unaware of cartographic grammar. In other words, there is no guarantee that the maps will be effective. These cartographic rules could also be applied in the analysis phase, but the necessity to do so would be less strong here. When cartographers and analysts discuss this matter, the second group would always claim 'Who cares about mapping rules, as long as one understands one's own maps?' And because the analysts knew their own data, they probably would understand their own maps, but when showing their maps to others communication problems would start. In a data exploration environment, it is likely that the user does not know the exact nature of the data and therefore might not be able to apply the relevant cartographic rules.

At this moment, the terms 'private visual thinking' and 'public visual communication' should be introduced (DiBiase, 1990). Private visual thinking refers to the situation where users explore and analyse their own data, and public visual communication refers to the situation where users present their results in the form of maps to a wider audience. The first describes the exploration circumstances,

and the second presentation circumstances. Analyses can be found somewhere in the middle along a line between the two. This becomes more evident when it is realized that private versus public map use (i.e. maps tailored to an individual versus those designed for a wide audience) is just one of the axes of the so-called map-use cube, first introduced by MacEachren (1994). Along the two other axes, the revelation of the unknown versus the presentation of the known, respectively, and high versus low user interaction are plotted, which are shown in Figure 1.2.

Most chapters in this book concentrate on maps that should communicate geospatial information (the lower-left front corner of the cube). However, recent developments in cartography and other disciplines handling geospatial data not only require a new line of thought, but also create one. This can be illustrated by plotting the evolutionary stages of the development of electronic atlases in the cube along the diagonal from the corner 'wide audience, presenting knowns and low interaction' towards the corner 'private use, presenting unknowns and high interaction' (Figure 1.2b). Possibilities for interaction are boosted by the advent of the Internet and its potential for querying the databases behind the maps presented there. Early electronic atlases were, in effect, sequential slide shows, but today's electronic atlases have high interactive multimedia mapping capabilities, and allow users to combine

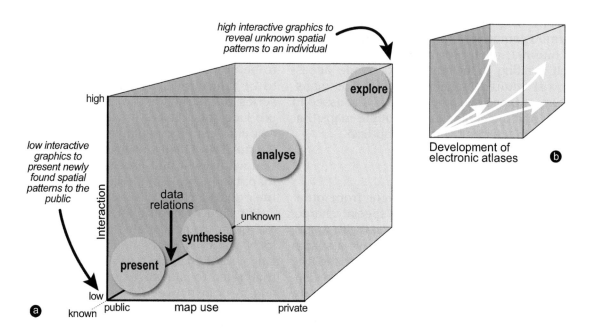

FIGURE 1.2 The map-use cube (adapted from MacEachren and Taylor, 1994): (a) the four main situations to visualize data in a GIS, (1) to present, (2) to synthesize, (3) to analyse and (4) to explore; (b) the evolution of the electronic atlas since 1987 plotted in the map-use cube

their own data with atlas data. Each category of map use in Figure 1.2 cube asks for its own visualization approach. New cartographic tools and rules have to be found for these approaches. They are probably not as restrictive as traditional cartographic rules, but on the other hand not as free as the technology allows either.

The demand for sophisticated geospatial data presentation is further stimulated by developments in scientific and information visualization, multimedia, virtual reality and exploratory data analysis. In each of these external developments influencing GIScience and maps, it would appear that from a technical point of view, there are almost no barriers left. The user is confronted with a screen with multiple windows displaying text, maps and even video images supported by sound. Important questions remain. Can we manage all the information that reaches us? The ever more detailed satellite imagery available, the increasing number of sensor networks and new techniques for analysing textual sources with spatial references like geoparsing all lead to highly varied 'big data', characterized by large volumes of data, coming available with high velocity. In order to make sense of them and derive meaning or trends, cartography, with its capacity of generalizing data in order to fit their purpose, plays an important role. What will be the impact of these developments on the map in its function to explore, analyse and present geospatial data? This book tries to provide an overview of the role that maps will play both today and in the near future in the world of geospatial data handling. There is an enormous amount of geospatial data out there, on the Internet, useful for any kind of geospatial research, waiting to be harnessed, made accessible and structured by being visualized as maps. The nature of these geospatial data is discussed in Chapter 2, and the characteristics of the maps that visualize them are dealt with in Chapter 3.

1.2 GEOSPATIAL DATA

Geographical information is different from other information in that the data, as a special characteristic, refer to objects or phenomena with a specific location in space and therefore have a spatial address. Because of this special characteristic, the locations of the objects or phenomena can be visualized, and these visualizations – called 'maps' – are the key to their further study. Figure 1.4 shows how objects from the real world that can be localized in space (such as houses, roads, fields or mountains) can be abstracted from the real world as a digital landscape model (DLM), according to some

predetermined criteria, and stored in GISs (such as points, lines, areas or volumes) and later, after being converted into a digital cartographic model (DCM), represented on maps (with dots, dashes and patches) and integrated in people's ideas about space. When stored in a database, these geospatial data are usually divided into locational data, attribute data and temporal data. The first refers to the geometrical aspects (position and dimensions) of the phenomenon one has (geometric) information about, and the second refers to other, non-geometrical characteristics. The temporal data refers to the moment in time for which both the locational and attribute data are valid. These three aspects are linked to the elementary questions 'Where?', 'What?' and 'When?' (see Figure 1.3a), and define the nature of an object (Figure 1.3b). An object's location, attribute or time can have multiple characteristics, such as different coordinate systems, multiple variables and even different kinds of time (Figure 1.3c). Apart from these three questions, one might also ask 'Why?' or 'How?' Answering these last two questions requires further analysis of the data. It might require more attention from one of the data's components, resulting in a perspective from what could be called 'location space', 'attribute space' or 'time space' (Figure 1.3a). Chapters 6–8 will, respectively, deal with questions related to these three spaces.

The stored geospatial data on a specific study area is called its 'digital landscape model' (DLM). Of course, it is an abstraction: selected characteristics have been measured or assessed and incorporated in this DLM. As soon as this DLM is considered suitable for communication to other persons, and has to be produced in hard copy form, this model has to be converted into a DCM, which consists of a series of instructions to the printer or screen, to produce dots, dashes or patches, in different sizes, colours, etc. for multiplication and distribution (Figure 1.4). Finally, users of the mapped information will view it and process it into their cognitive map, the mental construct of space they will base their decisions on.

For data to qualify for the tag 'geometric data 'or 'georeferenced data', information about their location would be required. This can be geographical or reference grid coordinates, code numbers that refer to statistical areas, topological terms (e.g. A is in between B and C) or nominal terms, as in street addresses and postcodes. The geospatial nature of the objects can be expressed in their shapes, with which one represents objects from the real world. There is a basic subdivision into point-, line-, area- or volumetrically shaped objects

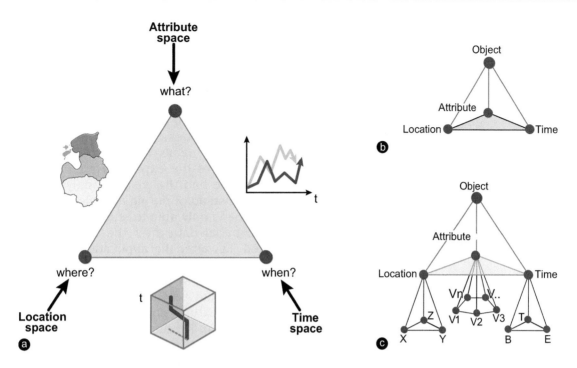

FIGURE 1.3 The characteristics of geospatial data: (a) its components' location, attribute and time, and their related elementary questions 'Where?', 'What?' and 'When?'; (b) the object view; (c) detailed characteristics of the data components

(see Figure 1.5), and this can be further subdivided into, for instance, elongated, triangular, irregular or convex-shaped objects. In a sense, this is scale- or resolution-dependent, as a populated settlement will be rendered by a point in a national context and as a built-up area in a municipal context.

Whether the objects or phenomena from the real world are abstracted as discrete or continuous is very important for subsequent storage and mapping procedures. Discrete objects can be bordered on all sides, and the coordinates of these borderlines can be made explicit. These can either be the locations of tactile objects (houses, streams, etc.) or be the locations of predetermined areas (states, enumeration areas or distribution areas). Continuous representations are abstractions of those phenomena that are considered to change non-incrementally in value. They can be tactile or measurable (like precipitation data or gravity field data) or be based on models (like isochrones, i.e. lines linking points that can be reached in an equal travelling time from a given starting point).

For later visualization procedures, it is essential that the nature of the attribute information be established. These attributes can refer to visible characteristics (e.g. deciduous trees) and invisible characteristics (e.g. temperature). When attempting to define these attribute values of objects, one usually tries to measure or categorize them, and

then, it will appear that these characteristics are either qualitative or quantitative. One may distinguish a number of measurement scales (see also Chapter 5), on which the values for these characteristics can be assessed:

- *Nominal scale*: Attribute values are different in nature, without one aspect being more important than another (e.g. different languages or different geological formations).
- *Ordinal scale*: Attribute values are different from each other, but there is one single way to order them, as some are more important/intense than others (e.g. warm, mild, cold).
- *Interval scale*: Attribute values are different and can be ordered, and the distance between individual measurements can be determined. Temperature is a good example: because the respective zero points of their measurement scales have been selected at random, it is impossible to say that, for instance, a temperature of 64°F is twice 32°F. This is plain when the values are converted into Celsius and become 18°C and 0°C, respectively.
- *Ratio scale*: Attribute values are different and can be ordered. Distances between individual measurements can be determined, and these individual measurements can be related to each other. If, for instance, the gross domestic

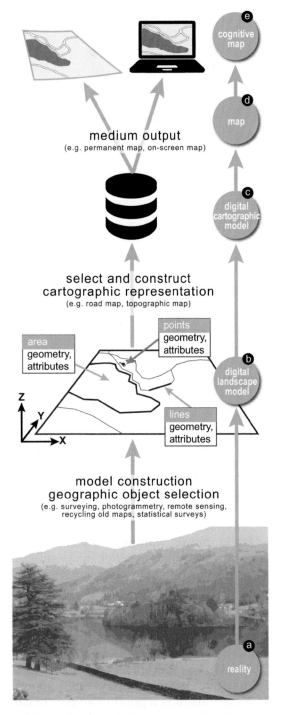

medium output
(e.g. permanent map, on-screen map)

**select and construct
cartographic representation**
(e.g. road map, topographic map)

area
geometry,
attributes

points
geometry,
attributes

lines
geometry,
attributes

**model construction
geographic object selection**
(e.g. surveying, photogrammetry, remote sensing,
recycling old maps, statistical surveys)

FIGURE 1.4 The nature of geospatial data: from reality (a), via model construction and selection to a digital landscape model (b), followed by selection and construction of a cartographic representation towards a digital cartographic model (c), presented as a map (d), which results in the user's cognitive map (e)

income per capita in Sri Lanka in 2017 was $13 000 per annum and in Bangladesh $4300, then one can say that the amount in the former was thrice the value of that in the latter.

All the geospatial data will be subject to changes over time: the attribute information on an object can change over time (like the composition of the population of an area), and even the object's location itself may change (for instance, through continental drift). The data's timestamp is seen as the third major component, next to geometry and attribute values. Especially these days, the interest in the data's temporal component increases because of the expanded number of time series available and the wish to analyse processes over time instead of during a single time slice.

One is only able to study or analyse or interpret geospatial data after the data have been visualized, i.e. after the application of the cartographic grammar to render these objects and their relationships. Here, symbols and signs are used, i.e. dots, dashes and patches, and these can vary in size, shape, texture, colour, value and orientation (see Chapter 5). These signs are linked to the objects or relationships they represent, and by doing so, one is able to convey geospatial relationships between point, line, area or volumetric objects, in a number of dimensions, to the map user.

If only one dimension is available, then the location of geospatial data can be expressed, for instance, by their distance from a central market, or from a point of origin – represented as a straight line. Two-dimensional representation with these dots, dashes and patches will result in a planimetric map, and Figure 1.6a showing a contour map of a hill in the south of the Netherlands is a good example. In order to have a true three-dimensional representation, a physical model of this hill could be produced from cardboard, or a virtual model could be created, which, by rotating it, or through anaglyphs, using red and green glasses, could be seen from all sides on a monitor screen. By drawing the model in perspective, and using a hidden-line algorithm, this 3D aspect could be simulated (represented in Figure 1.6b). The current description of this type of rendering is '2.5 dimensional', because it is a projection of 3D reality on a 2D plane but still gives the map user a three-dimensional impression. Maps with hill shading (see Figures 6.30 and 6.31) are another example of this.

If one adds the time dimension, the representation will become four-dimensional (Figure 1.6c), when, through juxtaposition of two states of this object, for instance, in 1950 and 2020 respectively, the change in its geometry or attributes during the intervening period can be ascertained.

Through their visualization in maps, the geospatial relationships of objects can be made visible. These geospatial relationships will usually refer to

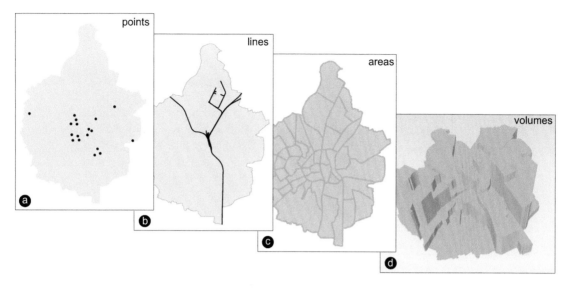

FIGURE 1.5 The representation of geographical objects in a digital environment as (a) points, (b) lines, (c) areas and (d) volumes

relations to some specific location on the Earth's surface, or can be those of objects to one another, and these relations can have many forms. Primary geospatial relationships are those between objects and their location on Earth, or between these objects and their attributes, such as the type of vegetation occurring at that location and the type of road classification. By visualizing object categories from a data file (e.g. car factory locations or stream networks, or fields used for horticulture), relations between the elements for an object category will be made clear, and one will be able to perceive patterns or geospatial trends. By combining geometric and attribute data, one would be able to perceive how the locations of elements from different object categories might influence each other.

Next to these primary relationships, it is possible to perceive secondary types: relations of objects to linear or area reference units, such as that of inhabitants to surface area, the number of cars to the length of the highway network and the relative amount of horticulture in all agricultural areas. One could go further and introduce other dimensions (like height or time), so that tertiary or even higher-order relationships would emerge.

1.3 GEOGRAPHICAL INFORMATION SYSTEMS

For most disciplines working with geospatial data, one of the first uses of the computer was to create an inventory of discipline-dependent data. In this period, cartographers worked to build a database from which they could produce the maps that were previously created manually. In the next phase, spatial analysis of the collected data was emphasized. Forestry scientists, for instance, would apply statistical methods to the maps' attribute data. For cartographers, this meant the possibility of creating different derived products from the existing database. Nowadays problems are approached in an interdisciplinary way. In physical planning processes or in environmental impact studies, data from many different fields are needed. The need to combine them led to the development of GIS. Cartographic knowledge is used in GIS to create proper visualizations. GIS offers the possibility of integration of geospatial data sets from different kinds of sources, such as surveys, remote sensing, statistical databases and recycled paper maps. Its functionality allows one to manipulate these data, or to set up geospatial analysis operations in conjunction with application-based models, and they allow for the visualization of the data at any time during this process. The core functionality of a GIS is provided by the disciplines such as geography, geodesy and cartography that are used to work with spatial data. To this core, functionality from database technology and computer graphics is added. Currently, GIS is used in virtually all disciplines and professions that require geospatial data to execute their tasks or solve their problems.

Why is GIS unique? Because it is able to combine geospatial and non-geospatial data from different data sources in a geospatial analysis operation in order to answer all kinds of questions. The fact that those questions can be answered is quite remarkable if one realizes that geographical phenomena

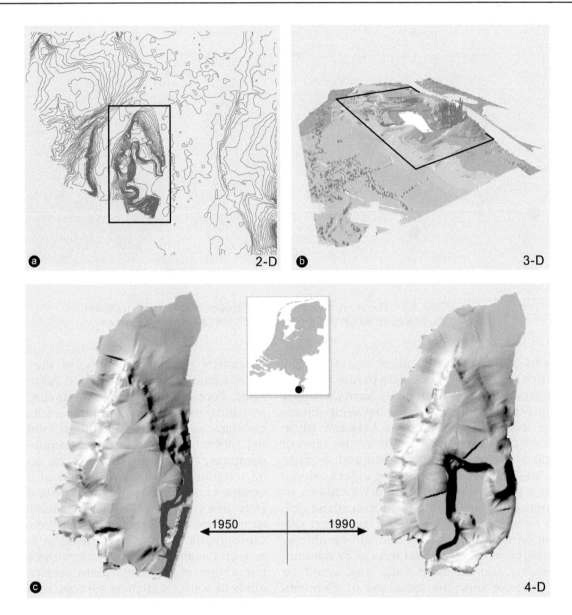

FIGURE 1.6 The dimensionality of geographical objects: (a) (2D), (b) 3D; (c) 4D/time. In Figure 1.6c, part of the hill has been excavated since 1950 (Rijkswaterstaat AHN (CC0))

are almost never homogeneously distributed. And added to this observation, one can add a quote from Tobler (1970) who said: 'all things are related, but nearby things are more related than distant things'. What type of questions can be answered by a GIS (Figure 1.7)?

→ *What is there ...? Identification*: By pointing at a location on a map, a name, or any other attribute information stored on the object, is returned. This could also be done without maps, by providing the coordinates, but this would be far less effective and efficient.

→ *Where is ...? Location*: This question results in one or more locations that adhere to the criteria of the question's conditions. This could be a set of coordinates or a map that shows the location of a specific object, or all buildings in use by a certain company.

→ *What has changed since ...? Trends*: This question includes geospatial data's temporal component. A question related to urban growth could result in a map showing those neighbourhoods built between 1950 and 2020.

→ *What is the best route between ...? Optimal path*: Based on a network of paths (e.g. roads or a sewage system), answers to such queries for the shortest or cheapest route are provided.

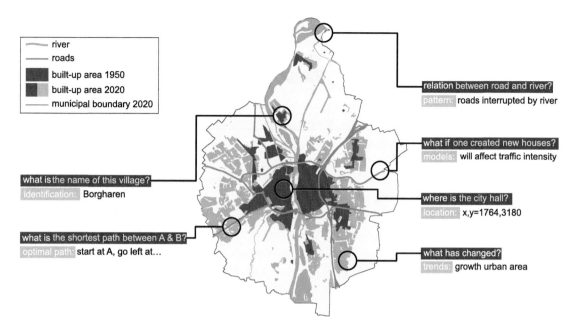

FIGURE 1.7 Typical GIS questions answered by maps such as those used to identify, locate or find geospatial patterns

→ *What relation exists between ...? Patterns*: Questions like this are more complex and often involve several data sets. Answers could, for instance, reveal the relationship between the local microclimate and location of factories and the social structure of surrounding neighbourhoods.

→ *What if ...? Models*: These questions are related to planning and forecasting activities. An example is: 'What will be the need to adapt the local public transport network and its capacity when a new neighbourhood is built north of the town?'

Of course, one does not only query a GIS but also uses it interactively, for instance, in physical planning procedure, through manipulation of designs, etc.

GIS development was stimulated by individual fields such as forestry, defence, cadastre, utilities and regional planning (see Chapter 4). Since they all have different backgrounds and different needs, the functionality of the software GIS initially used was different as well. It ranged from statistical analysis packages to computer-aided design packages. Functionality was added, and each of these groups started to call their software a 'GIS package'. This resulted in different meanings for the same term. Next to GIS, literature offers wordings such as land information systems, geo-based information systems, natural resources information systems and geodata systems (Longley et al., 2015).

The multidisciplinary background of GIS led to a multitude of definitions. In general, they can be split into two groups: those with a technological perspective and those with an institutional/organizational perspective. An example of the first is the definition by Burrough and MacDonell (1998): 'a powerful set of tools for collecting, storing, retrieving at will, transforming and displaying spatial data from the real world'. An example of the second is that of Cowen (1988), 'a decision support system involving the interaction of geospatially referenced data in a problem-solving environment'. So, it is the potential combination of different data sets that is paramount. A working definition for this book is derived from a combination of the two above: a GIS is a computer-assisted information system to collect, store, manipulate and display spatial data within the context of an organization, with the purpose of functioning as a decision support system.

In order to manipulate geospatial data, to procure added value, a GIS consists of software, hardware, geospatial data and people (the organization). These components communicate via a set of procedures. In Figure 1.8, which summarizes the view of GIS adopted in this book, the central schemes present the GIS components: the problem-solving production line (from exploration to presentation), the potential for geospatial analysis (with application-based models) and integrating geospatial data sets, the unique and basic ingredients of the system. The outer shell renders the organization in which GIS functions. The configuration of the

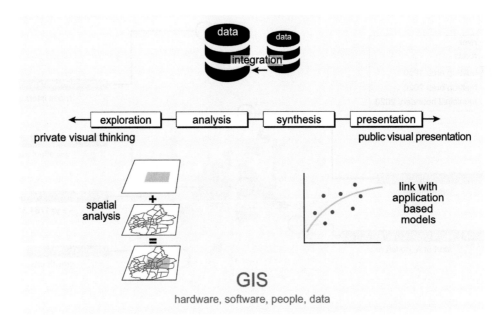

FIGURE 1.8   View on GIS: its characteristics and relation to visualization

scheme stresses the need for a proper user interface and management of the system.

Each organization involved in geospatial data will require a GIS with emphasis on a specific set of functions, depending on its task. In general, functions are needed for data input and encoding (e.g. digitizing, data validation and structuring options), data manipulation (e.g. data structure and geometric conversions, generalization and classification options), data retrieval (e.g. selection, spatial and statistical analysis options), data presentation (mainly graphical display options) and integrated data management.

For the purpose of this book, also written for GIS analysts who have to learn to use the methods, cartography will be regarded as an essential support for nearly all aspects of handling geographical information for the following reasons:

➤ Maps are a direct and interactive interface to GISs, a sort of graphical user interface with a geospatial dimension;

➤ Maps can be used as visual indexes to phenomena or objects that are contained in the information systems;

➤ Maps, as forms of visualization, can help in both the visual exploration of data sets (also the discovery of patterns and correlations) and the visual communication of the results of the data set exploration in GISs;

➤ In the output phase, the interactive design software of desktop cartography is superior to the output functions of current GISs.

These should be enough reasons for cartography to have an important place in GIS, but there are more reasons, if one looks at the context in which GISs are being used: they are aimed at decision support, and as this regards decisions about geographical objects, it should be visual decision support, in order to take into account the geospatial dimension as well. In order to correctly use these visual decision support aids (the maps visualized on the computer screen or the hard copy output of these systems), the users should adhere to proper map-use strategies (see Chapter 11). This ability to work with maps and to correctly analyse and interpret them is one very important aspect of GIS use. Strangely enough, not one GIS manual gives any clarification in this field, assuming that all GIS users are aware of the ins and outs of map use.

But there is another important decision support aspect to the information that is processed in and presented by a GIS: data quality. GISs are very good in combining data sets; notwithstanding the fact that these data sets might refer to different survey dates and different degrees of geospatial resolution, or might even be conceptually unfit for combination, the software combines them and presents the results. Cartographers, in compiling maps, have worked with different data sets for centuries and have some experience in the transformations that are necessary in order to combine data sets with different resolutions, projections, reference systems, geoids and dates of survey. They have developed transformations and modelling procedures (like generalization) that take account of these differences and will allow for real data integration. They have developed documentation techniques that will describe all relevant data characteristics (metadata) necessary for proper integration to take place. They have also lobbied for decades to standardize these documentation techniques so that the data sets can be easily exchanged (Figure 1.9). So much of the methodology for the determination of data quality is potentially available for GIS users from cartography. Finally, cartographers have strived to have access to geospatial data improved, as is the object of improving the SDI (see Section 1.5).

As to the assessment of data quality, this can be defined, as was shown in the preceding section, as a measure of the suitability of data for specific applications. So, one can, for instance, determine the precision x to which objects (like parcels) in a data set have been localized, as well as the probability p with which these objects have been correctly classified or categorized (e.g. regarding their land cover). Now the GIS will allow the combination of this land cover information (e.g. surveyed in 2018) with precipitation information (e.g. surveyed for the period 1990–2010, for five points in the area, and interpolated for all other locations), with a planimetric accuracy y of the rain gauge locations and a representativity factor z_{1-5} of these measurements for their surrounding areas. Well, what will be the value of correlations between precipitation data and land cover data, taking into account these accuracy, probability and representativity values?

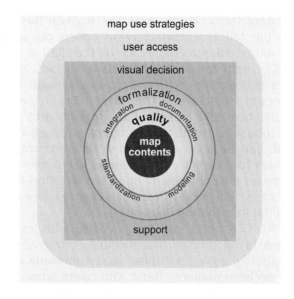

FIGURE 1.9 Visual decision support for spatiotemporal data handling. Keywords in the GIS cartography approach are map-use strategies (how people make decisions based on maps), public access (whether and how people can work with the information), visual decision support (showing what the quality of the information is) and formalization (building expert systems)

Until recently these have been disregarded in GISs, but in a mature GIS that really functions as a decision support system, these values should be indicated to properly inform the decision-maker.

Figure 1.10 shows the classification of remote sensing imagery of urban areas in a part of the

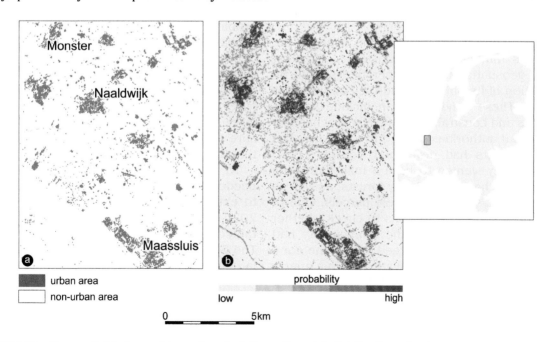

FIGURE 1.10 Geospatial information and meta-information: (a) distribution of urban areas over a part of the western Netherlands; (b) probability map for the classification 'urban' (from Hootsmans and Van der Wel, 1993)

western Netherlands (near The Hague): the image on the left is a classification based on spectral qualities, whereas the image on the right is a probability map for the classification 'urban'. It takes into account the potential confusion with related spectral signatures. The probability of a pixel being correctly classified as 'urban', and not as 'hothouse' or industrial complex or beach or bare soil, can be computed, and visualized probability values for correct classifications like these should form an essential element of the decision-making process, which takes place everywhere where geospatial information is involved.

So, GIS users can be provided with essential tools in all phases of collecting, processing and analysing geospatial data, and communicating it to decision-makers. Those GIS users who able to use maps are provided with the infrastructure for a correct decision-making procedure, and with the necessary information (meta-information) on the quality of the data contained in those maps.

1.4 GEOSPATIAL ANALYSIS OPERATIONS

Geospatial analysis operations are the unique processes GIS has to offer to the geospatial data handling community. However, its principles have been known since the 1950s from fields such as quantitative geography and statistical geography (Hägerstrand, 1967). This section will explain the principles of these operations, illustrated with a simple example that demonstrates the strength of GIS and the prominent role maps can play (Figure 1.11).

The first example deals with an issue in the Netherlands municipality of Maastricht. The first step in a geospatial analysis operation should be the definition of its objective and the conditions to adhere to. These conditions can include specific restrictions and constraints. In the Maastricht case, the municipal authorities wanted to know how the municipal forests had developed between 1950 and 2020. They wanted to have a map indicating obliterated, unchanged and new forests, as well as a table with the size of the forest parcels. One of the reasons for this analysis was to check whether a large private company in the municipality had adhered to the conditions agreed upon. The company had a concession to excavate marl in a quarry in the south of the municipality, but it had to plant new trees in those areas that had been affected by its operations.

The second step in an analysis is to prepare the geospatial data. Usually, not all data are available in a format that fits the requirements of the geospatial

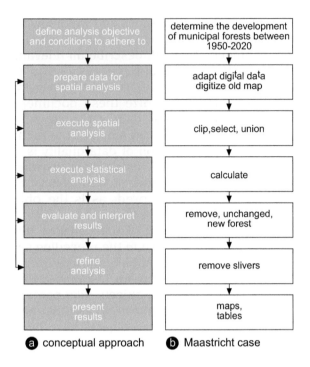

FIGURE 1.11 Spatial analysis: (a) conceptual approach; (b) Maastricht case

analysis. Some data have still to be collected in the field, whereas other data must be brought from external sources or brought in from other municipal departments, while there is a fair chance that some of the data are available on paper maps or in tables only. When available digitally, they could still be in a different coordinate system. Other problems that are likely to occur are the data being available on a different aggregation level and the density of coordinates in one set being too low to be compared with another data set; an example of the former problem is statistics being available at a neighbourhood level instead of at a street block level. The main GIS task at this stage is to make sure all data can be integrated and formatted so that they will be fit for use.

To answer its query, Maastricht municipality needed data on land use from both 1950 and 2020 to extract information on the forest parcels. The 2020 data were obtained from a database that contained data from the topographic map 1:25 000. The 2020 municipal boundary was taken from the large-scale map database available. However, it was too detailed to be used in conjunction with the 2020 topographic map data. It had to be generalized first. The 1950 data were not available in a digital form. The original topographic map had to be ordered from the municipal archives and had to be converted into the digital form. An interesting problem that occurred during the conversion was that part of the area for which the data were needed, within the 2020 municipal boundary, was not within the municipal boundaries on the 1950 map, because the area of Maastricht municipality had been enlarged considerably since, as can be seen in Figure 1.12a. The generalized 2020 boundary was used as a digitizing mask to extract the correct information from the 1950 map (see Figure 1.12b). The same 1990 boundary was used to select the land use data from the 2020 topographic map database. This database contained, in several layers, all the data of a topographic map sheet that covers a large area in the southern part of the Netherlands. From it, only data related to Maastricht were needed. Working with the whole database would slow down the execution of the operation because of the large amount of data.

The third step in the operation was the actual execution of the geospatial analysis. Most packages have all kinds of operations available, which somehow operate on the geometric and non-geometric components of the data. In general, one distinguishes three major types of geospatial analysis operations: overlay and buffer operations, network operations and surface operations. The first category often combines several data sets based on certain criteria, the second category uses an infrastructure network to find optimal paths, and the third category determines all kinds of (terrain) surface characteristics. Most packages allow the

combination of these operations. The Maastricht case was limited to a simple overlay operation. The data sets from 1950 to 2020 were combined, which revealed those forest parcels unchanged, those removed and the new ones. Figure 1.13 demonstrates this process. From the 1950 to 2020 land use data sets, the forests were selected to create two new data sets, forests 1950 and 2020. It was those two sets that were combined in the overlay operation. This calculated all possible intersections between the 1950 and 2020 forest parcels. In this process, all attributes were inherited and saved in the final data set. Based on these attributes, the map in Figure 1.13e could be drawn. The command 'draw all forest parcels with an attribute year 1950 and not year 2020' would result in all the parcels obliterated since 1950 being drawn.

The statistical analysis of the results is the step after the geospatial analysis. It is executed to fulfil the conditions set when the objective was determined. When, in the Maastricht case, one criterion would have been that the municipality is only interested in those parcels over a certain area or perimeter or when, in addition, it would like to know the average size of the new parcels, some basic statistics could be applied here. In complex geospatial analysis operations, it would be likely that more sophisticated statistical methods are needed. The next step in the analysis was the evaluation and interpretation of the results. In general, when executing a geospatial analysis, one has certain

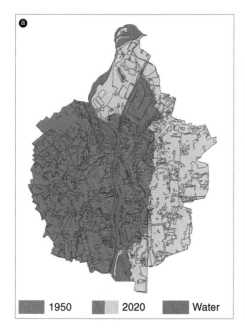

1950   2020   Water

FIGURE 1.12 Some of the available data in the Maastricht case: (a) land use data derived from a topographic database – the 1950 and 2020 municipal boundaries are given for reference; (b) a detail from the 1950 topographic map to be digitized (Kadaster Geo Informatie)

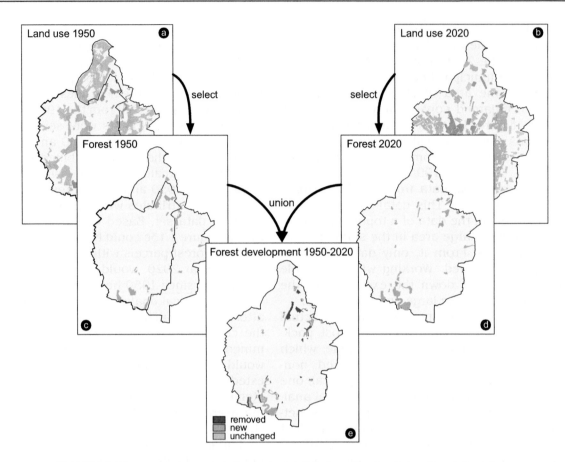

FIGURE 1.13 An overlay operation: (a) and (b) show the basic land use data from 1950 to 2020; (c) and (d) show the selected forest parcels from 1950 to 2020; (e) shows the overlay result: removed, unchanged and new forest parcels

expectations when there is some familiarity with the data. If the map revealed large new forest parcels in the city centre, it would be safe to conclude that something went wrong. If this were the case, one would have to correct certain steps in the analysis or refine analysis conditions.

Looking at the result in the map in Figure 1.13e, everything seems to be right. However, if one takes a closer look at the result, it can be seen that the result is not quite perfect. Figure 1.14a shows why. It reveals lots of small polygons indicating change. However, if the enlargement in the same figure, together with the two map details, is analysed (Figure 1.14a–c), it is obvious that in reality nothing has changed at all, although the GIS operation created 11 new polygons. These polygons are called 'sliver polygons'. A comparison of the basic statistics of the resultant data set with the original data set would have caused suspicion as well. The original data sets have 32 and 36 polygons, respectively, while the new data set has 100 mainly small polygons. The main reason for their occurrence is that the same feature in both the 1950 and 2020

data sets has a different geometry. The digitizing of the 1950 and 2020 maps was not performed by the same operator. Even if it had been the same operator, it would have been unlikely that the same points would have been selected during both digitizing sessions. For a problem like this, most GIS software packages offer simple solutions. One can delete all polygons with a size smaller than a certain threshold or calculate an average polygon boundary. Both approaches have disadvantages. But if the sliver polygons are not removed and the results are used in future geospatial analysis, the errors will propagate into the future results, and the database might grow unnecessarily.

During the collection of geospatial information, many types of error can be made: errors in measuring, classifying or categorizing data, localization errors, mistakes in data entry, etc. When these data are not directly incorporated into a GIS during the collection process, but are, for instance, mapped first because the new technology had not been applied yet, other kinds of error will emerge. Among these are generalization errors

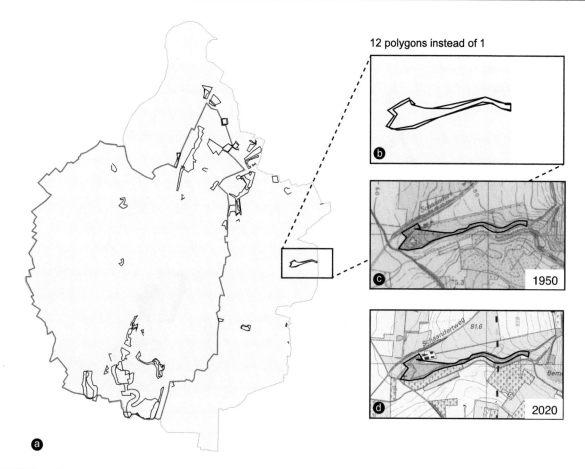

FIGURE 1.14 Overlay results and sliver polygons: (a) the forest parcels; (b) a detail: from 1 to 12 polygons; (c) and (d) the original 1950 parcels and the original 2020 parcels (Kadaster Geo Informatie)

or misrepresentations due to data amalgamation, reproduction errors and errors due to deformation of the printing paper. When these map data are subsequently digitized or scanned for input into the database, these errors are at least duplicated in it, but more probably the digitizing process itself will be another source of error as the Maastricht case demonstrates. At this moment, it is not quite clear, however, how these errors (for errors and their propagation, see Heuvelink, 2019) may affect the geospatial analysis results, i.e. whether they would lead to uncertainties in the results of analysis operations that would exceed some critical level. Not only are there errors in the input values, but also errors can be caused by analysis operations themselves and by the application-based models used. Examples are geospatial computational modelling techniques that forecast groundwater flow or polluted air diffusion and try to approximate reality but might in fact 'misrepresent' it. The combination of input error and these geospatial modelling techniques might lead to other error types. It is therefore very important to make sure that the data quality (i.e. suitability for specific applications – its fitness for use) is sufficient before basing decisions on maps that represent the results of geospatial operations executed on these data.

The results of a geospatial analysis operation are often presented in a report with maps, diagrams and tables to emphasize certain points or illustrate the conclusions. Most GIS packages do have a basic cartographic functionality to create the graphics. However, dedicated desktop packages have a more extended cartographic functionality and are better suited to producing the final maps. An example of these are the maps created by the municipality of Maastricht to illustrate its final report on the development of municipal forests. Some of these are shown in Figure 1.15. Included is a qualitative map that shows forest developments as well as a shaded topographic map with the forest development map draped over it to show relations with the terrain surface. Chapters 5 and 7 discuss the characteristics of the most existing map types.

It would be possible to execute the geospatial analysis discussed above without maps. For instance, one could ask for the area of forest stands in both years, compare them and answer the question without ever

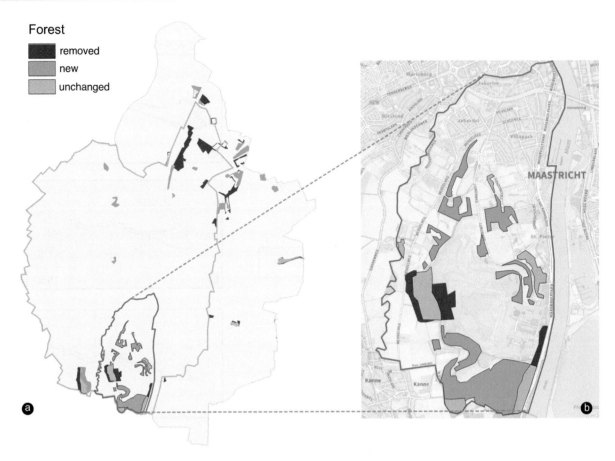

Forest
- ■ removed
- ▨ new
- ▨ unchanged

MAASTRICHT

ⓐ ⓑ

FIGURE 1.15 Presenting the results: (a) a qualitative display of forest developments in Maastricht; (b) relation between forest development and the terrain 1950–2020 (Source: J.W. van Aalst, www.opentopo.nl CC-BY)

visualizing them. But by doing so, one would deny oneself the opportunity of obtaining additional information from the process, as any geospatial trends or patterns in the answer may never come to light. The same size of forest stands in both years would not necessarily mean that they were at the same location. The information transfer without maps (e.g. through tables) would be more cumbersome as well. For instance, an increase is not necessarily due to the activities of the company active in the marl quarry, while the map would give a clear answer. It would be wise to apply the 'cartographic method', i.e. to visualize the geospatial relationships between the objects using abstraction techniques and the transformation based on the graphical grammar explained in Chapter 5; in other words, to map it. Observing the geospatial connections, relationships and patterns is only possible through the abstracting capacity of maps. As an example, another practical case study will be elaborated, showing the role that maps can play.

This case deals with the location of the TGV or high-speed train in the Netherlands 'Randstad' area when it was decided to extend the Paris–Brussels

TGV link to Amsterdam. Consequently, as the existing rail links were already overburdened with traffic, and as extra foundations had to be constructed because of its high speed, a new route for the rail link had to be selected, through the 'Green Heart' of the Randstad area, i.e. the non-urbanized centre of the urban agglomeration in the western Netherlands. The proposed route should spare the environment as much as possible (nesting birds should be disturbed as little as possible, and there should be no polluting influence on groundwater or on vegetation). Moreover, no valuable geoscientific monuments should be affected.

An environmental information system (EIS) built for the Netherlands was therefore consulted. This EIS contains data on soils, groundwater, vegetation and fauna and even on rare geological outcrops (geoscientific monuments). These data had been collected for the EIS on a grid-cell basis. For each grid cell (1 × 1 km), dominant soil types (by putting the grid over a soil map), dominant vegetation types, the number of different vegetation types found per grid cell, the total number of vegetation type units, the types of wild animals that occurred, etc. were ascertained. Because of the fact that this information was stored in the EIS, the effect of the proposed routes could be easily estimated. In order to select the best route from a number of alternatives (Figure 1.16a), the susceptibility of the soils to water-table lowering (Figure 1.16b), the susceptibility of mammals to fragmentation of their habitat (Figure 1.16c) and the effects of disturbances and pollution on bird life (Figure 1.16d) for all the affected grid cells were determined. Subsequently, the computer was used to calculate how many of these grid cells would be affected, and to what degree, for every proposed route. In other words, one could use the computer to define the sum of the environmental values, which, because of the construction of the TGV along the various routes, would be affected or nullified. This created the opportunity to select the route that would create the least damage.

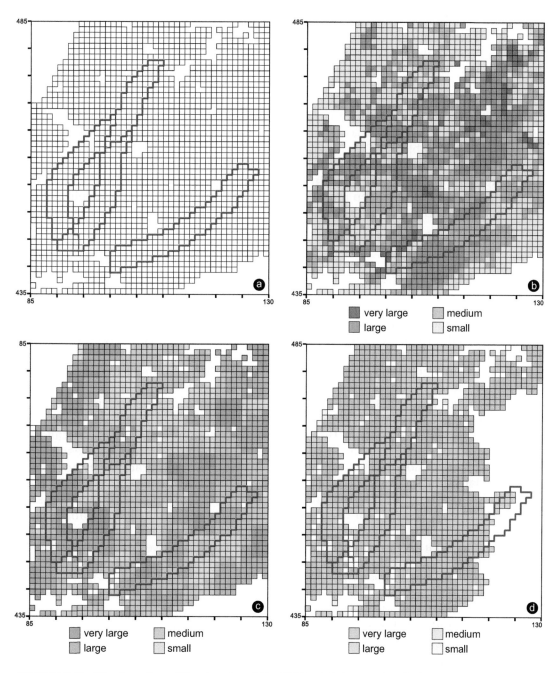

FIGURE 1.16 The location of the tracks of the TGV or high-speed train in the western part of the Netherlands: (a) alternative routes; (b) susceptibility to water-table lowering; (c) susceptibility of mammals to habitat fragmentation; (d) susceptibility of birds to traffic intensification

Digital maps are used outside of GIS as well, as will be discussed in Chapters 9 and 10. Electronic atlases are one example. Also called 'electronic atlas information systems', their function is less one of information processing than of answering specific questions, providing the support to integrate the answers in the mental map of the atlas user. This requires specific scenarios for a gradual immersion of the user into the new information environment. These atlas information systems can be extended to contain drawings, photographs, text and sound, and so become multimedia systems.

GISs are not yet well enough equipped to handle multitemporal information, and it is here that animated cartography comes in. Animation techniques are being developed that show the geospatial effects of developments at every stage. This presents extra potential for analysis and is one of the avenues for advanced data exploration that will, in future, also be available in a GIS. The potential for analysis is already greatly enhanced by the possibilities of applying GIS processing on the Web to all the data that have currently been made available there through the SDIs that have been constructed all over the world.

1.5 THE SPATIAL INFRASTRUCTURE AND MAPS

Companies and government departments at different levels (municipal, provincial/county, state or national) create and use information, for inventorying objects they administer or manage; for monitoring the state of the environment, or crime; for forecasting the weather, sales or schooling needs; or for reacting quickly to emergencies. In executing their day-to-day task, they offer part of this information to others and often need data from other organizations as well. Location is often the glue to link the different data sets.

Now this location component (street addresses, postcodes or ZIP codes, geographical coordinates or other geographical reference systems) allows one to combine the data from various information systems on the basis of common locations: it would be possible to link items from different files with the same postal address or coordinates. These possible linkages provide an enormous added value, but it would only be possible to realize this added value when a number of conditions have been met. Not only do the programs or packages used for the information systems have to be compatible, but also the structure of the files as well. When it comes to coordinate-based files, from

which, for instance, maps of the phenomena could be created, joint ellipsoids and projection systems would be a necessity as well. Similar resolution is a prerequisite and preferably the data should be surveyed in a similar time frame as well. There seems to be little point in combining population data from the 1990 census with sales data from a 1999 sales drive.

The stimulus for spatial infrastructures originates from the above need to exchange data smoothly and is based on the motto 'collect once, use many times'. A GDI can be defined as (Groot and McLaughlin, 2000) 'A set of institutional, technical and economical arrangements, to enhance the availability (access and use) for correct, up-to-date, fit-for-purpose and integrated geo-information, timely and at an affordable price, with the goals to support decision making processes related to countries' sustainable development'. In Europe, the GDI implementation is guided by the EU INSPIRE initiative, which, based on legislation, will implement the GDI concept throughout the European Union (http://inspire.jrc.ec.europa.eu/). In their guidelines, INSPIRE emphasizes several basic principles:

- Data stewardship and data security, meaning, e.g., that data should be collected once and maintained at the level where this can be done most effectively;
- Data accessibility and data interoperability must be possible to combine seamlessly spatial data from different sources and share it between many users and applications;
- Reusability and data synchronization are necessary: it must be possible for spatial data collected at one level to be shared between all different levels, e.g. detailed for those performing exhaustive investigations, but more general for strategic purposes;
- Data availability, e.g. spatial data needed for good governance at all levels, should be abundant and widely available under conditions that do not restrain its extensive use;
- Data discoverability, data validity and data rights, meaning that it must be easy to discover which spatial data is available that fits the needs for a particular use and under what conditions it can be acquired and used. By nature, an atlas has several relevant facilities for data discovery, such as an index for geographical names, a topical index and index maps;
- Data usability, meaning that spatial data must become easy to understand and interpret

because it can be visualized within the appropriate context and can be selected in a user-friendly way.

INSPIRE follows the standards established by the Open Geospatial Consortium (OGC – http://www.opengeospatial.org/). This is a non-profit, international, voluntary consensus standards organization that is leading the development of standards for geospatial and location-based services and has companies, government agencies and universities as members. The standards and protocols describe interfaces and encodings that allow the so-called geoservices to operate. This means that data providers can offer their products in a standard way.

In order to enable data users to find out whether data sets from different sources can be combined at all, clearing houses are being set up. Such clearing houses provide metadata on data files. Metadata ('data on data') refers to information on the quality, time frame, accuracy, fitness for use, lineage and completeness of the data and on the way the data have been collected. The major function of metadata is that it allows us to check whether specific data are available, suitable and accessible. One of the possible functions of a clearing house is to check whether spatial data are not collected more than once, by different government institutions, for example. If data sets are found, the Open Geospatial standards guarantee the data can be used.

For many years, the Web (we will use this term instead of the 'World Wide Web' or 'Internet') has been the medium to acquire and disseminate geospatial data. Today's Web-induced revolution has even further increased the number of people involved in making maps. Should mapmaking via GIS still involve geo-professionals, the Web includes potentially everyone having access to this new medium to create maps. This has led to potential situations where the organizations offering maps via the Web will never know what their map products will look like, because the mapmakers on the other end have so many interactive options (to render reality through symbols that represent feature or phenomenon categories).

In the process of acquiring and disseminating geospatial data, maps are often used to represent the data. Maps can disclose or reveal geospatial patterns and help to increase insight into geospatial relationships. However, a specific design approach is required because of the nature of the Web (see Chapter 5), especially since the Web is highly interactive and users expect maps to be clickable. This interactivity and the possibility to use the Web to link distributed databases also make web maps good instruments to explore the different databases.

Maps can play different roles too (see Figure 1.17). They can function as an interface to other (geo)-information in cyberspace or can guide the surfer navigating the Web. In providing access by clicking objects, the map will bring the surfer to other web pages, which again can contain maps, photographs or text. In this respect, the map could play a prominent role in a country's SDI as being part of a search machine. In its role of guiding the surfer through parts of cyberspace, maps are used as a metaphor to keep track of paths through, for instance, a single website.

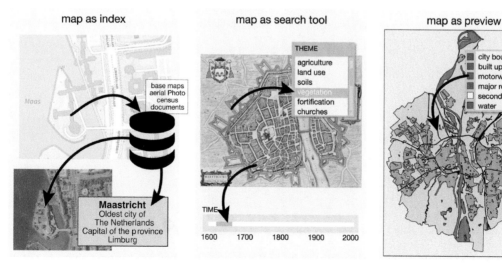

FIGURE 1.17 Maps used on the Web as index to other (geo)-information, as part of a search engine and preview of geospatial data offered

Why is the Web an interesting medium for maps? The answer is that information on the Web is virtually platform-independent. Also, many users can be reached at minimal costs and it is easy to update the maps frequently, although this last argument is only valid when the data provider is geared for it. Furthermore, the Web allows for a dynamic and interactive dissemination of geospatial data. This results in new mapping techniques as well as new possibilities for uses not seen before with traditional printed maps and most on-screen maps.

A true revolution was caused in 2005 by the introduction of the programs Google Earth and Google Maps. For everyone with Internet access, these programs were made available and could display satellite data and maps for free on a level of detail that was not heard of before. With an interface that is very intuitive in its operation, people could visit any location on Earth. Google Earth even allows three-dimensional flight through the landscape and cities. Not only could one visit places but it was also possible to add one's own data such as photographs, GPS (Global Positioning System) tracks and even maps to the Google environments to share these with others. This new but already very popular development was coined 'neo-geography' by Turner (2006). Figure 1.18 shows how Google Maps accommodates the sharing of collected photographs and other information. In Figure 1.19, it can be seen how users can drape their own maps on top of the Google Earth terrain and imagery data.

1.6 CARTOGRAPHIC EDUCATION

This book has been written to assist in carto-graphic education and intends, as a first objective, to provide an overview of the role that maps will play both today and in the near future in the world of geospatial data handling. To realize this, the nature of geospatial data is described as well as the characteristics of maps and the ways in which they can be put to use. Teaching map design is a second objective, which is shared out over top-ographic, statistical (or thematic) and temporal maps. Showing the manner in which maps func-tion, either independently or combined in atlases, and can be analysed and interpreted is the third main objective.

The field of cartography is vast, and this book can only provide a selection of all subject matters relevant for the discipline. As such, it provides a subset of the Cartographic Body of Knowledge (BoK), i.e. all knowledge relevant for the art, sci-ence and technology of producing or using maps (a BoK presents the scope, boundaries and structure of a discipline, and it helps to establish expecta-tions in professional practice and specialist skills (Fairbairn, 2017)). Currently, the International Cartographic Association (ICA) is engaged in the compilation of this Cartographic BoK. It has

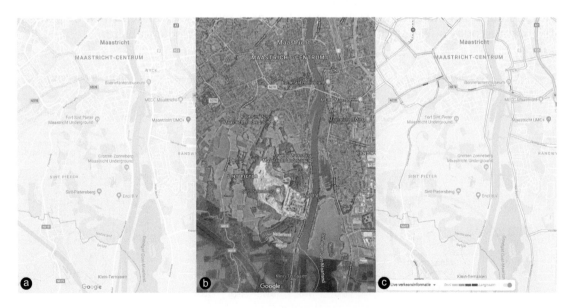

FIGURE 1.18 Google maps and several of its optional appearances: (a) topographic, (b) satellite image, (c) with traffic information (©2019 Google LLC, used with permission)

FIGURE 1.19 Google Earth: a three-dimensional view of part of the Lake District, south of Keswick and Derwent Water, with user-added data (©2019 Google LLC, used with permission)

instituted a special Working Group on Body of Knowledge on Cartography to realize this task. A closely related subject, Geographical Information Science and Technology, has already developed its own BoK, primarily directed to curriculum development (https://Gistbok.ucgis.org), and this BoK also contains a section on Cartography and Visualization, with learning objectives for some (2020) 32 subsections. For the time being, until ICA has completed its own BoK, this will serve as a well-informed substitute BoK for part of the field of cartography. The relevant ICA working group's endeavour is also based on the Strategic Plan of the International Cartographic Association 2019–2027, to be found https://icaci.org/files/documents/reference_docs/ICA_Strategic_Plan_2011-2019.pdf and on its research agenda for the coming years (https://icaci.org/files/documents/reference_docs/2009_ICA_ResearchAgenda.pdf). For those interested in pursuing a career in cartography, the items in this research agenda should be indicative for the direction we assume the field of cartography is moving in.

FURTHER READING

DiBiase, D. 2015. The nature of geographic information. An open geospatial textbook (E-education.psu.edu/natureofgeoinfo/) In State College: Department of Geography, Pennsylvania State University.

Groot, D., and J. McLaughlin, eds. 2000. *Geospatial data infrastructure – concepts, cases and good practice.* Oxford: Oxford University Press.

Heywood, I., S. Cornelius, and S. Carver. 1998. *An Introduction to Geographical Information Systems.* Harlow: Longman.

Kraak, M. J., and A. Brown, eds. 2000. *Web Cartography – Developments and Prospects.* London: Taylor & Francis.

Longley, P. A., M. F. Goodchild, D. J. Maguire, and D. W. Rhind. 2015. *Geographic Information Systems and Science.* 4th ed. Chichester: Wiley.

Rhind, D. 1998. *Framework for the World.* Cambridge: Geoinformation International.

Slocum, T. A., R. B. McMaster, F. C. Kessler, and H. H. Howard. 2008. *Thematic Cartography and Geovisualization.* 3th ed. Upper Saddle River, NJ: Pearson.

2

Data Acquisition

2.1 THE NEED TO KNOW ACQUISITION METHODS

In GISs, it is usual for many files to have been combined in order to boost the potential for analysis of the geospatial data. In an ideal situation, all the data combined have been collected, identified and measured on the same date, with the same geospatial resolution, according to identical procedures, and consecutively entered into the GIS using the same method. It is only then that the users would be sure of an adequate quality of the results of the analytical operations for which the files are being combined.

The reality is nowhere near this ideal: data are collected at different moments, are valid for different spans of time and have a different geospatial resolution; some might be collected in the field, while others were taken from existing maps that were generalized to an unknown degree. Some files might have been entered after they have been made compatible using some rubber-sheeting technique; others have been subjected to numerical transformations from other projections. Some might be based on random samples, while others on complete surveys. If it is numerical data collected at regular sample points, some might have been interpolated on the basis of linear, while others on the basis of geometric types of progression.

So at least the potential situation exists that the data, when compatible at face view, might not warrant the conclusions drawn from their analysis. It could be the case that the analyst or the GIS user in general should be warned about the results, for instance, that these should be interpreted with care. Traditional topographic maps used documentation helps, like reliability diagrams to show that navigation in certain areas on the basis of the map would be hazardous, as the producer could guarantee neither the accuracy nor the completeness of the data. In the more complex world of GIS, one needs numerical aids to be informed about the quality of the data files, in order to be able to decide on the validity of analysis results.

It should be kept in mind that in all these cases, one is collecting coordinates with which to describe locations of objects, with attributes the nature of which is determined either at the same time (e.g. during terrestrial topographic surveys) or later, in a laboratory (as is the case, e.g., for soil surveys), or through field checking (as is the case for remotely sensed data).

The various geospatial data acquisition methods used in GIScience (geographical information science) can be divided into the following types (see also Figure 2.1).

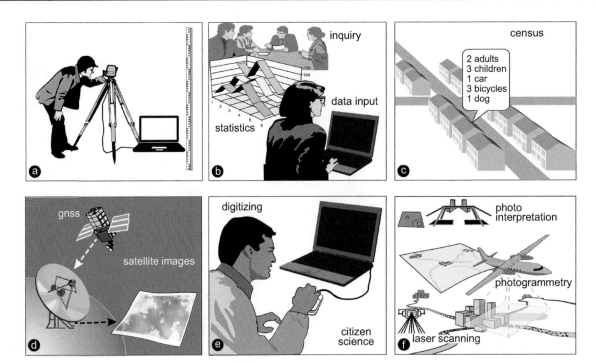

FIGURE 2.1 Various types of data acquisition methods: (a) surveying; (b) enquiries and statistics; (c) collecting census data; (d) remote sensing – GNSS (Global Navigation Satellite Systems); (e) digitizing maps – citizen science; (f) photogrammetry and laser scanning

2.1.1 Terrestrial Surveys

Large-scale topographic data can be acquired through terrestrial surveys. Such surveys immediately lead to digital files that can be imported into a GIS. When surveying new extensions for telephone companies' or cable companies' networks that dispose over digital files of their networks, surveyors would use 'total' stations with which the survey data are immediately edited and transformed into files that are extensions of existing files, and can be added to these main files at will.

2.1.2 Photogrammetrical Surveys

From aerial photographs produced from manned aircrafts or close-range cameras (drones), object coordinates can be determined in digital stereo-plotters and imported directly into information systems. The attribute information required could be determined either through interpretation or through field checking.

2.1.3 Lidar or Laser Altimetry

Lidar or laser scanning equipment sends out a beam of light which is returned by the surface it touches upon. The result is a large point cloud of data, which after corrections can be used to reconstruct the objects or the terrain. The scanners can be mounted on a tripod, a backpack or car for terrestrial surveys, or on plane, helicopter or drone for airborne surveys. GPS (Global Positioning System) equipment at all times keeps track of the location of the scanner to be able to determine the accuracy of the measurements that are within centimetre range.

2.1.4 Satellite Data

Satellites have been built that contain scanners with sensors susceptible to radiation emitted or reflected by the Earth's surface. These scanners operate in such a way that the sensors measure radiation sequentially from patches or grid cells along paths perpendicular to the line of flight. These radiation data are later put in their proper geospatial relationship, and by doing so, a map is simulated. Here, the data accuracy also depends on a number of correction techniques, both for the radiation values and for the geometric accuracy. After these corrections, the data frequently have to be resampled in order to fit specific grids, or to be comparable with other data sets. By being collected for grid cells (with, for instance, 5×5 or $20 \times 20\,m$ resolution), the data are generalized from the start.

2.1.5 GPS Data

A special set of satellite data is provided by the GPS, which, on the basis of 24 satellites, is able to pinpoint one's position three dimensionally with an accuracy of some centimetres. GPS is the name of the positioning system developed by the United States. In 2020, GPSs developed by the European Union (Galileo), Russia (GLONASS) and China (BeiDou) are also operational. GPS will become a generic term for positioning systems. These GPS recordings are used to increase the accuracy of existing georeferencing methods, or can be used directly in data surveys; they can be used for both point surveying and linear surveying. The data are recorded on the basis of a global reference system, which can be transformed to a local reference system.

2.1.6 Digitizing or Scanning Analogue Maps

Digitizing refers to the conversion of analogue images into digital representation. Initially, this was done manually by registering with cursor sequences of characteristic points belonging to lines on a map. Through this action, the coordinates of the positions touched were recorded digitally. Nowadays this conversion is mostly effectuated through scanning (see Section 2.4.3): optical records of the existence of specific colours at specific positions on a sheet are transformed into files with information on positions with attributes (hue or colour value). If digitizing or scanning could be effectuated with 100% correctness, the results would still depend on the accuracy of the original maps.

2.1.7 Using Existing Boundary Files

From open data domains, governmental data portals and commercial organizations boundary files (digital geometric descriptions of administrative units) can be acquired for applications that occur more frequently or for a larger number of uses: topographical files, files for car navigation, boundary files to be used in conjunction with statistical data (like for marketing applications). For these existing files, it is essential that they are compatible with one's software and therefore are based on the same standards as used by the buyer.

2.1.8 Socio-economic Statistical Files

National statistical services are increasingly publishing their (census) data online, with the appropriate software to query the files or even to visualize them in map form. These files are mostly presented in standard formats that make them compatible with most current mapping packages. One of the items contained in these socio-economic data files is the standard area code that relates the data to the appropriate areas for which they have been collected. This link between data and area should also be preserved when entering the data into other packages. It is through these code links that these packages will know what data to link to specific areas (see also Figure 6.48). On a global level, the United Nations Sustainable Development Goals indicator database provides current statistics (https://unstats.un.org/sdgs/indicators/database/).

2.1.9 (Geo)physical Data Files

Earth scientists have been producing data files since the late 1980s and were among the first to make them available to the public at large. It was to provide base maps for these global physical data files that the Digital Chart of the World (DCW) (see below) was first conceived (a base map is either a map showing topography that enables one to understand the location or distribution of a phenomenon (such as in Figure 3.7) or the (relatively) large-scale topographic map of an area from which all other, smaller-scale maps of that area are derived (such as in Figure 3.3)).

2.1.10 Environmental Data Files

Some examples are the Data Distribution Centre (DCC http://www.ipcc-data.org) of the Intergovernmental Panel on Climate Change (IPCC), which provides climate, socio-economic and environmental data, and the Corine Land Cover data from the European Union (https://land.copernicus. eu/pan-european/corine-land-cover).

2.1.11 Volunteered Geographical Information; Citizen Science

As fieldwork becomes increasingly expensive, mapping or nature conservation agencies become more and more dependent upon the information received from citizens that report on, e.g., changes in the topography or toponymy, and on the number of bird or vegetation species counted or found. For areas without detailed topographic maps, in emergency situations, volunteers might develop plans and maps during mapathons, sessions organized under the Missing Maps formula, to provide humanitarian agencies with the maps they need, on the basis of current satellite data and the information from local informants.

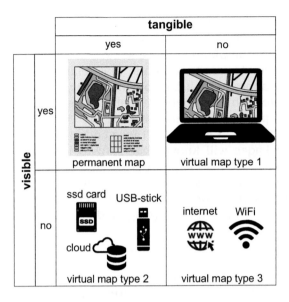

FIGURE 2.2 Analogue and virtual maps (after Moellering, 1983)

Before they are entered into the GIS, data from these sources are stored in different ways: some in the form of paper maps and some in the form of files (e.g. commercially produced files well protected by copyright (like boundary files) and published on online; remote sensing files; socio-economic, environmental or (geo)physical files; files of oceanographic surveys; photogrammetrically plotted data; or GPS data that are distributed mainly through the geodata infrastructure portals).

It is in relation to the various types of storage or procuring of geospatial data that one discerns between analogue maps and various types of virtual maps (after Moellering, 1983). Virtual maps of the first type can be seen, not touched: these are the maps made visible on monitor screens. Virtual maps that are not visible but tangible are the ones that can be downloaded from the Internet. These are dubbed virtual maps type 2. The third type of virtual maps is neither visible nor tangible and can be procured through Internet. As soon as one has queried them from the Internet, they can be viewed on a monitor screen (type 1) or printed as analogue maps (Figure 2.2). The virtual maps type 3 can be equated to the digital cartographic model (see Sections 1.2 and 2.5.1).

2.2 VECTOR FILE CHARACTERISTICS

Though actual file structures will be dealt with in Section 2.5, Sections 2.2 and 2.3 will focus on other aspects of files used as base maps (boundary files) for GIS maps that are available via portals. Before doing so, vector files and raster files have to be differentiated. In vector files, lines or boundaries between areas are defined by a series of point locations and their connecting links; in raster files, boundaries, or any other relevant background information, are defined as strings of picture elements (pixels) in regular grids that have been activated with specific values (see Figure 2.3).

To use these vector files as geographical reference frames for the geospatial information one wants to visualize on a monitor screen, one has to take account of the following aspects: resolution (the relation between the area as represented by a pixel and the same area in reality; see Section 2.3), digital scale and the possibility to separate the files into different categories or layers.

Vector files, to be used as boundaries, can be produced by scanning maps (see Section 2.4). This procedure concerns maps of a specific scale – of course, once the information is digitally available (through digitizing), scale seems to become less important issue, as it will be possible to zoom in or out and thus change the scale at will. But whenever the scale is increased beyond the original scale, the ensuing image will look poor and coarse, and cannot be used further for precise referencing. This is because the original map that was digitized will have been subjected to generalization (see Section 6.4). Apart from decreasing the scale, the projection of the map can be changed easily; as soon as the proper transformation formula from one projection to another has been determined, the digitized data can be displayed in any other projection system.

An early global product was the DCW (also called 'VMap0') (a data file containing all the linear elements (coastlines, rivers, contour lines, state boundaries, major roads and railways and cities) was produced mainly by the US military mapping agency National Geospatial-Intelligence Agency (NGA) in 1990 from Operational Navigation Charts (ONCs) at the scale 1:1 million). These ONCs are ultimately based in most cases on topographic maps 1:50 000 or 1:250 000, produced from aerial photographs. Both in producing the 1:50 000 maps from the photographs and in producing the 1:1 million charts from the 1:50 000 maps, generalization has been applied with its ensuing simplification, exaggeration, displacement and selection (see Chapter 6 for generalization). Objects that would still be retained at a scale 1:100 000 or 1:250 000 are omitted on the scale 1:1 million. So when the DCW is zoomed in on, and enlarged to a scale 1:100 000, it would not show all objects one would expect on this scale.

It would be otherwise if large objects and minor objects were stored in different layers or files and

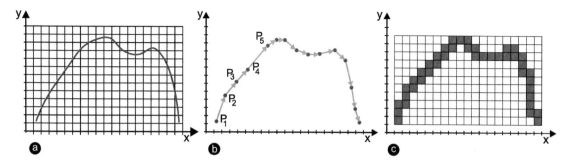

FIGURE 2.3 Representation of a line (a) in vector format (b) or raster format (c)

could be activated whenever a specific threshold scale value was passed: by zooming in beyond the scale 1:250 000, minor rivers could be activated and be made visible, etc. So it is important that (boundary) vector files have their objects stored in at least as many layers from which a selection can be made for display. An example of such a subdivision of objects into categories would be:

Hydrography: (a) Major rivers, lakes; (b) minor rivers, lakes.
Territorial boundaries: (a) National boundaries; (b) state/provincial boundaries; (c) county boundaries.
Coastlines: (a) Coastlines and major islands; (b) minor islands.
Administrative names: (a) Names of countries; (b) names of states/provinces; (c) names of counties/departments.

The various administrative areas should be supplied with the codes assigned to them, for the purpose of being able to match them to the statistical files (see Section 2.1).

This approach is currently followed by global map products such as OpenStreetMap, Google Maps and Bing Maps, which have organized their maps in 20 tiled layers, each with content appropriate for that particular scale level. Tiles are pre-rendered maps which speed up the map rendering process of user's area of interest. However, most of these are raster tiles and only useful as base map and not as layer to be used in a spatial analysis.

The DCW project has been replaced by Global Map, an initiative led since 1996 by Japan, to make available topographic files on a country basis with a similar resolution as DCW but more recent data and a more extensive thematic coverage. These geographical data sets are composed of the following thematic layers: elevation, vegetation, land cover, land use, transportation, drainage systems, boundaries and population centres. The major reason to develop this global data set was to enable the monitoring of changes in the global environment. After being transferred to the United Nations, the project ended in 2018. The original DCW files can still be downloaded from the Pennsylvania State University library site (https://psu.box.com/v/dcw) or be obtained from ESRI, be it in an updated form.

Other examples of vector files are EuroGeographics' EuroBoundaryMap files (see Section 6.6.2) and the Electronic Chart Display and Information System (ECDIS). Member states of the International Hydrographic Organization (IHO) developed regional databases for ECDIS (using a Digital Data Transfer Standard (S-57), with a common object code and exchange format (DX-90)). Though preference is given to the production of regional databases covering all the routes used by international shipping, this will take time. Such data has not been generally available worldwide before 2002. In the meantime, hydrographic chart raster files, like the one developed by the United Kingdom, will be used.

2.3 RASTER FILE CHARACTERISTICS

In a scanner, optical records of the existence of specific dots, dashes or patches in black and white or colour at specific positions are transformed into files with information on positions with attributes. As the registration of these dots, dashes and patches takes place along regular parallel scanning paths, in incremental temporal steps, the output of these files consists of regular grids, built up from picture elements (pixels), each representing specific attributes (see Figure 2.3c). The higher the resolution of the scanning device (and its price), the smaller the pixels will be, and the more detail of the original image will be rendered.

To revert to the case at the end of Section 2.2, existing hydrographic charts are now being scanned and stored digitally, to provide charts in a form that can already be used in ECDIS, in order to bridge the gap before they will be made available in vector format. The first to do so was the British Hydrographic Office, with its ARCS (Admiralty Raster Chart Service) package. ARCS consists of DVD files produced by scanning existing charts, with an update option that allows one to combine data from weekly DVD notices to mariners with the original raster chart to produce new charts on screen that are fully corrected. Figure 2.4. shows a detail of the OpenNauticalChart (http://opennauticalchart.org/).

The resulting charts when displayed on the monitor screen can be overlaid with radar images and also visualize the ship's position as well as data on its course, speed and planned track. The electronic navigation systems using ARCS DVDs will allow for the selection of the area to be viewed, and also allow for zooming in or out and for adding or omitting a number of additional data layers. When the information contents of a paper map are available on different films that can be scanned separately, the raster file can consist of different layers – in this case with contour lines, names, buoys, bathymetric colours, etc. that can be activated either separately or together.

Apart from updating, the raster image itself cannot be queried or otherwise electronically analysed – it forms a backdrop picture against which geospatial processes can be visualized. Important aspects of these backdrop raster files are the same as for the vector files: the scales at which they have been scanned and the existence of information in different layers.

Another example of such a raster file are the raster versions of the topographic map series of the many countries. These files, available online, contain all the information of the paper topographic maps. These maps can be used as georeferenced background in GIS or in apps on mobile devices for leisure purposes. It is for products like

FIGURE 2.4 OpenNauticalChart (http:// OpenNauticalChart.org/ (CC-SA))

this one that the resolution is important: scanners (see Section 2.4) register with a specific number of dots per inch, indicating the smallest areal units for which information is being determined. A resolution of 250 dpi (dots per inch), e.g., refers to the fact that areas with a size of $0.01\,mm^2$ are represented independently (see Figure 2.5). Equally important as the geometrical resolution is the radiometric one. It refers to the number of colours that can be differentiated between by the scanner and to its display capabilities as well.

In contrast to vector files, transforming raster files to other projections is very difficult. Coordinate systems with which the images are overlaid cannot easily be changed for other systems. Additionally, the pixel is the basic unit of the image structure, and it might not be relevant as a reference unit for the theme mapped. A pixel is a representation of the smallest area where electromagnetic radiation is collected from individually. The larger these areas, the less information there will be to store and the easier it will be to handle the resultant files. On the other hand, the smaller these areas on the Earth's surface, the more accurately the resulting raster image will model the original data.

Satellite imagery is made available as raster files. Resolution here is not a function of the scanner measurement device's size but of the sensor's field of view: by collecting the radiation within a specific angle and from a specific distance from the Earth, the sensor registers and measures radiation from a specific patch of the Earth's surface. The size of this patch, the dominant radiation of which is being registered, will determine the usefulness of the imagery for constituting a model of the objects whose radiation values are being registered. These element areas on Earth, represented as pixels on the satellite imagery, have decreased in size since the first civil satellites began sending their information to the Earth in the 1970s: Landsat registered $80 \times 80\,m^2$ first and later (Landsat Thematic Mapper) $30 \times 30\,m^2$; SPOT in 1986 $20 \times 20\,m$ and even $10 \times 10\,m$ for its 'panchromatic' applications. Today, resolution goes beyond the one-metre resolution. Many countries as well as commercial organizations have satellite programmes. Some of these offer the imagery for free like the Copernicus programme of the European Union with their Sentinel satellite family. Up-to-date information on satellites, and their sensors and resolution, orbit height and repeat cycles can be found at ITC's satellite sensor database: https://www. itc.nl/research/research-facilities/labs-resources/ satellite-sensor-database/.

When one compares raster files with vector files, the latter have a greater overall resolution. This is

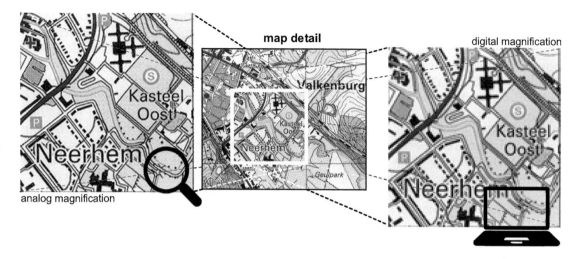

FIGURE 2.5 Tenfold enlargement of photographs taken from Top25 Raster (the raster-based screen representation of the topographic map of the Netherlands at scale 1:25 000 (right)) next to a tenfold enlargement of the analogue version of this map (left) (Kadaster Geo Informatie)

because in the raster technology, the input device provides the information divided into discrete pieces ('pixels') with a finite size. When the pixels are large, much information is lost. On the other hand, this grid structure of satellite and scanned images allows for analytical operations that are much easier and much less time-consuming than is valid for vector images. It also provides for relatively easy combination with other files, as well as for a multitude of image processing possibilities. As raster technology is also behind the monitor screens, vector images are simulated in reality on these screens, being built up from activated raster cells.

2.4 DERIVING DATA FROM EXISTING MAPS

The technical procedures for deriving data from existing maps will be described in Chapter 5. In this section, the requirements that existing maps have to answer to, in order to be suitable sources for digital files, will be indicated. In particular, it is documentary aspects that are important here. Such a plethora of paper maps has been produced that finding the proper one can sometimes be extremely difficult. Furthermore, the organization and aims will be covered, while the principles of the hardware used will be described. The data to be derived from existing maps are complex in the sense that they have both locations and (other) attributes. So as a first requirement, these locations and attributes should be unambiguous on the source documents: the definitions used or implied in the map legend should be clear and consistent, the period of time during which the data

were gathered should be mentioned (in order to be copied and stored somewhere in the digital files as well). The data quality aspect is important and will be covered in the next section.

2.4.1 Finding the Proper Map: Documentation

Finding the proper map will depend on the selection of a map of the proper area, the proper theme and the proper time period in which the data have been collected. Most map libraries will contain map descriptions in their catalogues that refer to the area and theme of a map, and also have data on the map's date of publication. They will rarely have information in the catalogue files on the time period, in which the data for the maps have been collected, i.e. of the period for which the map is valid. For topographical maps, this is used to be about 2 years before the map has been published; for thematic maps, this might take longer.

2.4.2 Preparation

Finding out about the way in which locations have been indicated on the paper map is essential, as this will be a guide to the manner in which the information to be digitized or scanned will have to be transformed in order to fit in one's GIS. When preparing an existing map for scanning, the method according to which the data are geospatially referenced should be taken account of. This georeferencing (providing a geospatial address) can have been done by using either geographical coordinates, grid coordinates or no coordinates at all. In the latter case, the map cannot be further fitted into or integrated with existing files, unless by rubber sheeting (see Section 6.3). Using geographical coordinates, the Earth is regarded as a sphere; however, using grid coordinates, the area represented is regarded as a flat plane (see Section 6.1). The UTM (Universal Transverse Mercator) grid coordinate system combines the latter two in a sense, as it consists of a number of planes that together cover the Earth (see Section 6.2).

If, e.g., a file digitized at scale 1:250 000 of North-west Europe has to be extended or updated, topographic maps 1:50 000 of the Netherlands have to be digitized. This means that all coordinates implied in the file produced when digitizing the map will have to be transformed from the stereographic azimuthal projection in which the maps of the Netherlands are rendered to the UTM projection of the 1:250 000 file. Digital files are scaleless in principle, but their resolution is still governed by the scale at which the data have been digitized. Now, in order not to have discrepancies in the resolution, for the data from these maps of the Netherlands to fit in the new 1:250 000 North-west Europe file, they have to be generalized as well (see Figure 2.6).

2.4.3 Digitizing

As indicated in Section 2.1, digitizing refers to the conversion of analogue images into digital representation. The main means of converting analogue data into digital data are scanners. Before scanners were used, manual digitizers were employed, consisting of a tablet in which wires were embedded, located along Cartesian axes. A cursor was linked to the tablet and, when a cursor button was pushed somewhere on the tablet, the electrical charge generated was picked up by those wires directly underneath in the tablet, and the wires activated would then provide their specific codes

as x- and y-coordinates. Additional buttons on the cursor would allow one to join attribute information to the locational information. For entering topological information (information about logical relationships between geospatial objects, such as about relative positions, adjacency and connectivity) such as stating whether a node belongs to a specific line, or is the beginning or end point of a line or is located somewhere in between, specific areas (called a 'menu') on the digitizing tablet could be activated and used as legend boxes: by pressing these areas by hand or button, the link between the digitized location and specific attribute information was forged and entered into the computer memory.

This hardware is hardly used any more, but it still serves well in order to discuss what the main characteristics are one should want to hold on to during this process of converting geospatial information from analogue into computer-readable form. It is to preserve the relationships that were visualized on the map. This means, e.g., (as shown in Figure 2.7) that

Existing links between points on the map should be retained in the digital file;
Parallel lines should remain parallel;
Relative locations should be preserved;
Absolute locations (as expressed in coordinates) should be preserved;
Adjacency should be preserved;
Lines that merely touch each other should not intersect each other.

Many digital files in current use have still been based on manual digitizing, and this might have resulted in the kind of mistakes as described above. These are avoided by using scanners (see Figure 2.8a). The operating principle of high-resolution scanners can be that of a rotating drum to which a paper

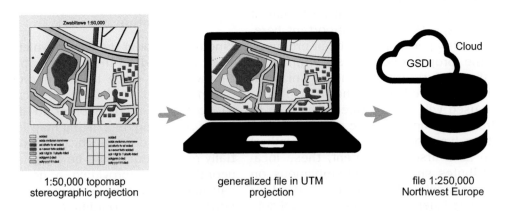

| 1:50,000 topomap stereographic projection | generalized file in UTM projection | file 1:250,000 Northwest Europe |

FIGURE 2.6 Conversion of an analogue map into a digital file

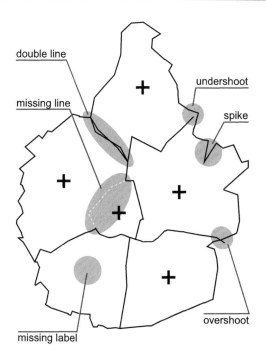

FIGURE 2.7 Potential errors as a result of manual digitizing

The working principle of low -to medium-resolution table scanners used in desktop environments is that of registering the characteristics of an image on a page, put facedown on an A4 or A3 tablet, line by line (Figure 2.8b). All data points or pixels on a line are registered sequentially before moving to the next line. It subdivides images into discrete data points in which the sensor (ccd = charge-coupled device = electronic light sensor) measures their light value or colour value. In this way, an analogue map is turned into a description consisting of geospatial addresses (grid cells) and their characteristics (light values).

This is the easy part of scanning. After scanning, there is still the operation to make sense of the scanned data, and to change it back into a vector file, if necessary. If possible, it is not the printed maps themselves that are scanned, but the original colour separates prepared for the map's reproduction. On these, there is already a separation of functions that should then be further edited. On scans from the separates to be printed in blue, one should indicate whether lines refer to rivers, coastlines or lake shores; codes can be added to the rivers that should indicate their importance; this is relevant should the scanned image subsequently be generalized and represented on a smaller scale.

The currently (2020) optimal procedure is considered on-screen digitizing. It is a combination of scanning and digitizing. For this procedure, the map as a whole is scanned first and displayed and enlarged on a monitor screen. The lines to be scanned can be highlighted, and the operators will then follow/trace and enter with the cursor the relevant lines or point locations on the screen.

map has been attached. While rotating, a sensor will scan the map in narrow (e.g. 0.1 mm wide) contiguous bands and will register the light intensity – or colour value – of small squares (0.01 mm²). These light measurements are then transformed to digital values, and can be digitally represented on screen, thus reconstituting the scanned original. By splicing the signal picked up by the sensor over measurement devices susceptible to red, green or blue light, the original colours can be reconstituted.

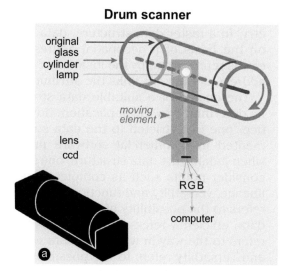
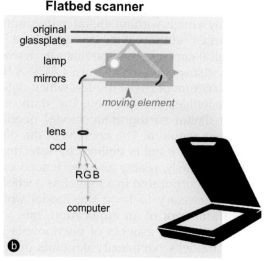

FIGURE 2.8 Working principle of a scanner: (a) drum scanner; (b) flatbed scanner

FIGURE 2.9 On-screen or heads-up digitizing

The parts of the lines entered will be displayed in a different colour; the image can switch between that of the scanned map, partly digitized, and that of the digitized information only, allowing the operator to better check the consistency of work done (see Figure 2.9). This approach is also used by volunteers who want to update map data for humanitarian purposes after disasters such as hurricanes, flooding or earthquakes (in the so-called Missing Maps projects). The volunteers use recent satellite imagery to, for instance, indicate where the damage is most severe, and the new maps, often part of the OpenStreetMap, can be used to restore the basic infrastructure (see also Section 6.6.2).

2.5 WORKING WITH DIGITAL DATA

2.5.1 Modelling the World

Producing today's maps without digital cartographic models is almost unthinkable. As explained in Chapter 1, digital cartographic models are derived from digital landscape models (Figure 1.4). It is the content and structure of these models which determine the possibilities for querying the data and for defining a digital cartographic model needed to draw maps as required. The content of the digital landscape model itself is defined by selections from reality. Obviously, reality as experienced outside cannot be incorporated in a model as a whole. Selectivity is necessary to keep the model workable. In the framework of an application, one will try to process as many aspects of phenomena as possible. The model's complexity depends on the nature of the application and the intended manipulation in the GIS database or the maps required.

How an application influences the contents of a digital landscape model can be illustrated when considering the road concept. To an environmentalist and a traffic manager, a road seems to be the same object. However, the two viewpoints may differ considerably. An environmentalist will look at the road as a barrier to wildlife migration patterns. From this perspective, he or she wants to know its width, how busy it is, the width of the verge, whether there are any crash barriers, the level of noise pollution, etc. A traffic manager will look at the road from a safety and transport perspective. Questions he or she will have are related to the capacity of the road, the number of accidents, traffic lights, flyovers, etc.

Figure 2.10 depicts the modelling process. The illustration is a broadening of Figure 1.3. The steps in the modelling process correspond with the approach suggested by Peuquet (1984). Next to reality, where geographical objects and their characteristics can be found, she distinguishes a data model, a data structure and a storage structure. For each application, selections are made which adhere to conditions defined in a data model. This data model is a conceptualization of reality without conventions or restrictions regarding its implementation. It contains a defined set of geographical objects and their relationships. The example data model comprises the six districts of Cumbria (Figure 2.10a). The next step in the modelling process is to structure the data. In GIS, this implies the representation of the data model in a vector or raster data structure as illustrated in Figure 2.10b. In a vector data structure, the data are organized according to the objects. Geometric characteristics of the data are represented by sets of coordinates, which, in the map image, are connected by lines (the vectors). Labels link the attributes to the geometry. In a raster data structure, data are organized on the basis of a geospatial address. The geometry is represented by the location of grid cells. The address of the cells links the attributes.

The choice of a suitable data structure should be determined by the application. However, in practice, one is restricted to the data structure implemented in commercial software packages. Still, when judging the data structure, one should at least consider points such as completeness, efficiency, lineage, versatility and functionality. Completeness refers to the possibility of representing all selected data, efficiency refers to data accessibility, lineage refers to the way in which the data were collected, and versatility refers to the possibility of adapting the structure to new circumstances. Functionality refers to the operations that can be executed with

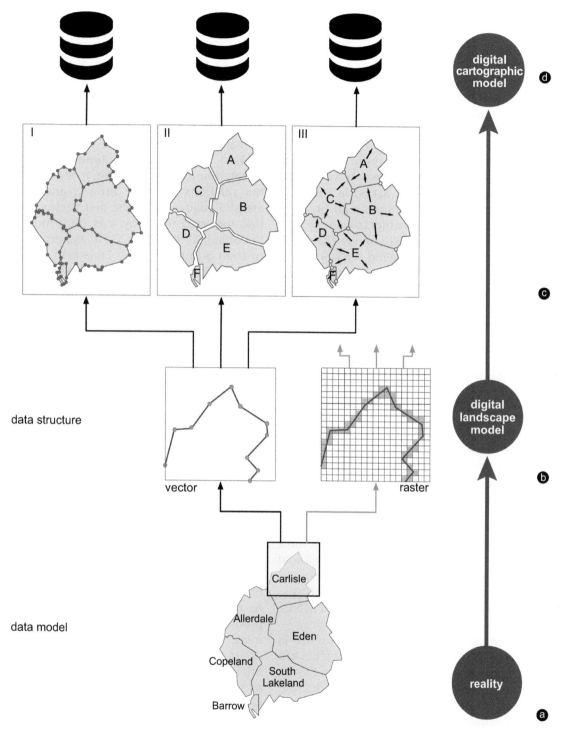

FIGURE 2.10 Organizing geospatial data: (a) selections from reality are based on a data model; (b) the selected data are often structured in a vector or raster format (digital landscape model); (c) the nature of the data structure defines the query level; (d) determination of what maps can be drawn (digital cartographic model)

the data; in other words, it refers to the kind of query that can be answered by the structure. The query level will define the structure's suitability for certain tasks. This is illustrated in Figure 2.10c by three examples of some possible vector data structures in relation to their cartographic use. Map I represents just the individual lines, map II represents the areas as well, and in map III, lines, areas and topology (i.e. mutual relationships) are also known to the system. If one would like to have a simple map with just the outline of the districts, all three data structures will function. If the map to be drawn is based on the request 'draw only the largest and smallest area', the structure represented in map I will fail. However, the structure represented by map II will also fail when the request would be 'draw only those areas bounded by district C'. The type of request will, for raster and vector structures, influence the response times.

The nature of the data to be represented will strongly influence the choice between vector or raster data structures. If one is active in the field of utilities, the vector approach seems an obvious choice, because of the type of objects one is dealing with (e.g. pipes, networks). Whenever the organization depends on remote sensing data, a raster data structure is advisable, because it is suitable for both interpreted and uninterpreted data. It should also be realized that a vector data structure is only suitable for data that have been interpreted fully.

2.5.2 Vector Approach

The vector structure is one of the oldest structures in use. This is partly because the vector approach is close to the traditional cartographic drafting techniques. Look at an analogue map and it is possible to 'see' lines constructed from nodes and arcs. Another reason is the limited computer technology available at that time. The small computer memories were unable to deal with the vast amount of data involved with raster structures, while vector data structures are relatively small. The basic unit of the vector data structure is the geographical object. Several kinds of vector data structure exist. For an extensive elaboration on these structures, see Laurini and Thompson (1992). Figure 2.11 illustrates two of these structures.

A non-topological or spaghetti data structure is the simplest type of vector structure. All objects are defined as single items. As can be seen in Figure 2.11a, the line between points 6 and 7 is defined by a set of coordinates $(x6, y6)$ and $(x7, y7)$ and is labelled 37. However, the data structure does not include any reference to other objects, such as lines 38 and 34, which are connected to line 37. No reference is made to the areas bounded by line 37. It is similar to map I in Figure 2.10c, and it does not refer to any geospatial relationships between the objects defined in the structure. It is unable to answer questions that are not related to drawing its content. Checking its consistency can only be done visually. This approach was introduced at the beginning of the 1970s.

More advanced are those vector data structures that contain topology. Topology defines the mutual relations between geospatial objects and can be used to check consistency among point, line and area objects, or help in finding answers to more complex queries. Topology can also be described as the highest level of generalization possible. Graphically, this can result in different images of the same area, as can be seen in Figure 2.11b. Area C in Figure 2.11b I has its 'natural' boundaries. In II and III, these boundaries have been strongly generalized, but relations between area C and its neighbours are still valid. One of the first vector data structures that included topology was Dual Independent Map Encoding (DIME). This DIME system was developed by the US Bureau of the Census to deal with census data. An important difference to the spaghetti structure is the incorporation of geospatial relationships between the objects registered by the structure. In the 1980s, the Bureau of the Census replaced DIME with the more advanced TIGER (Topologically Integrated Geographic Encoding and Referencing) system.

More complex and advanced is the georelational data structure. Similar to topology, it includes links to a database system containing attribute data. This results in an efficient and flexible structure, and can be found in many of today's GISs, among them ArcGIS. Figure 2.11c shows the principles of this approach. The basic unit is the line segment. From each segment, the beginning and end points (the 'from' node and the 'to' node, respectively) are known. Those two points give direction to the segment and make it possible to define a left and a right area. In Figure 2.11c, the line labelled 17 has point 8 as a begin node and point 6 as an end node. The left and right areas are OUT and A, respectively. When applicable, the number of points between both nodes is registered. The segment labels can be used as pointers to refer to other tables with information on points, areas or attributes. They function as a link to the database as well. The georelational data structure allows for flexible search operations, area aggregation, linking attribute data and consistency checking. The popular file formats GeoJSON and ESRI's shapefiles are kind of georelational data structure, but not quite.

2.5.3 Raster Approach

Since 1990, the use of raster data structures has increased. Although it is difficult to code during input, the speed of the scanner, as well as that of output equipment such as laser and electrostatic

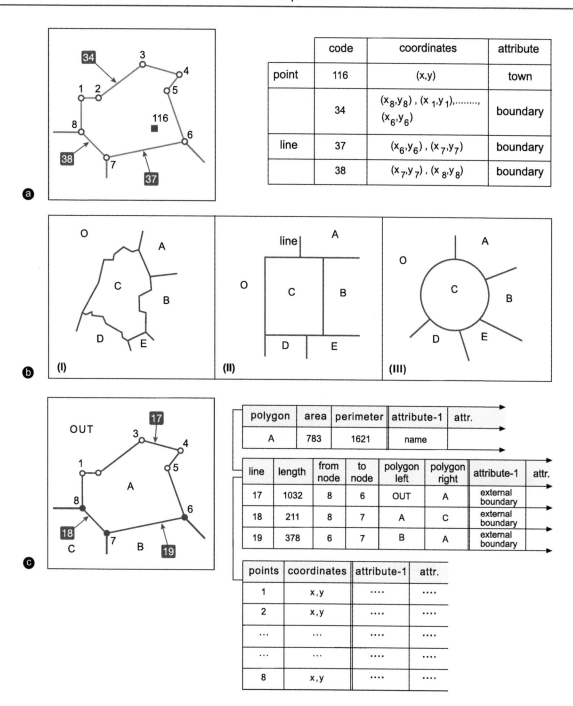

FIGURE 2.11 Vector data structures: (a) a spaghetti structure; (b) topology; (c) a georelational structure

plotters, offers advantages. Physical storage is simple, and stored data are easily accessible. A wealth of processing techniques is available from the remote sensing and image processing disciplines. In the GIS environment, this has led to the development of raster modelling techniques (Tomlin, 1990).

In the raster approach, geospatial units function as basic reference units, instead of as geographical objects (as is the case with the vector data structure). Squares are most common (Samet, 1989) since manipulations with squares are easily performed by the hardware (e.g. pixel on a screen or paper). Figure 2.12a shows how a geographical feature is registered in a simple raster structure. The grey squares define Cumbria, and the white squares the area outside Cumbria. Introducing more grey values allows for the registration of more different objects (for instance, different grey values for each of Cumbria's districts). However, it is not possible to register more than one attribute for each square. If there is a need to have more attributes, which occurs especially in a GIS environment, one has to store more raster layers. The size of the squares (the resolution) will define how well a raster structure can represent geographical reality. A small size will give a better representation, but will result in a very large data set. Sometimes the data source defines the resolution. An example is the resolution of the satellite's scanner, which sets conditions for data collected by remote sensing techniques.

Another advantage of the square above a triangle or a hexagon is that it can be split into subunits of the same shape and orientation (Figure 2.12b).

The quadtree data structure is based on this approach and is used in GIS packages. Considering the quadtree, the whole map is seen as a square and is subdivided into four smaller squares (see A, B, C and D in Figure 2.12c). Splitting the squares continues until each single square has a homogeneous content. In practice, this goes down to six or seven levels. The example in Figure 2.12c is four levels deep. The tree below the map image shows how it is done. Each solid circle represents Cumbria, and each open circle the area outside Cumbria. When no branches leave a circle, it is considered homogeneous (black or white). Several variations on this basic approach exist.

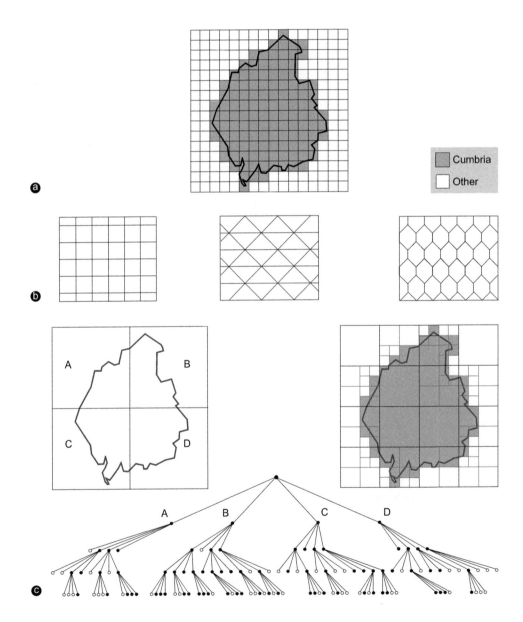

FIGURE 2.12 Raster data structures: (a) normal; (b) basic raster types; (c) quadtree

2.5.4 Hybrid Use of the Database

Both raster and vector data structures are used via database management systems. A database management system often has one of the following structures: hierarchical, network or relational (Figure 2.13). The first has a fixed tree-like structure, and questions can only be asked along the tree's branches, as can be seen in Figure 2.13a. Here, Cumbria is divided into districts, each of which is further subdivided into wards. Because of its fixed structure, it has relatively short response times to queries. It is applied mainly in those information systems with management tasks. Here, the nature of the questions is known beforehand.

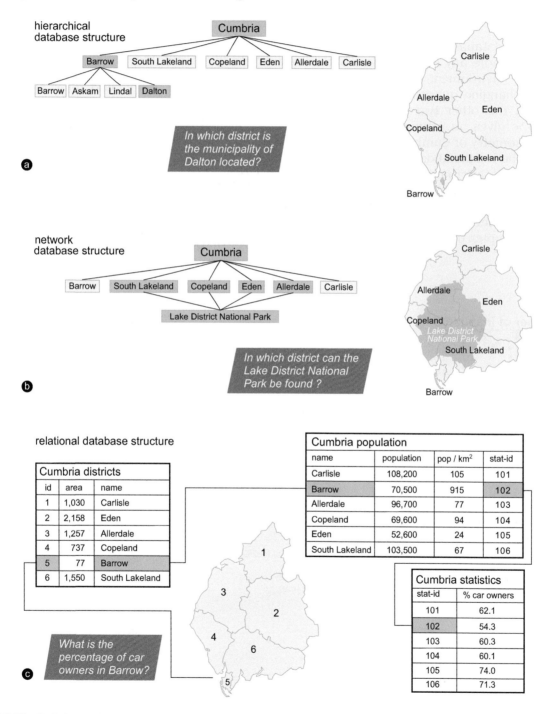

FIGURE 2.13 Database management systems: (a) a hierarchical structure allows for fixed queries along the tree's branches; (b) a network structure offers greater flexibility and allows a combination of items on the same level; (c) relational structures are most flexible and allow any kind of question on the data stored in related tables

When a more flexible approach is needed during the query process, a network structure can be used. As Figure 2.13b demonstrates, this structure is not limited to queries along branches. Database elements of the same level, like the districts in the figure, can be combined in the query. These databases can be found at utility companies. However, as with hierarchical databases, the type of question is fixed at the moment one defines the database structure.

A relational database structure is even more flexible. It can handle any kind of query and is very useful in an environment with unpredictable and constantly changing queries. However, it is often slower than the other two database structures. Figure 2.13c gives an example. In the GIS database, the geometry of Cumbria's districts is stored in a table. The relational database principle allows one to link this information with any other table when a common variable is available. This figure shows the district's name and its statistical identifier. The links between the different tables allow one to ask complex questions.

2.6 CONTROL AND ACCURACY

The advent of GIS has accelerated and simplified the process of information extraction and communication. Combining or even integrating various data sets was made possible on a large scale. The ease with which operations can be effected provides a danger as well, as technical possibilities will also allow for irrelevant or inconsistent data integration. On the other hand, the present storage potential in the cloud (storing and accessing programs and data in the Internet instead of in one's computer) will allow one to store the original data and not the derived or aggregated data. Figure 2.14 provides an example of the difference between using original and aggregated data. Here, population distribution data (visualized as a dot map, showing locations of specific numbers of inhabitants) and population density data (visualized as a choropleth map, showing densities for enumeration areas) are both used as a starting point for an analysis of the average distance the inhabitants of a region have to walk in order to reach specific municipal facilities. Therefore, both data sets are combined with

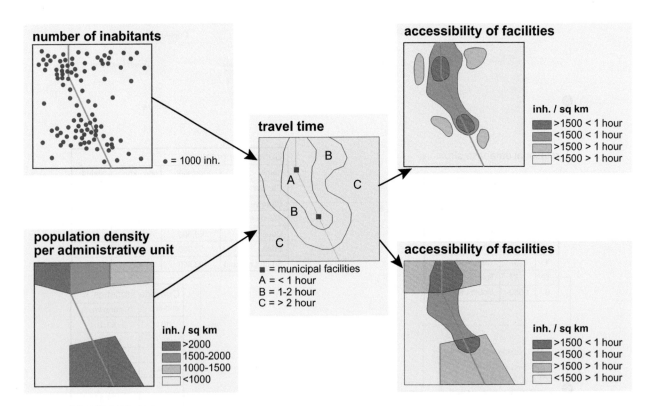

FIGURE 2.14 Results of data integration. If the aggregated data set is used (bottom), the densities of the population able to reach the facilities within a specific time will be visualized differently as compared with the map based on distribution data (top) (after Hootsmans and Van der Wel, 1992)

data on the average travelling time (here expressed as an isochrone map). If the data set visualized by the map at the lower left is used, a distinctly different image is provided when compared to the image which results from starting with the data set visualized at the upper left. If decisions about siting new facilities are based on either of these maps, the decision-maker should be able to ascertain the quality of both maps.

The end product of every type of data integration will inevitably have a certain degree of uncertainty, because of mistakes in the original data and because of the data processing. The result will only be fit for use when a certain level of reliability is reached. That is why the user has to be presented with information on the quality of the original data and of the various processing steps. Users might depend on the data, in which the end product quality is determined by the available data, or the user might require end products of a specific quality, which will determine the data quality of the original data to be collected and of the processing methods to be followed.

Traditionally in cartography, the quality of the original data was expressed in the form of reliability diagrams. These consisted of statements that described the manner in which the data for specific parts of the area mapped had been collected, and did not differentiate within these parts (see Figure 2.16b). Thematic equivalents of these topographic procedures were statements such as 'soil samples have been taken for every hectare' and 'map unit boundaries should be accurate within ± 100 m'. But for correct decision support, data quality information is not only needed on an overall level; local/regional deviations from the overall accuracy have to be provided as well. Overall accuracy of information will serve to relativize the importance of data sets for decision-making, while local anomalies in accuracy will be used to restrict the decision-making process to specific areas.

In the literature (Guptill and Morrison, 1995), the following aspects of data quality or accuracy are differentiated: lineage, positional accuracy, attribute accuracy, logical consistency and completeness. Lineage will indicate when the data have been collected, and what kinds of processing they have been subjected to. Logical consistency will only refer to data sets as a whole, while the other aspects might deviate locally. This will allow one to visualize their geospatial variation in map form.

In satellite data, various land cover categories or classes can be differentiated on the basis of the spectral characteristics of the radiation measured

from their pixels and from field checking them in the terrain (training). On the basis of a number of trained categories, for every element (pixel) of a satellite data set, a probability vector can be determined, defining the probability of this pixel being assigned to each of the categories discerned. The category with the highest probability value will be selected finally for representation.

So far, so good. But one can imagine that it will mean a big difference to the user, whether the highest probability value (called the 'maximum likelihood') of a pixel to be assigned to the 'bare rock' category is 0.9 (probability vectors vary between 0 and 1) or 0.3 (the second likelihood being 0.28). In the latter case, the certainty of assigning this pixel to 'bare rock' is much lower than in the case of it being 0.9 (see also Figure 1.10).

Uncertainty information can be visualized in one of the following map forms:

Probability images, such as maps of the maximum likelihood and maps of the second likelihood;
Maps of the difference between maximum likelihood and second likelihood;
Maps of likelihoods for each category discerned.

Apart from being visualized, it can also be rendered by sound: shrill sounds generated when moving a cursor over a less certain area might, for instance, indicate low data quality.

A measure for uncertainty or ambiguity could be the confusion index, in which the maximum likelihood is compared to the second likelihood:

$$\text{Confusion} = 1 - (m_{max} - m_2).$$

Small differences between m_{max} and m_2 will lead to high values for confusion, which can also vary between 0 and 1.

If classes cannot be sharply defined, it is conceivable that objects have not only a specific certainty (possibility) to belong to one category, but also another certainty (possibility) to belong to other classes/categories as well. This is expressed by the fuzziness index, which compares possibilities (which can vary from 0 (not a member of a category) to 1 (a member of a category)) to Boolean possibilities (which can only take values of 0 *or* 1), as in Boolean logic the objects are defined in an either/or way, leading to crisp boundaries between objects. So if a location – because of overlap between category or class definitions – may be assigned to different categories, in a situation

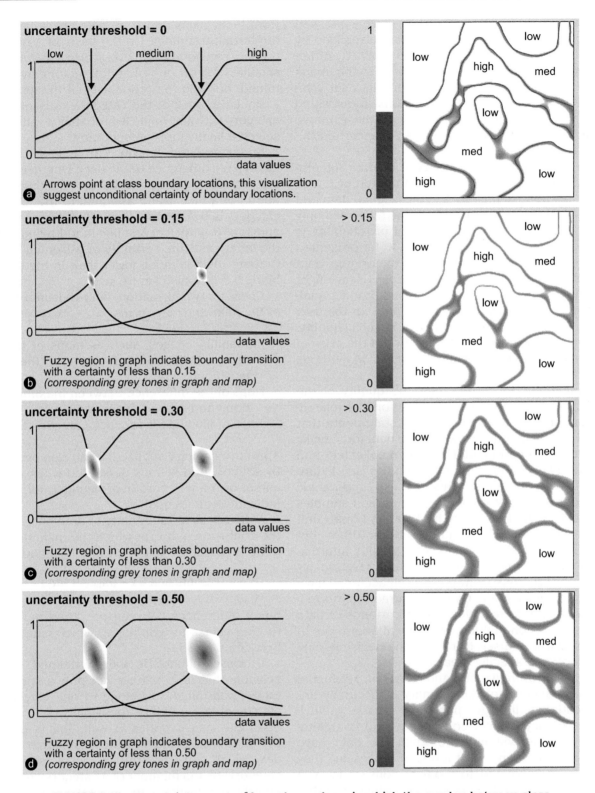

FIGURE 2.15 Uncertainty zones of boundary values, in which the overlap between class definitions is expressed with shades of red in the map. Setting higher uncertainty threshold values will increase the areas where the classes overlap in the map: a) threshold 0; b) threshold 0.15; c) threshold 0,30; d) threshold 0.50 (after Hootsmans and Van der Wel, 1993)

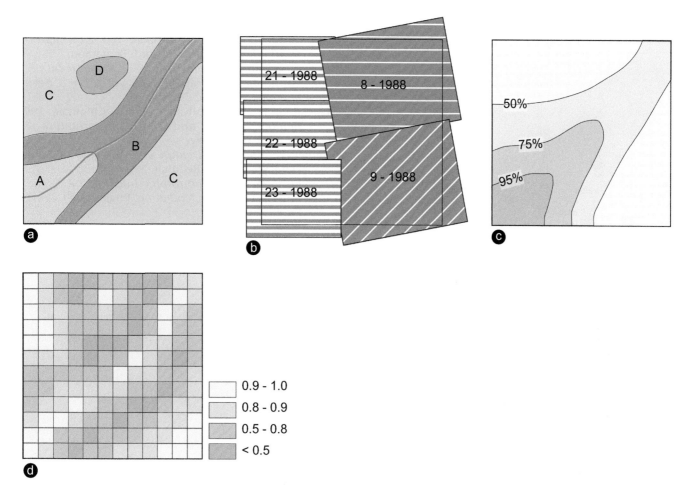

FIGURE 2.16 Examples of the visualization of meta-information (from Hootsmans and Van der Wel, 1992)

where it is not imperative that it be assigned to one class only, possibility vectors can be determined.

In Figure 2.15, maps are shown in which only the uncertainty zone of the class boundary is visualized for different uncertainty threshold values as an alternative to conventional 'crisp' boundaries.

Next to probability and possibility factors, certainty factors are assigned on a more arbitrary basis, i.e. by experts who estimate the validity of data and express these values also on a scale from 0 to 1. All the different methods of representation can be expressed as shown in Figure 2.16. It is the ideal that it would be possible to ask for information as contained in Figure 2.16, to toggle it with the visualized data themselves, in order to allow for proper decision-making on the basis of the data.

Figure 2.17 shows the three options available to display with that topic. The simplest option is to display the uncertainty information in a map next to the original map. In an interactive environment, pointing in one of the maps will highlight the same area in the other map keeping the two maps directly linked (Figure 2.17a). An alternative is to display the uncertainty information as a separate layer on top of the original map. Figure 2.17b shows an example. It does show a direct relation but might 'blur' each of them depending on the spatial distribution in both maps. Finally, it is possible to completely integrate the original topic with the uncertainty information into a single map as shown in Figure 2.17c. This results in a so-called bivariate map, which has two variables integrated into a single legend.

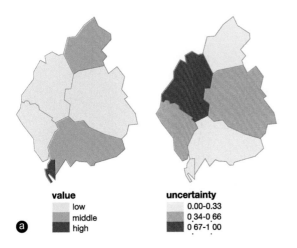

value
- low
- middle
- high

uncertainty
- 0.00-0.33
- 0.34-0.66
- 0.67-1.00

(a)

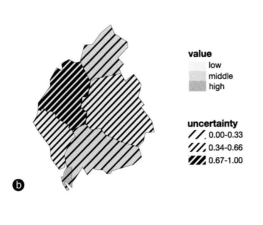

value
- low
- middle
- high

uncertainty
- // 0.00-0.33
- /// 0.34-0.66
- /// 0.67-1.00

(b)

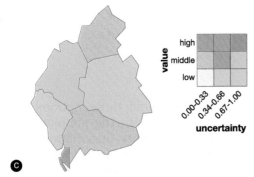

(c)

FIGURE 2.17 Uncertainty visualization options:
(a) uncertainty information in a separate
map; (b) uncertainty information as overlay;
(c) uncertainty information integrated

FURTHER READING

Guptill, S. C., and J. L. Morrison. 1995. *Elements of Spatial Data Quality*. Oxford: Pergamon.

Laurini, R., and D. Thompson. 1992. *Fundamentals of Spatial Information Systems*. Vol. 37, *APIC Series*. London: Academic Press.

Lillesand, T. M., R. W. Kiefer, and J. Chipman. 1999. *Remote Sensing and Image Interpretation*. 7th ed. New York: Wiley.

Samet, H. 1989. *The Design and Analysis of Spatial Data Structures*. Reading, MA: Addison-Wesley.

Tomlin, C. D. 1990. *Geographic Information Systems and Cartographic Modelling*. Englewood Cliffs, NJ: Prentice Hall.

3

Map Characteristics

3.1 MAPS ARE UNIQUE

It is not possible to get an overview of an area in any way other than by consulting a map. A map places geospatial data, i.e. data about objects or phenomena of which one knows their location on the Earth, in their correct relationship to one another. A map can be considered as a geospatial information system that answers many questions concerning the area depicted: the distances between points, the positions of points in respect of each other, the size of areas or the nature of distribution patterns. The answers can be read off directly from the map image most of the time, without the need for a keyboard or the loading up of some files.

Theoretically, GISs would be able to work out conclusions to problems set to them without maps, just on the basis of geospatial information in digital form, which can have been collected as such (though in most cases, it would have been digitized from a map). But performing such a task would be questionable in practice, as without maps one would hardly be able to formulate the relevant geospatial problems that can be set to a computer.

The term 'map' is used in many areas of science as a synonym for a model of what it represents, a model that enables one to perceive the structure of the phenomenon represented. Thus, mapping is more than just rendering; it is also getting to know

the phenomenon which is to be mapped. By 'cartographic method', one understands the method of representing a phenomenon or an area in such a way that the geospatial relationships between its objects and its geospatial structure will be visualized. When representing geospatial information in map form, one has to limit oneself, on account of the available space, to the essentials, among which is the information's structure.

One of the most important moments for cartography was when the first satellite pictures became available. This created the opportunity to check whether the mapping activities of previous centuries, and especially the generalization from large-scale detailed maps to small-scale overview maps of larger areas, had been done in the right way. Comparison of those first Apollo satellite photographs with existing maps generally showed a great deal of agreement, which proved the correctness of the applied techniques.

A second important moment for cartography was the introduction of the computer. Initially (in the period 1960–1980), the computer was used to automate existing mapping tasks such as the calculation of projections and the plotting of the grid or graticule on the map. It proved to be feasible to map an area according to different projections, based on the combination of the same digital file with different transformation parameters. Gradually, cartographers also came to realize the potential for analysis inherent in the digital (digitized) data which computers offered. It then became clear that with the aid of the computer, one would be able to do calculations with the digitized map data and one could have the computer determine the distances, areas and volumes much more precisely than could be done by paper maps.

As soon as a link was made between these cartographic (boundary) files and statistic files, it also becomes possible to evoke numbers of inhabitants, average income data or agrarian production figures, and combine them digitally with the cartographic files in map form; the same could be done for the relation between certain socio-economic and physical phenomena and topography. This has developed into cartographic information systems, which operate similarly to the GISs defined in Chapter 1, but are more geared to visualization than to analytical functions. We must continue to realize that it is the abstracting capacity of the map that allows us to observe geospatial connections, patterns or structures; this also applies when looking at the screen of a GIS.

Maps are nowadays regarded as a form of scientific visualization (see Figure 3.1), and indeed, they existed before visualization developed into a distinct field. Its objective is analysing information about relationships graphically, whereas cartography aims at conveying geospatial relationships. Visualization consists of graphics (with which symbols and lines are indicated) and geometry, which refers to their relative positions. In cartography, these relative positions are usually defined on the basis of a geospatial grid – Cartesian or geographical – which refers these locations to real positions on the Earth's surface. The emphasis in scientific visualization (Hearnshaw and Unwin, 1994) is more on its analytical power ('explorative analysis') than on its communicative aspects: it is primarily directed at discovery and understanding. In cartography, equal emphasis can be placed on analysis and communication. Information visualization has similar objectives for non-geographical data (Keim et al., 2010). Information graphics or infographics for short pertains to the graphic presentation of information, to be clearly and quickly understood by large audiences.

An example of the difference between the analytical and communicative aspects of cartography is shown in Figure 3.2. Here, on the right, the distribution of the precipitation classes has been visualized, and this has been done correctly, as it is possible to see in one glance the distribution of every class in relation to all other classes. But getting an idea of the distribution of the various precipitation classes is not the same as conveying to the reader a proper idea of the distribution of the phenomenon 'precipitation' itself. If it is the communication objective to show the effect of a particular thunderstorm over the Netherlands, then one would want to perceive the geospatial trends, the increases and decreases in its effects over the area, and that is something different from providing an idea of the location of the various precipitation classes. The wrong graphic variables have been chosen to answer that objective in Figure 3.2b. It is in Figure 3.2a that correct graphic variables have been selected, so that the increase in the (grey) value is proportional to the increase in precipitation values, which makes the map fit for communication objectives, as it portrays the phenomenon as a whole. Figure 3.2b is only fit for analysis, independent of

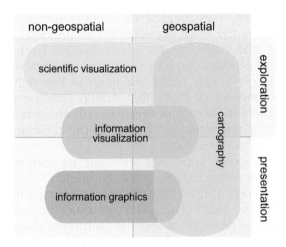

FIGURE 3.1 Relation between scientific visualization, information visualization, information graphics and cartography

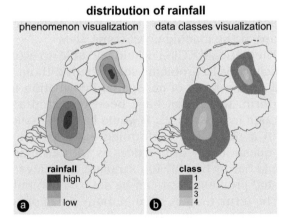

FIGURE 3.2 Distribution of rainfall in the Netherlands: example of map design geared to either (a) data communication or (b) data analysis objectives

the phenomenon. This would be acceptable in a data exploration environment, where the user is in command of the display time and would be able to adjust the class boundaries at will.

One of the most important aspects of exploratory analysis in cartography is the change of perspective that is provided through transformations. In cartography, these transformations are effected by deviating from traditional map projections, by deviating from Euclidean geometry by representing other than geographical distances and areas, by deviating from real life by exaggerating the values or measurement data to be represented (as happens, e.g., in 3D models, where the vertical scale might be ten times the horizontal scale), by separating local from regional trends, so as to make the latter stand out, or by experimenting with class boundaries and the hues or values assigned to them. Other ways of changing perspective are by deviating from traditional geographical map frames (through scrolling in electronic maps), by changing the orientation of maps (by not having the north at the top) as well as their time frames (by monitoring) so that the effects of a random survey activity or a random data gathering activity are offset.

3.2 DEFINITIONS OF CARTOGRAPHY

The meaning of the term 'cartography' has changed fundamentally since 1960. Before this time, cartography (Van der Krogt, 2015) was generally defined as 'manufacturing maps'. The first change in the definition originated from the fact that the subject was put in the field of communication sciences. Under the influence of the former, cartography nowadays is seen as 'the conveying of geospatial information by means of maps'. This results from the view that not only the manufacturing of maps but also their use is regarded as belonging to the field of cartography. And it is indeed evident that only by investigating the use of maps and the processing of the mapped information by their beholders is it possible to check whether the information in the maps was represented in the best way.

The International Cartographic Association defined cartography (2011) as 'the art, science and technology of making and using maps' (http://icaci. org/strategic-plan). Taylor (1991) defined cartography as 'the organisation, presentation, communication and utilisation of geo-information in graphic, digital or tactile form. It can include all stages from data preparation to end use in the creation of maps and related spatial information products'. These 'spatial information products' already refer to database-derived products; the second change in the

definition came about by the rise of the computer and GISs in the field of mapping, and new definitions emerged like: 'Cartography is the information transfer that is centred about a spatial database which can be considered in itself a multifaceted model of geographical reality. Such a spatial database then serves as the central core of an entire sequence of cartographic processes, receiving various data inputs and dispersing various types of information products' (Guptill and Starr, 1984).

For this book, we define cartography as: 'making accessible spatial data, emphasizing its visualization and enabling interaction with it, aimed at dealing with geospatial issues'.

The unsatisfactory aspect in most of the above definitions is that the concept 'map' has not yet been defined. The elements that belong in a definition of maps are geospatial information, graphic representation, scale and symbols. A possible definition of a map runs as follows: 'a graphic model of the geospatial aspects of reality'. According to French cartographers, the map is 'a conventional image, mostly on a plane, of concrete or abstract phenomena which can be located in space' (ICA, 1973). By 'conventional', it is meant that one works with conventions, such as the fact that the sea is represented in blue, that the north is at the top of the map or that some graded series of circles denotes settlements with increasing population numbers. By 'image', the graphic character of a map is stressed. But not all maps are printed on a sheet of paper: relief models and globes are also considered to be maps. It is, of course, also possible to map phenomena that are not physically tangible, such as political preferences or language borders. And it is obvious that it must be possible to locate the phenomena in space.

The International Cartographic Association defines map as follows (2011b): 'a symbolised representation of a geographical reality, representing selected features and characteristics, resulting from the creative effort of its author's execution of choices, and designed for use when spatial relationships are of primary relevance' (see http://icaci.org/). For practical purposes, we will simplify this to: *a map is a visual representation of an environment* (Kraak and Fabrikant 2017), while we imply that it is a product of art, science and technology and the object of scientific study in a digital information society.

The usefulness of a map depends not only on its contents but also on its scale. The map scale is the ratio between a distance on a map and the corresponding distance in the terrain. There are several possibilities to indicate the map scale, as can be seen in Figure 3.3. Next to a verbal description

FIGURE 3.3 Large- and small-scale maps: (a) detail of the base map of Maastricht at a scale of 1:1000 (courtesy municipality of Maastricht); (b) Maastricht at a scale of 1:500 000

(like one inch to the mile), a representative fraction (like 1:1000) or a graphic representation can be used. The representative fraction in Figure 3.3a means that 1 cm on the map corresponds to 1000 cm (10 m) in the terrain. The scale bar in Figure 3.3b represents a distance of 10 km. When the representative fraction is small, a map is considered to have a large scale. Figure 3.3a shows a large-scale map of the city centre of Maastricht, scale 1:1000. The map reveals many details on the level of individual houses. Figure 3.3b shows a detail from a small-scale map with a scale of 1:500 000. Again, the map displays some data on Maastricht. Here, the whole urban area of the city is shown as a small polygon, however.

Confusion exists about the concepts *large scale* and *small scale*. In everyday linguistic usage, small scale is linked with small areas; in mapping, this is used in the reverse sense: small scales in cartography are linked with large areas that are represented on a small map area (see Figure 3.3b). A large scale in cartography is connected with a small area, about which detailed data are presented on a relatively large map area (see Figure 3.3a). Technically, the linguistic usage in cartography is correct: a large scale represents a fraction, which has a relatively small figure in the denominator, whereas a small scale represents a smaller value, which has a bigger figure in the denominator.

In Figure 3.4, the effect of changes in scale is demonstrated. When scales are changed, generalization becomes necessary, as not all the information from large-scale maps can be incorporated in small-scale maps in a legible way. In socio-economic thematic maps, this generalization usually takes the form of aggregation: smaller enumeration units are taken together to form larger units. In Figure 3.4, the enumeration units in map (a) are grouped together to form enumeration wards in (b) and to form municipalities in (c) and nodal regions in (d), i.e. groups of municipalities oriented on the same large towns. This aggregation has a direct effect on the data, in this case the population density: as the figures are counted for larger areas, excessive local scores are averaged, and differences become less notable. If instead of nodal regions, economic regions (groupings of municipalities on the basis of common economic characteristics, e.g. agricultural versus industrial regions, with the ensuing political preferences) would have been demarcated, the differences in population density would have been more outspoken than in the case of nodal regions. This illustration perfectly shows that it would be incorrect to assume, for an area on a map with a higher aggregation level (such as b, c or d), to which a population density of say, 1500–3000 inhabitants per km^2 is assigned, that that value would

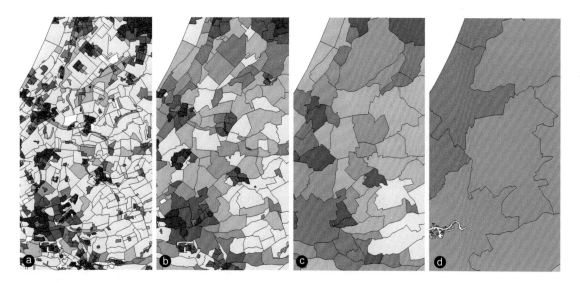

FIGURE 3.4 Thematic map and the influence of aggregation on socio-economic phenomena: the population density in the central part of the Netherlands in 2009 in (a) neighbourhoods; (b) wards; (c) municipalities; (d) nodal regions. Produced by Ton Markus

be valid all over that area, in other words that the areas are homogeneous regarding the phenomena mapped. This misinterpretation of choropleth maps is called the 'modifiable areal unit problem'. Changing boundaries like grids would also modify the distribution image.

Traditionally, the main division of maps is between topographic and thematic maps. Topographic maps supply a general image of the

Earth's surface: roads, rivers, houses, often the nature of the vegetation, the relief and the names of the various mapped objects (see Figure 3.5a). Thematic maps represent the distribution of one particular phenomenon. In order to illustrate this distribution properly, every thematic map, as a basis, needs topographic information; often this would be provided by a thinned-out version of a topographic map. In Figure 3.5b, one can find a

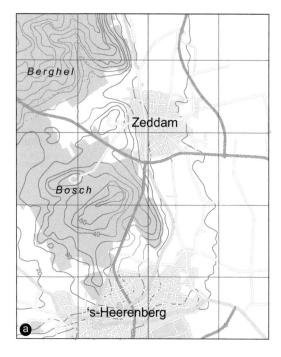

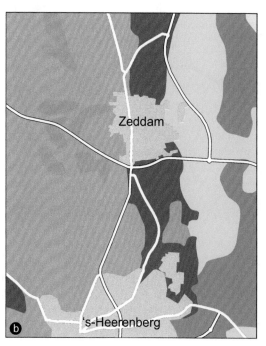

FIGURE 3.5 Topographic and thematic maps: (a) the topography of Montferland in the eastern part of the Netherlands; (b) a soil map of the same area (after Netherlands soil map 1:50 000)

detail of a soil map as an example. Such a map is only usable if one could perceive, with the help of this base map, where the types of soil which have been differentiated can be found.

A thematic map would also emerge if one aspect of the topographic map – such as motorways or windmills – is especially highlighted, so that the other data categories on the map could be perceived as ground. An example of this is presented in Figure 3.6.

In a digital environment, the differentiation between topographic maps and thematic maps becomes less relevant, as both map types consist of a number of layers – a topographic map would be a combination of separate road and railway layers, a settlement layer, hydrography, a contour line layer, a geographical name layer and a land cover layer. Each of these would be a thematic map in itself; a combination of these layers in which each data category had the same visual weight would be a topographic map. When one category is graphically emphasized or highlighted, and the others thereby relegated the status of ground, it would again change into a thematic map.

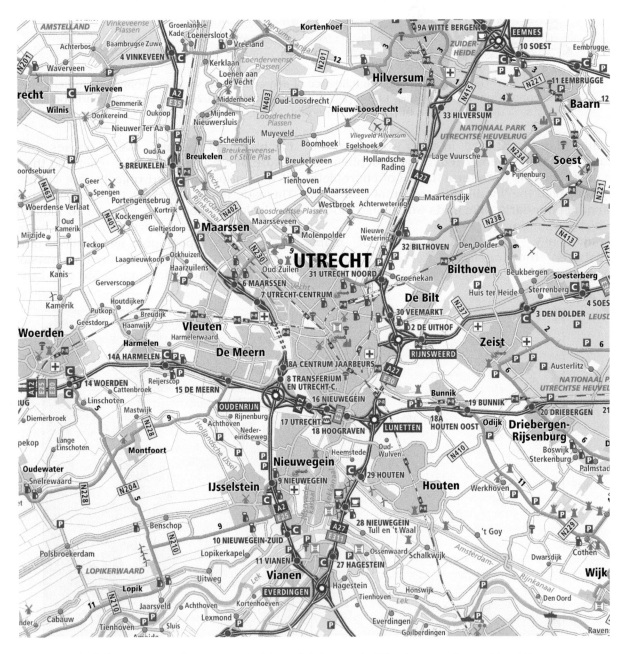

FIGURE 3.6 Thematic map created by subduing part of the topography and by doing so highlighting a specific information category (highways) (© ANWB B.V., Den Haag)

The topographic base of a thematic map can even be far more schematic than an excerpt from a topographic map. In the representation of socio-economic phenomena, the data are gathered for enumeration areas instead of at individual locations (as would be the case for physical phenomena), the boundary files are usually strongly generalized, so as to distract from the map theme as little as possible. An example is shown in Figure 3.7. In Figure 3.7, the base map only consists of national boundaries and coastlines. Such a presentation (the combination of thematic data (Figure 3.7b) plus schematic base map (Figure 3.7a)) leads to a clear map image (Figure 3.7c) – which can only then be interpreted and memorized correctly when the map user also recognizes the area concerned. To arrange for this recognition, it may be necessary to add the names of the (most important) enumeration units.

People form for themselves a mental model of reality. For example, after living in a village for some time, one would operate, while traversing the village for one's daily chores, on the basis of this mental construct, which operates like a *virtual map* (see Figure 3.8); when someone else asks for directions, one would consult this mental construct in order to provide an answer. Answering the request could also be done by drawing a sketch map, a 'mental map', which is a permanent print-out from our mental construct of reality, designed for answering a specific request for directions from and to random locations. Such a mental construct (we could also call it a 'cognitive map') can be generated not only from one's contact with reality but

also through consulting a proper, tangible paper map (a 'permanent map', or hard copy map in computer talk). When one is an experienced map user, one would be able to grasp the essential information from a paper map and store this in one's mind. This process can also take place from a map displayed temporarily on a monitor screen (a 'temporal map', which is visible but not tangible). This temporal map can have been generated from a geospatial database stored in the computer, from which a specific selection has been made in order to answer specific requirements or objectives. Therefore, this geospatial database, which can also be used to produce other maps, functions as a *virtual map*, in the same way as different sketch maps can be produced from one's mental perception of reality (see also Figure 2.2).

Maps can be said to show three dimensions of the phenomena represented: the nature or the value of the objects and their location. The location is defined at the hand of the x- and y-coordinates; the nature or value is considered as the z-coordinate. It is one of the tasks of cartography to have the z-coordinate stand out sufficiently to provide map readers an adequate vision of the mapped area's relief.

3.3 THE CARTOGRAPHIC COMMUNICATION PROCESS

In Section 3.2, cartography is described as 'the conveying of geospatial data by means of maps'. In order to illustrate this, a model of cartographic information transfer is presented in Figure 3.9.

The starting point of the cartographic communication process is the data or the information (I), usually collected by third parties (geodesists, photogrammetrists, geographers, statisticians). When producing a map, one has to study this information, as one will have to get acquainted with its characteristics as well as with the purpose of the information transfer in order to be able to represent the information correctly in map format. Often, the resulting map will not contain every particle of information with which the mapper was supplied; generalizations or classifications may have been applied in order to present a clearer picture of the phenomenon.

The map user or map reader who sets eyes upon the map will derive certain information from it. The information retrieved (I') will, however, never completely overlap or coincide with the original information (I) as during the communication process, data may have fallen out or been left out on purpose, or because mistakes may have been made.

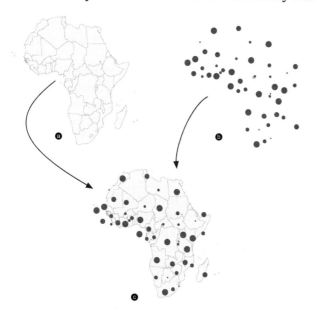

FIGURE 3.7 A thematic map (c) is constructed from a base map (a) to which symbolized data (b) are added

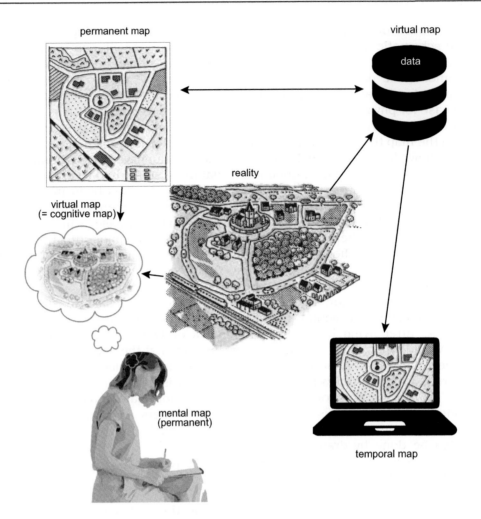

permanent map

virtual map

data

reality

virtual map
(= cognitive map)

mental map
(permanent)

temporal map

FIGURE 3.8 Examples of permanent, virtual, temporal and mental maps

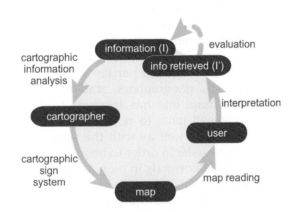

information (I)

evaluation

info retrieved (I')

cartographic
information
analysis

interpretation

cartographer

user

cartographic
sign
system

map reading

map

FIGURE 3.9 Model of the cartographic
communication process

Those that produced the map may have interpreted the original information incorrectly, and even if they did interpret it in the right way, then they may still have made errors in the process of mapping, when representing the information. The map reader may read out the data in the wrong way, or may draw the wrong conclusions from the right data. Thus, there are ample possibilities for the derived information not to coincide completely with the original information.

Cartography aims for the elimination of these various sources of errors, and for the provision of a correct transfer of the data, by means of such a graphic presentation that the map reader is able to draw the correct conclusions. A feedback of I' to I in Figure 3.9 is necessary, because in this way the map producer is able to check the effect of the cartographic products and, depending on this evaluation, may adjust the image of the map. During the process of evaluation, moreover, one keeps on learning about the depicted phenomenon: whether the correct structures have been transferred, whether the most relevant characteristics have been selected and whether the most recent quantitative data have been represented; therefore, the cartographic representation is also a cognitive process: one has to get to the essence of a geospatial phenomenon if it is to be represented adequately.

Of course, in principle, maps show us the situation of the Earth's surface; by viewing a map, we visualize the Earth's surface, and we try to match this mental vision with the real world. The overlap referred to earlier should also apply to the real world but this overlap can never be complete: there will always be a time gap between the moment the data were gathered or surveyed and the moment they were made available to the users. In some cases – as in recent military conflicts – it may only take a couple of hours before the data are made available, but even then this time gap may be crucial as the very planes stationed on a photographed airfield or tanks on the ground may have left the area in the meantime.

The cartographic information process, as presented in this book, thus starts with the recording. The traditional, cartographic 'recorder' is the topographer, who surveys the terrain for the sake of the production of the topographic map, who describes functions, distinguishes road categories, collects geographical names, measures the location of new objects, etc. The topographer's work has already been referred to in Chapter 2, and the results of this work will be discussed in Chapter 6.

Topographers (together with geodesists) supply mainly the *x*- and *y*-coordinates of the geospatial information; the *z*-coordinates are defined (with the exception of altitudinal data) by others – by census-takers who collect information by means of socio-economic questionnaires, by soil scientists who take samples of the soil, by traffic enumerators, etc. Since 1930, the aerial photograph, as a data source, has held a very important place, while since 1970, the satellite image has also become increasingly important.

From all these sources, traditional and new, a lot of information is directed at the cartographer or map producer, most of it not in the right shape for cartographic presentation. Thus, some kind of (cartographic) processing is necessary first. The elaboration of the survey data, in many cases, takes the shape of a classification. Individual characteristics of observations are then replaced by group characteristics. Therefore, the quantity of information decreases, and the map resulting from this becomes better ordered. Other ways of elaboration are the correlation with other data, the expression of the degree of conformity or difference in the distribution of different data sets, or the comparison of absolute enumeration data taken at different times for the same area, in order to be able to show the development in the intervening periods.

The design of maps is mainly concerned with making choices: the choice of mapping method, the choice of graphic variables (such as differences in size, value, grain, colour, direction and shape, see Chapter 7) to be used. In Figure 3.10, a number of selection moments are shown, which play a role in the design of maps, as well as the

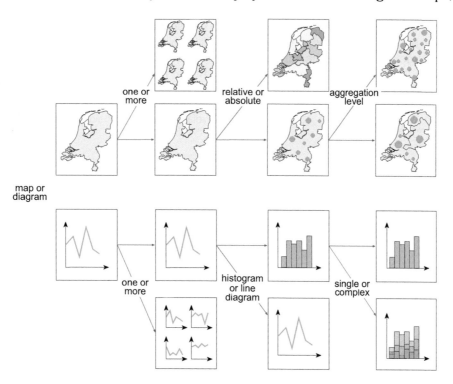

FIGURE 3.10 Examples of choices during the visualization process

cartographic results of these choices. Here, the development of the population of the Netherlands in the period 2000–2020 is the concern. The first choice that has to be made is the one between map and diagram. Should the diagram be selected, then the opportunity to show relations to other geospatial phenomena is lost. Even if a geospatial component is introduced which shows the developments in the various regions separately, it is still far more difficult to draw geospatial conclusions from this complex image than in the case where the same information had been mapped. Of course, there are various possibilities for the design of a diagram, e.g. a line diagram or histogram, a representation with index values or a logarithmic presentation. Some possibilities are shown in Figure 3.10. Should a map be selected, then one first has to select the aggregation level on which the information has to be depicted: on the level of statistic enumeration areas, municipalities or even higher-order areas. The resolution (level of geospatial discrimination) at which the information is presented depends on the space which is available for the map and the goals of the map author. On the level of local enumeration areas or municipalities, one would be able to show the local trends, concentrations and dispersions. On the basis of aggregations of municipalities or on county level, only regional patterns can be shown.

After the selection of the aggregation level of the data, a choice can be made between an absolute or a relative representation. The results of both choices may be contradictory: a map of the absolute unemployment in the Netherlands shows a concentration in the central Randstad area, whereas the unemployment problem seems to lie in the marginal areas in the south, east and north of the Netherlands, when judging from the image of the relative unemployment (Figure 3.11). Selection of either absolute or relative representation also implies the selection of a specific cartographic method that is for proportional symbol maps or choropleth maps, respectively. The same information could also be presented as an isoline map, a grid map or a map with diagrams. This will also depend on the ulterior motive or objective of the communication process.

Supposing data are available relating to the participation of a city's inhabitants in public transport, and that it is known through questionnaires or otherwise how many passengers were transported over each stretch of the bus routes between bus stops, where the passengers boarded and descended again, and where they came from and went to. Similar data have been collected for a number of major cities. If one wanted to show the existing geospatial relationships between the inhabitants, then the movements by public transport from each part of the city could be shown through sets of arrows (Figure 3.12a). If the transport authorities were more interested in the routes followed by the inhabitants to get to their destinations, these could be plotted (Figure 3.12b), and the frequency with which they were used could be shown by proportional widths (not indicated in Figure 3.12). By comparing origins and destinations of all the trips recorded in the questionnaires, the places where people changed buses could be traced and a hierarchy could be portrayed, showing these bus stops with circles proportional to the number of changes (Figure 3.12c).

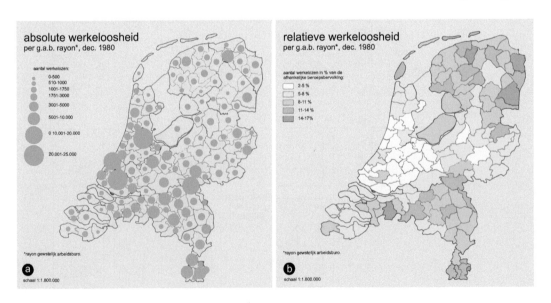

FIGURE 3.11 Absolute (a) and relative representations (b) of the same information (unemployment data)

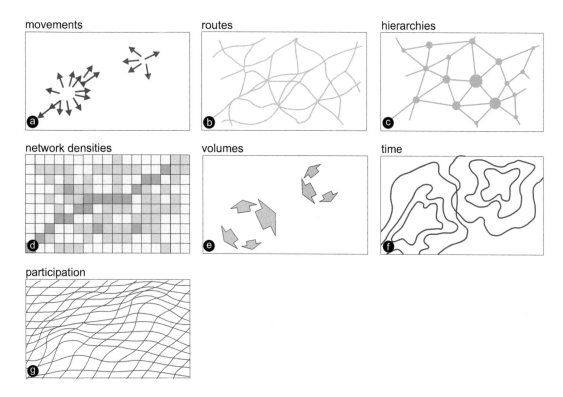

FIGURE 3.12 Visualized results of changes in definition of attribute data from a public transport database

Accessibility of the public transport network could be shown by portraying the number of bus stops per regular area unit (Figure 3.12d). The data could be aggregated for the various wards the city would be divided into, showing the volumes of passengers travelling between the major wards by proportional arrows (Figure 3.12e). By comparing the destinations of the passengers from each city ward with the public transport timetable, the average travelling time could be determined and portrayed through isochrones (Figure 3.12f). The degree of participation of city inhabitants in the public transport, computed either as a percentage of all inhabitants or as the number of people travelling by bus per areal unit, could be expressed three dimensionally (Figure 3.12g). So depending on the actual aims or objectives, the database built up from the questionnaires could be used for the production of quite different types of maps, each conveying the answers to different questions.

The possibilities for its distribution and the available financial resources determine whether the map will be made available on the Web or reproduced, in black and white or in colour. In the latter case, it might have to be very much generalized because it is printed on poor-quality (newsprint) paper. The manner of distribution, the way in which it will be documented for search engines or the management in libraries, will determine who will see the map and who will therefore be able to use it. The role of publishers or webmasters is essential here, as it is usually they who are able to control the distribution, through assigning specific prices to the cartographic products, through adding ISBNs (International Standard Book Numbers) or DOI (digital object identifiers) to the maps, through the use of their distribution outlets and channels, through marketing campaigns, through the distribution of review copies to cartographic journals or by providing links to the websites where they occur. The map's potential for use will also be determined by the skills of the users and their foreknowledge about the theme that is mapped out. Of course, the first requirement here is that map user and map producer speak the same graphical language, i.e. they use a common set of symbols or associations, so that the signs in the map are understood by the map users, and so that they can communicate. The addition of legends is another essential aspect of map documentation, and it is not always automatically generated, not even in the newest GISs.

Increasingly, we live in a society where maps become ubiquitous, i.e. are available anywhere, anytime, of any location we want and with every possible topic of theme. This total availability may emerge because we store our own maps on the Internet in a way in which we can access them anywhere where we have Internet access, for instance, through the Microsoft OneDrive or, more general, in the cloud. Cloud storage implies that users upload their data to an Internet storage pool provided by a hosting company which keeps the data available and accessible. Alternatively, people, companies or institutions provide their map data via the Internet where it can be accessed freely or under passwords. Examples are the town plans or topomaps we have on our mobile devices. Availability of geographical information is also discussed by Buckley and Rystedt (2015).

Map use is also the subfield where cartographers do research on the map's effectiveness. There are currently two different views on improving cartographic communication: one is to improve map-reading expertise by user training and the other consists of improving the maps. Both routes of course have to be followed simultaneously. In many countries (cf. Great Britain with its map-use standards), high school students during geography lessons are confronted with standard tasks they should be able to perform with the help of maps in specific grades (e.g. between the ages of 13 and 16, pupils should normally be able to correlate information on two or more thematic maps in an atlas such as on relief and vegetation).

The other route, towards improved maps, has been boosted by the computer. It has not only allowed mapmakers to experiment freely with map design options, but also allowed them to customize maps, i.e. to adapt them to specific requirements. A good example is the hydrographic chart. These are rather complex documents, as they have to cater for many different types of map use. They do contain information about shipping hazards, depths (isobaths, soundings, wrecks, and reefs), currents, tides, navigational lights and buoys. They contain several grids for locating purposes (geographical coordinates, Decca grids), boundaries, one-way shipping lanes in densely navigated waters, warnings, munitions dumping areas and drilling platforms, and also contain information on the nature of the underwater sediments, etc. When navigating, one never needs all this information at the same time, and some of the information categories one never needs (see Figure 3.13).

Now that electronic charts are emerging, with all these information categories on layers, only those layers that are needed for specific purposes can be visualized on the monitor screen, thereby avoiding the clutter of symbols on paper charts. If a ship draws only 3 m, then all the depth information beyond the 3 m isobath can be omitted. During daylight, information would not be needed on navigational lights, while at night, one would only need information on the location of those buoys that also had lights. So, the map could be adapted and displayed on the monitor screen to the specific navigational requirement of a specific time and place.

In order to check a map's effectiveness in conveying geospatial information, test subjects are asked to perform tasks on the basis of the map. When testing out various versions of a town plan, test subjects could, for instance, be asked to locate addresses, to describe the shortest route from one address to another, or to describe the optimal link through public transport between two locations. The use of maps with the same informational contents but with different contrasting colours, type sizes, degree of generalization or grid reference systems might result

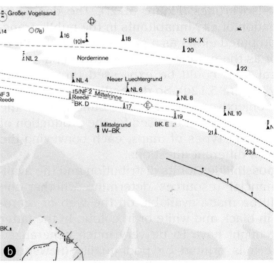

FIGURE 3.13 (a) Detail of a traditional hydrographic chart and (b) detail of a partial rendering of this chart in an electronic navigation display (right) (courtesy Bundesamts für Seeschifffahrt und Hydrographie Hamburg/Rostock)

in different performances regarding the time it took to answer these test questions correctly.

On the basis of map-use tests like these, the best possible design for answering specific tasks can be selected – but here one has to take account of not only the functions the map has to perform (see next sections) but also the circumstances in which the geospatial information will be used (constraints in space, lighting, etc.) and the constraints imposed by the target audience (age, school experience, etc.), which might result in a specific colour selection, map complexity and appropriate wording of the marginal information.

Usability, according to the ISO (International Organization for Standardization), is: 'the extent to which a product can be used by specified users to achieve specified goals with effectiveness, efficiency and satisfaction in a specified context of use'. The study of usability of maps starts with the users, which may range from professionals such as geographers, soldiers or planners to tourists, children and mobile device users. Each of these user categories deserves its own approach regarding information presentation as they all have different use characteristics. That is why map design always should be user-centred (see also Section 11.6).

User-centred design usually consists of the following steps: (1) a problem definition phase, (2) the development of a functional specification, (3) a building phase and (4) an iterative testing or evaluation phase. Applied to atlas production, this could be translated as: (1) For whom do we develop an atlas with this content, and what information do we want the users to be able to extract from the atlas? (2) What specific tasks should the intended users be able to perform with the atlas, and what functionality should be needed to make those tasks possible? (3) How do we build the prototype? (4) How do we test whether the prototype answers the criteria we set under (2)? In Figure 3.14, the last two phases have been combined under (3), and a fifth phase has been added, the user satisfaction feedback phase, during which the results of the previous phase are checked.

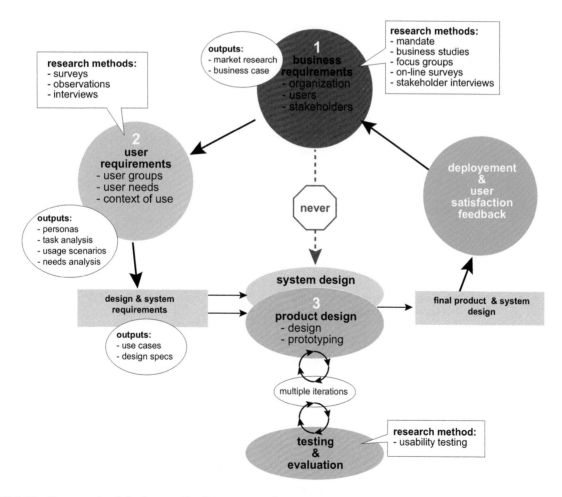

FIGURE 3.14 User-centred design methodology used for the production of the Atlas of Canada (Kramers, 2020)

3.4 MAP FUNCTIONS AND MAP TYPES

The most important function of maps is probably the function of orientation or navigation. In any case, most of the maps the general public comes across, with the exception of the weather chart, are produced to aid orientation and navigation (Figure 3.15). People use orientation maps (road maps – on paper, in a navigation system – topographical maps, charts) for getting from one place to another along a selected/plotted route, and want to be able to check against the map/chart on their mobile phone whether they are still 'on course' during their trip. These mobile maps (maps on mobile devices) have become the most consulted map category to date.

Maps that are used for town planning take second place to orientation maps, although it would be the other way around if it were the number of different maps and not the total number of printed copies that was the criterion. Town planning maps consist of maps that inventory the present situation, maps that define development processes and maps that contain propositions for a future situation, for instance, future land use. Generally, alternatives have been made for such plans as well, which are offered to the public before a public comment session. Up to the moment that the plan (development plan or regional plan) is codified in its ultimate form, hundreds of these town planning maps would have been produced (Figure 3.16). Also directed towards the future are maps that show forecasts: certain developments from the past are, on the basis of the development pace that is to be expected, extrapolated into the future. This applies to weather charts but may also hold good for other

FIGURE 3.15 (a) Car navigation planning example with (b) sequential instructions, from Google Maps (©2019 Google LLC, used with permission)

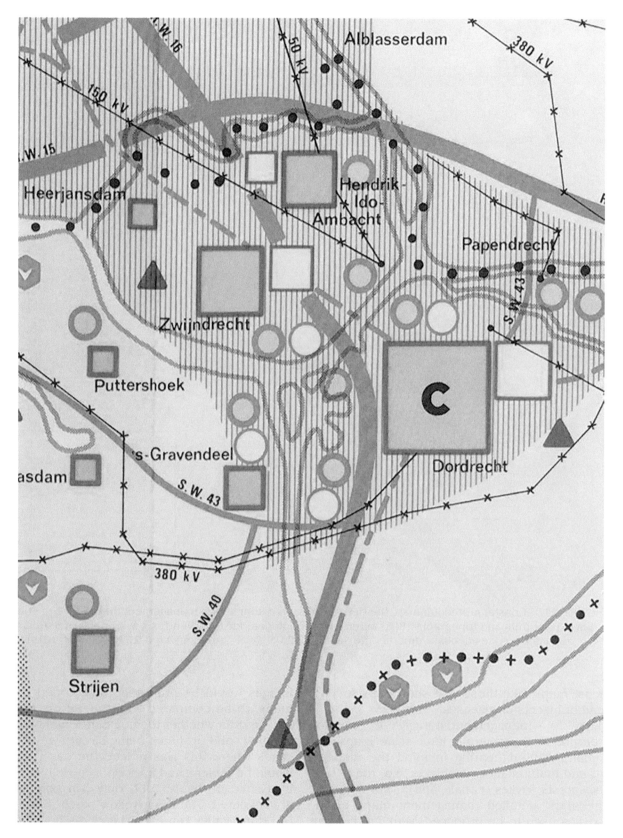

FIGURE 3.16 Physical planning map (courtesy National Atlas of the Netherlands, 1st edition)

FIGURE 3.17 Coastal protection map, the Netherlands, as example of a management map, with 5 m contour lines, layer tints and topography (Rijkswaterstaat, Hoogtedata kust 2019) https://www.rijkswaterstaat.nl/apps/geoservices/geodata/dmc/hoogte_kust_2019/05_DSM_2m/ (consulted 21 November 2019)

forecast maps, like those that show the expected spread of insects or diseases.

Maps for management/storage or monitoring purposes are generally large-scale maps that are manufactured bearing in mind the management and maintenance of objects, e.g. roads, railways, forests, dykes, canals and airports. In the Netherlands, detailed management maps of the sea dykes used to be produced, with 1 m contour lines. After every major storm, the dykes would be scanned again from the air with the LIDAR technique, in which pulsed lasers emitted from the aircraft measure distances to the Earth. Resulting images would be processed and turned into new maps, to be compared with the old ones in order to ascertain whether the coastlines had been damaged by sand or dunes being swept away. In those cases where this was indeed the case, the sand would be replenished as soon as possible. For clarity's sake in Figure 3.17, only 5 m contour lines are rendered. The three yellow patches in the central dune valley represent lakes, as heights below the water surface cannot be assessed by laser altimetry.

For educational objectives, special-purpose map material has been produced since around 1750:

FIGURE 3.18 Detail of an educational map of Africa (courtesy Georg Westermann Verlag, Druckerei und kartographische Anstalt GmbH & Co. KG)

school atlases, wall maps (Figure 3.18) and workbooks, which should provide the pupils with a geospatial frame of reference in order to understand national and worldwide developments. These educational maps are subject to more stringent generalization in order to provide for a better legibility.

Another map function is codification, e.g. showing the legal situation as it is, as, for instance, the situation of property rights. In continental European countries and indeed in an increasing number of countries worldwide, cadastral maps are being produced that have this function of codifying land ownership.

It is on these lines (orientation, town planning, forecasts, management, education and codification) that maps are divided functionally into different groups. These map functions should be discriminated from map types, which are groups of maps that have received their shape according to the similarity in the specific methods used, such as the choropleth method or the isoline method.

Map categories are divisions to which maps with the same theme are relegated: town plans, weather charts, geological maps, population maps, language maps, etc. From a map design or a map-use point of view, it is no use to discuss these map categories separately, since for different map themes, identical design, representation or interpretation, problems may occur. So in Chapter 7, the design problem will be the guideline of the discussion, not the theme which is depicted. The division into map categories as such is not of importance either for the discussion of the possibilities for the derivation of information; in Chapter 11, this discussion will take place along the lines of map types.

FURTHER READING

Harley, J. B. 2001. *The New Nature of Maps. Essays in the History of Cartography.* Baltimore, MD: John Hopkins University Press.

ICA. 2011. ICA strategic plan 2011–2019. https://icaci.org/files/documents/reference_docs/ICA_Strategic_Plan_2011-2019.pdf. International Cartographic Association.

ICA, Commission II. 1973. *Multilingual Dictionary of Technical Terms in Cartography.* Wiesbaden: Steiner.

Keim, D., J. Kohlhammer, G. Ellis, and F. Mansmann, eds. 2010. *Mastering the Information Age Solving Problems with Visual Analytics.* Goslar: Eurographics Association.

Kraak, M. J., and S. I. Fabrikant. 2017. "Of maps, cartography and the geography of the International Cartographic Association." *International Journal of Cartography* 2 (S1: Research Special Issue):9–31.

Krogt, P. v. d. 2015. "The origin of the word cartography." *e-Perimetron* 10 (3):124–142.

MacEachren, A. M. 1994. *How Maps Work: Representation, Visualization, and Design.* New York: Guilford Press.

4

GIS Applications
Which Map to Use?

4.1 MAPS AND THE NATURE OF GIS APPLICATIONS

To solve problems in the fields of Earth and social sciences, maps are often indispensable. Frequently, the researcher will have questions related to the nature and coverage of some particular phenomena. Finding answers will involve questions concerning what map type to use and which map scale to use. Map types and scales have been covered in Chapter 3, and one might be inclined to think that in a digital environment, map scale is no longer relevant, as it is possible to zoom in or out on the data at will. This misconception has to be countered here, as the scale at which geospatial data have been collected or digitized does indeed matter. Keywords here are clutter, resolution and being representative. When map data have been digitized at a specific scale and then rendered on screen at a much smaller scale, geospatial data on the map may be too dense, so that the overview is lost. Particular objects with a detailed character, like coastlines, might appear as a cluttered line when reduced to a smaller scale. On the other hand, line elements that have been digitized at a specific scale and are rendered much larger on the screen may no longer be

representative or characteristic, as straight lines and sharp angles might be acceptable at a small scale but will look unnatural at larger scales. Moreover, an increase in scale might lead to a map image with too little information. It is therefore wise either to use data that have been digitized in a similar scale range as the one needed, or to generalize existing files (for generalization, see Chapter 6).

4.2 CADASTRE AND UTILITIES: USE OF LARGE-SCALE MAPS

4.2.1 Cadastral Maps in Use

Land has been registered and mapped since ancient times because it is vital to humankind. We all somehow depend on land for our living. The cadastre, an information system that uses the land parcel as the basic geographical unit to register land ownership, was revived in continental Europe by Napoleon, who used it to collect taxes. The cadastral map plays an important role in this system since it shows boundaries that define the ownership. Cadastral maps are large-scale maps normally at a scale between 1:1000 and 1:2000. For orientation purposes, they sometimes include the outlines of important buildings, next to legal boundaries. On these maps, the parcels themselves are often numbered, and these numbers form the link to other components of the cadastral information system. These could be the registries of landowners or the original survey data. In their digital, online version, the parcels can be queried via mouseover in order to access this attribute information (Figure 4.1).

The nature of the relationship between cadastral and topographic objects differs, depending on the country's cadastral system. In some countries, like Great Britain, the general boundary system is in use. Here, the cadastral boundaries coincide with topographic objects; in other words, the boundaries are visible in the terrain. In the United States and the Netherlands, it is not necessarily the case that cadastral boundaries coincide with topographic objects. However, in the United States, these boundaries are marked in the land by pegs, while in the Netherlands, the original survey data have to be reconstructed to localize cadastral boundaries in the terrain.

Cadastral maps have had many uses during the last few centuries. Among these are land reclamation (in the 16th century in the Netherlands); evaluation and management of state land resources (in the 19th century in the United States and Canada); land redistribution and enclosure (in the 17th century in England); and colonial settlement, land taxation and land consolidation. Nowadays, similar tasks are executed by what is called a 'multipurpose cadastre', a parcel-based land information system. In that context, Dale and McLaughlin (2000)

provided an extensive overview of land administration (the process of determining, recording and disseminating information about ownership, value and use of land and its associated resources).

The process of rural land development is an example of the manner in which the cadastral map is used in a GIS environment (Figure 4.2). In some rural areas, the need to change the current historic pattern of land parcels arises because the parcel pattern no longer fits modern agricultural requirements or an economic approach to the use of land. In a land reconsolidation procedure, land ownership is a vital factor, but not the only factor. The land in the project area will not be of the same quality throughout and, because of this, reallocation based on ownership only would not be fair. The reallocation is therefore executed also on the basis of the land values. Land value is determined by combining the ownership data, the topography and a soil map. In a GIS, the value map is created by an overlay operation. The land value map is combined with a physical planning map that shows the new layout of the land (drainage system, road pattern, etc.). A model linked to the GIS will then, in an interactive iterative process, define the new parcels and ownership. The result will be a new cadastral map and often – because of the new infrastructure – a new topographic map as well.

4.2.2 Utility Maps at Work

The function of service area maps of utility companies is to keep an accurate and up-to-date record

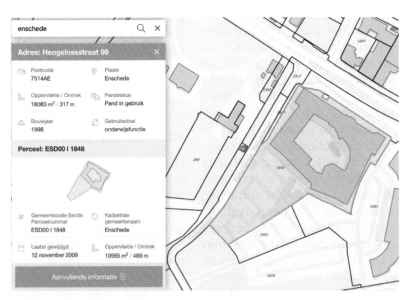

FIGURE 4.1 Cadastral map of the Netherlands (1:1000): a detail showing ownership boundaries (black lines), buildings (grey) and parcel numbers (black), plus the data (cadastral section, parcel number, size, date of sale and registration) accessed by clicking this digital map (Kadaster)

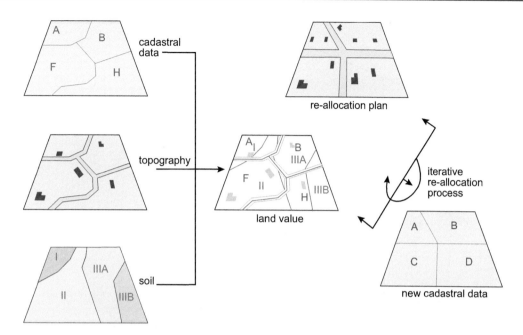

FIGURE 4.2 The cadastral map at work: its role in a land reconsolidation project. A land value map is constructed based on the cadastral map, a topographic map and a soil map. In an iterative process, the land value map is combined with a reallocation plan to create the new cadastral map

of their infrastructure for maintenance and planning purposes. Examples of utility networks are those of gas, water, sewerage, electricity, telephone and cable television. Next to geometric information related to the location and depth of the pipes and cables, the maps register attribute information such as pipe diameter, construction material, age and capacity. The types of maps needed are large-scale maps, which would, for instance, be used by a field engineer to locate a specific part of the network plan for maintenance, and small-scale maps to plan future demand.

The large-scale maps in use are often those supplied by official mapping organizations. They are used as a base on which to map the pipes and cables. When the utility companies have to survey their own maps, they show only selected topography. For instance, only the distance from the front of a building is needed to locate the cables or pipes under the pavement. Figure 4.3 shows an example for water. The coding along the pipes refers to their attributes. The location on the map is not exact. However, to avoid digging in the wrong place, with the possibility of damaging other cables or pipes, or even finding no cables or pipes at all, survey data like distances from buildings are included as well, as can be seen in detail in Figure 4.3. Maps of dense urban areas can become very cluttered because of the high density of lines and text representing cables or pipes. In these situations, the cables and pipes are often found above each other

in a special ditch. In a GIS environment, a solution to these graphic problems is the application of a layering technique. Although, in practice, the utility companies can solve their automated mapping and facility management problems, it is very likely that a water board will still need to consult an electricity company before starting to work in a particular area. Information is exchanged, often via local authorities, to keep each other informed on the whereabouts of cables and pipes. This exchange is related not only to the location of the network but also to the planning of works to avoid a street being successively opened and closed by different utility companies in a single month.

Schematically designed small-scale maps are used in the planning process of utility companies. Such maps contain only the network lines and selected attributes, and are used to plan future extensions or to play a role in emergency simulations. In the planning process, maps with geological and soil data are also used to determine the character of the subsurface for the underground pipes and cables. A database with customer information is also vital to the utility companies because it provides them with knowledge on the nature of the customers' connections as well as the addresses to which they should send bills.

Figure 4.4 demonstrates the role of utility maps for landlines in a telecommunication company's GIS. The monitoring system, tracking the status of the network, signals a problem in a certain circuit,

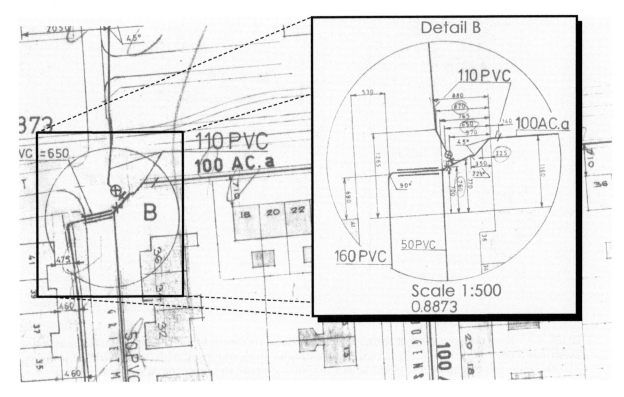

FIGURE 4.3   A utility map 1:1000: water conduits and pipes

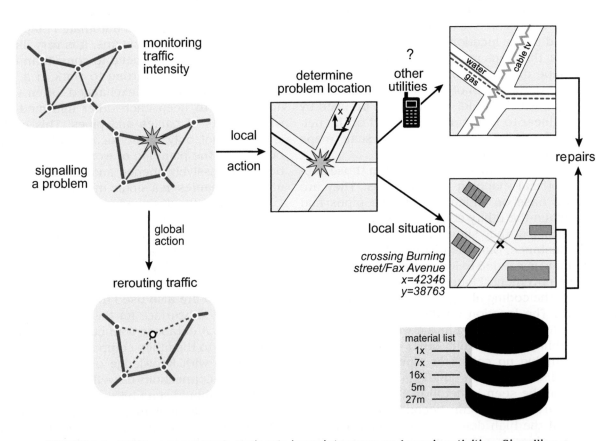

FIGURE 4.4   Utility maps at work, their role in maintenance and repair activities. Signalling a problem results in global (rerouting traffic) and local action (preparing and executing repairs)

which has damaged a certain part of the network. On an overall scale, the system tries to reduce the damage by rerouting most of the traffic along other paths of the network. Only local traffic is affected now. Meanwhile, field engineers have located the trouble and quickly produce a large-scale map that includes the necessary geometric and attribute information. A call to the local authorities provides them with a map that shows all other pipes and cables near the trouble spot. From the database, a material list is generated of those components that may be needed to repair the damage.

4.3 GEOSPATIAL ANALYSIS IN GEOGRAPHY: USE OF SMALL-SCALE MAPS

4.3.1 Socio-economic Maps

Themes represented in socio-economic maps are often derived from statistics related to topics such as census data, infrastructure, housing, employment, trade and agriculture. Figure 4.5 shows two examples. The map on the left shows how absolute values (here fish catches in Africa for 2016) are represented by proportionally sized circles. Interesting to see in this map are the high values for land-locked countries such as Chad and Uganda. Studying the map in more detail reveals that those countries have large inland lakes (Lake Chad and Lake Victoria). It is a good example of the need to add useful topographic information to thematic maps. It is necessary not only for orientation, as mentioned in the first section of this chapter, but also for a better understanding of the map. Adding the rivers to the map of Africa further contributes

to a better understanding of the theme. The figure on the right shows a typical choropleth map depicting the population density of East Africa.

Examples of socio-economic maps can be found throughout this book. Chapter 7 gives in more detail the characteristics of the different cartographic options available to map socio-economic data. To interpret maps correctly and retrieve relevant information from them is not always easy, as can be seen by looking at the map displayed in Figure 4.6. It shows the number of traffic accidents for each municipality in the north of the Netherlands. An insurance company created it in order to help it decide on regional differences in its rates. The data have been split into two classes. Graphically, this results in squares that can have two sizes. The smallest square, for instance, is drawn in those municipalities that counted between 1 and 100 accidents. When looking at the map, one's first impression is that it is much more dangerous to drive in the north-east than in the north-west since the map shows a greater concentration of symbols in the north-east. To base a rate increase on this visual impression would be completely unjustified, however. There are two main reasons for this. The first has to do with the number and size of the municipalities. The north-east has smaller and therefore more municipalities, each of which gets a symbol. However, the number of accidents represented by squares of the same size can differ considerably, as a comparison between selected municipalities shows when the original data are considered. When working in an interactive GIS environment, the original data are usually at hand.

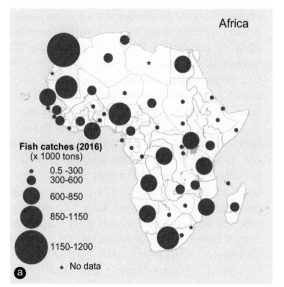
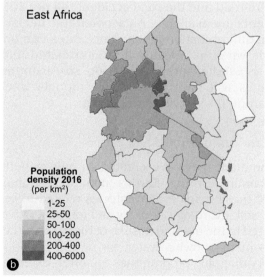

FIGURE 4.5 Mapping statistics: (a) fish catches in Africa in 2016; (b) population density in East Africa in 2016

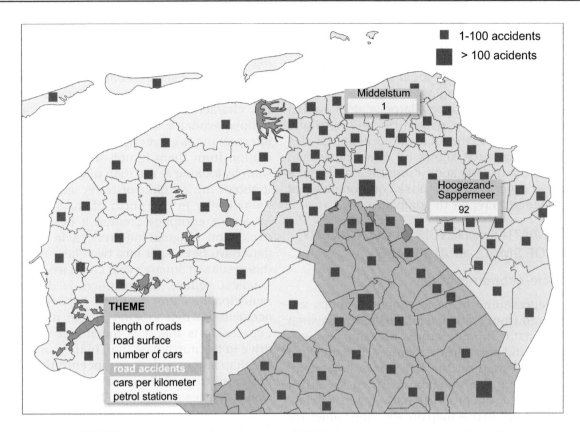

FIGURE 4.6 Socio-economic maps and GIS: road accidents per municipality

The user can point and click on the map to retrieve the data behind the map. It is also possible to quickly create other views on the subject by combining and visualizing other data or changing the class boundaries. The list on the left shows that, for each municipality, data on themes such as the total length of roads, the number of vehicles per kilometre road and the road accident toll per 1000 inhabitants are available. An accountable decision on local differentiation in insurance rates can only be made when all relevant and interrelated information is studied properly. To do so, insurance companies will have formulas or models where each parameter is given a weighting.

4.3.2 Environmental Maps

Environmental maps are created to gain a better understanding of the Earth's natural resources. Some of these maps are inventories related to vegetation, soil, hydrology, geology and forestry. Others are related to the use and misuse of these resources, such as maps showing water, air or soil pollution. Inventory mapping programmes are often extensive and time-consuming, especially in remote areas. To get up-to-date information on land use for those

areas through traditional fieldwork techniques takes too long. Here, the application of remote sensing techniques can be very helpful. Figure 4.7 provides an example. In a study for a rural development plan for this area in Eastern Bhutan a Sentinel 2 image (Figure 4.7a) was used. The radiation values measured by the satellite are rendered here by colours chosen to resemble the actual land cover; they have been combined with hill-shading in order to give an idea of the relief. The resulting map (Figure 4.7b) shows the actual land use for each pixel. Most of the region appears to be forested, alternated with shrubs and bare soils (landslides, bare rocks and moraines). It is an example of qualitative mapping.

For the urban area of Athens (Greece), a study has been carried out to judge the magnitude of air pollution (Koussoulakou, 1990). The study resulted in an air diffusion model, which could be used to predict the effects of different air pollution parameters. Maps and GIS played an important role in the model, which was capable of modelling a diversity of geospatial data. The model was calibrated with real-world data in order to be able to use it later to predict the level of air pollution under different circumstances. Parameters in the model were related to the topography (Athens is

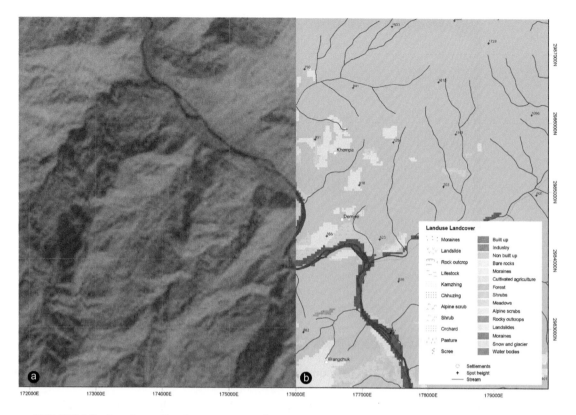

FIGURE 4.7 Land use/land cover map of Bhutan (b – Credits: National Lands Commission Bhutan) based on Sentinel 2 imagery (a) (https://scihub.copernicus.eu)

located in a bowl bounded by mountains and the sea), meteorology (wind, temperature, precipitation), urban conditions (land use, population, traffic density) and sources of emissions (industries, traffic, central heating). The individual parameters were each mapped in separate analytical maps using different thematic mapping techniques. The model itself, combining all parameters, resulted in several animated maps that showed air pollution concentration developments above the city during the day.

Figure 4.8 shows an example of an environmental map, displaying the concentration of carbon monoxide (CO) in the area. The quantity of CO is shown by tints of blue. Each represents a certain range of the data. The map itself is a single frame from a cartographic animation that shows the change in CO air pollution on 13 April 1985. The clock on the right shows the current time. Below the map, an interface allows the user to interact with the animation, and move forwards and backwards through time. More details regarding cartographic animations can be found in Chapter 8. In a smart-city environment with many different sensors streaming live data, a real-time map could be created.

4.4 GEOSPATIAL, THEMATIC AND TEMPORAL COMPARISONS

While working with geospatial data in a GIS environment, it is very likely that one will have to deal with three basic query types: 'Where?', 'What?' and 'When?' (see Figure 1.3). In a geospatial analysis operation, the queries will result in the manipulation of the geospatial data's geometric, attribute or temporal components, separately or in combination. However, just looking at a map that displays the data already allows one to evaluate how certain phenomena vary in quantity or quality over the mapped area. Often one is not just interested in a single phenomenon, but in multiple phenomena. For some aspects, analysis operations are required, but sometimes even a visual comparison will reveal interesting patterns for further study. Geospatial, thematic and temporal comparisons can be distinguished. A geospatial comparison means looking at different areas at the same scale, to see if certain patterns correspond or differ. A thematic comparison means looking at maps displaying different themes of the same area to see if the geospatial distribution of the themes is similar or different. Temporal comparisons are executed

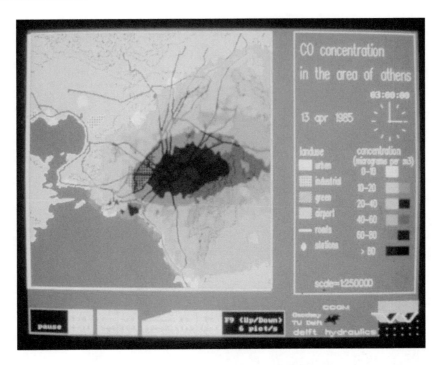

FIGURE 4.8 Air pollution in Athens, the CO concentration on a specific date and time (Koussoulakou, 1990)

by studying views of the same area and theme for different dates. The next subsections will look at each of these comparisons in more detail.

4.4.1 Comparing Geospatial Data's Geometry Component

Comparing two different areas seems to be relatively easy while focusing on a single theme, e.g. hydrology, relief, settlements or road networks. However, to make a sensible comparison, the maps under study should have been compiled according to the same methods. They should have the same scale and the same level of generalization or adhere to the same classification methods. If two areas are compared in order to get an impression of the population density based on the settlement density, both maps should show the same type of settlement. They should both show those settlements with, e.g., more than 10 000 inhabitants. If one is comparing the hydrological patterns in two river basins, the individual rivers should be on the same level of detail with respect to generalization and level of branches.

In Figure 4.9, the deltas of the Rivers Rhine and Scheldt (the Netherlands) and the Rivers Neuse and Roanoke (North Carolina/Virginia) are shown. At first glance, both areas look quite different. When one changes the orientation of the Neuse/Roanoke delta, this view suddenly changes. The shapes of the coastlines clearly have something in common. To

ease the comparison, the names in both maps have been left out. The links between the towns in the Neuse/Roanoke delta seem to be there just to connect the harbour with the hinterland. Links in the Rhine/Scheldt delta are more regular. On the basis of the more numerous settlement symbols, this area seems to be much more densely populated. The hydrographic pattern in the Roanoke/Neuse delta area is apparently much more of a barrier than in the Rhine/Scheldt delta area. So simply by comparing a map of an unknown area with one of a known area (produced to the same specifications), it is possible to learn quite a lot about otherwise unfamiliar areas. However, this is only possible when one has homogeneous geographical maps available. Traditionally, in the small-scale map range, the International Map of the World 1:1 million, produced up to the 1980s) or – as shown in Figure 4.9a – the 1:2.5 million Karta Mira (produced in the 1970s) could be used (see also Chapter 6). Standardization of their legend and the uniform level of generalization make these maps very suitable for small-scale comparisons. In Figure 4.9b, more current data (2018) from the Natural Earth data set has been used (https://www.naturalearthdata.com).

The comparison in Figure 4.10 is of a different nature. To study the similarity between the estuary of the River Thames near London and the estuary of the River Solent near Southampton and Portsmouth, not only the orientation of the area has been changed, but also both areas have been

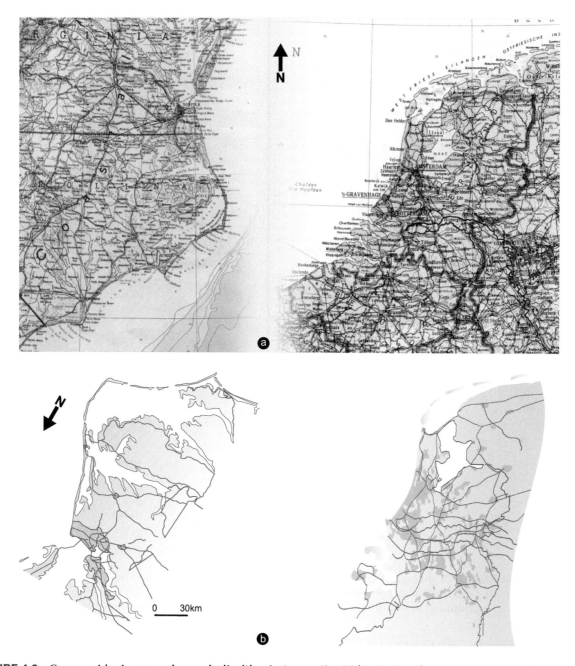

FIGURE 4.9 Geographical comparison: similarities between the Rhine Delta (right) and the Cape Hatteras area (left): (a) both areas north-oriented; (b) Cape Hatteras area rotated (map detail in a) from State Office of Geodesy and Cartography, Hungarian's People's Republic Budapest 1966 and VEB Kartographischer Dienst, Potsdam, 1964

transformed into simple geometric shapes. This example has been adapted from Cole and King (1968). To study the industrial development patterns of both areas (the docklands), the orientation has been altered so that comparisons are easier. The upstream regions of both rivers are now at the north and the open sea at the south of both maps. For the River Thames, this means a 90° rotation. In addition, the River Solent was also mirrored and its scale enlarged. To minimize the influence of the shape of the riverbanks and islands, the data were further generalized, as can be seen in Figure 4.10c. Actually, this type of generalization is called 'schematization', because of its rigorous approach. Comparing both schematic maps, it can be seen that the main urban areas can be found upstream. The docklands are to be found there as well. Both areas show oil refineries near to the open sea (large oil tankers), and a resort area very close to the open sea.

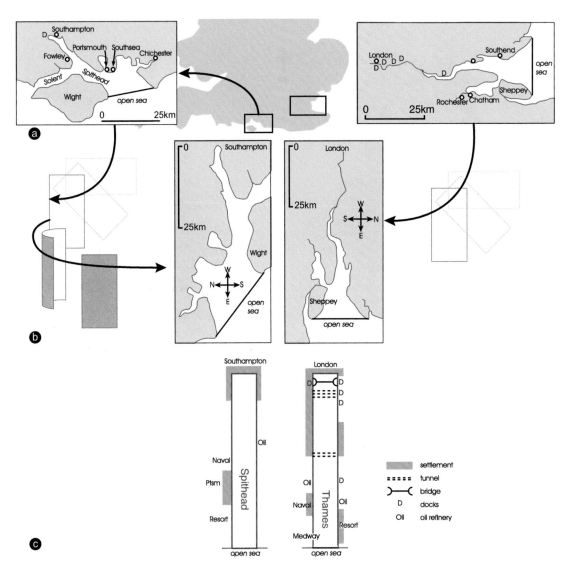

FIGURE 4.10 Topological comparison looking at the River Solent and the River Thames (after Cole and King, 1968): (a) both areas in correct orientation and scale; (b) after manipulation of orientation and scale; (c) both areas after schematization

4.4.2 Comparing the Attribute Components of Geospatial data

If two or more themes related to a particular area are mapped according to the same method, it is possible to compare the maps and judge similarities or differences. However, not all mapping methods are easy to compare. Choropleth maps are the most simple to compare, at least as long as the administrative units are the same in both maps. To be able to compare qualitative maps (e.g. Figure 4.7b), they must be converted into, for instance, isoline maps. Comparing isoline maps is a well-established technique. A classical example is the study by Robinson and Bryson (1957), who compared the precipitation and population density maps of America's Midwest.

The importance of their study is partly due to the fact that the authors considered the accuracy of the information during the comparison. Comparing isoline maps is executed by measuring and comparing values in each map at the same location. Figure 4.13c shows an example of this approach. Comparing maps with point symbols can only give a rough idea of similarities or differences.

The maps of south-east Britain in Figure 4.11, based on census data, show the distributions of children and persons over retirement age. Both age groups are mapped relative to their average value for the whole country. On average, children under 15 years of age make up 24% of the population and those retired 16%. The distribution of the children is closely linked to the distribution of

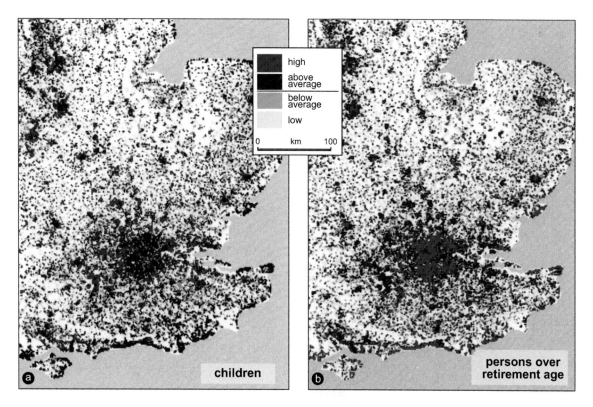

FIGURE 4.11 Thematic comparison of British population statistics: (a) distribution of children; (b) distribution of elderly people (courtesy HMSO, People in Britain, A Census Atlas 1980)

younger married couples. From the maps, it can be seen that they are concentrated in the suburban areas around most large cities as well as in the new towns around London. They are under-represented in the central London area, along the coast and in rural areas. The map in Figure 4.11b is a mirror image of the children's map. The elderly people are obviously concentrated along the coastline, especially in the south. People over retirement age are over-represented in the central urban areas and most rural areas as well. Some of these concentrations are popular retirement areas, like the south coast, while other concentrations, like those in the rural areas, are due to the younger people leaving these areas.

4.4.3 Comparing the Temporal Components of Geospatial Data

Users of GIS are no longer satisfied with analyses of snapshot data but would like to understand and analyse whole processes. A common goal of this type of analysis is to identify typical patterns in space over time. Change can be visually represented in a single map. A well-known example is the maps displaying the westward movement of the centre of population of the United States. It moved

from Maryland (around 1800) to its current location in Missouri (2010). The centre of population is calculated by summing for each census the centre coordinates of all counties, weighted by their population. The trend shows the growing importance of the west coast population. Understanding the temporal phenomena from a single map will depend on the cartographic skills of both the mapmaker and map user since these maps tend to be relatively complex. An alternative is the use of a series of single maps, each representing a moment in time. Comparing these maps will give the user an idea of change (see, for instance, the maps in Figures 1.12, 4.12 and 4.13). The number of maps is limited since it is difficult to follow a long series of images. Another alternative is the use of dynamic displays or animations. A well-known example is the animation shown during the daily weather report on television presenting the change in the atmosphere's meteorological conditions. The Athens map in Figure 4.8 is another example. Section 8.3 will explain more about cartographic animations.

Figure 4.12 shows how line patterns can be compared. To be able to do so, change has to be emphasized. The upper map in the figure is a representation of the Polish railroads at the beginning of the 20th century. The pattern reveals the political

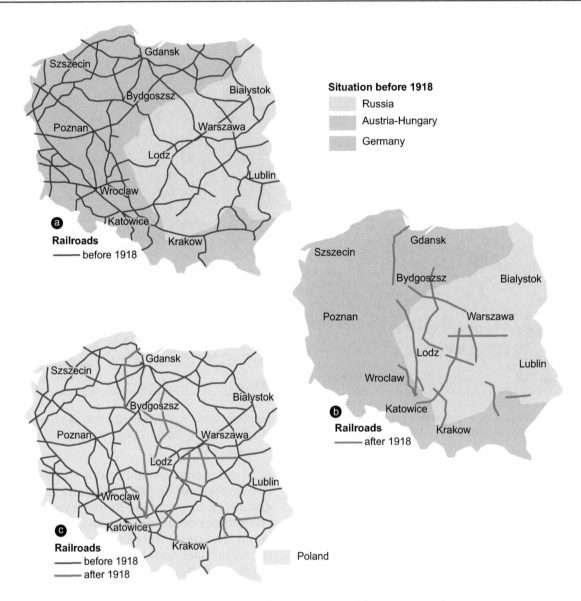

FIGURE 4.12 Temporal comparison of the development of the Polish railroads (after Grote Bosatlas, courtesy Noordhoff): (a) railroads built before 1918; (b) railroads built after 1918; (c) current railroad network

boundaries at that time. The lower map shows today's railroad network, which is clearly denser than the one in the upper map. However, to understand the changes, a map showing the changes has to be produced. From this map, which also includes the pre-World War I boundaries between Austria–Hungary, Germany and Russia, it can be understood that the railroads added were built to erase the boundary pattern and emphasize the importance of the new economic centre, Warszawa.

Interest in the behaviour of glaciers under today's environmental conditions led to the Greenland Ice Margin Experiment (GIMEX; Roelfsema et al., 1995). Part of this research project consists of monitoring change. One of the glaciers that

researchers focused upon was the Leverett Glacier on Greenland's west coast. By building three digital elevation models based on aerial photographs from 1943, 1968 and 1985, glaciologists expected to gain an impression of change in the volume and shape of the glaciers. Some results are displayed in Figure 4.13. To compare the change in the extent of the glacier, the glacier fronts in 1943, 1968, 1985 and 2015 can be compared (Figure 4.13a and c). An impression of changes in shape is given by the three perspective maps (Figure 4.13b). By adding the contour lines and the glacier margins, absolute comparisons are possible. Growth or decline could be suggested by the use of different symbols. The maps in Figure 4.13 give an impression of change.

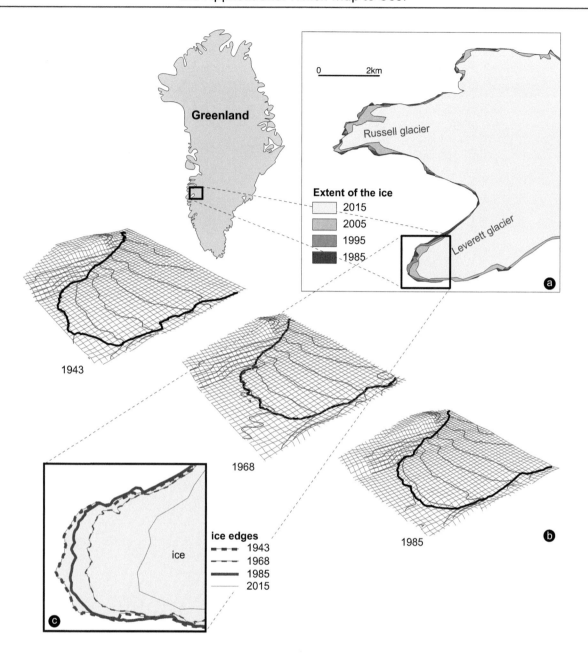

FIGURE 4.13 Decline and growth of the Leverett Glacier in Greenland (courtesy Roelfsema et al., 1995): (a) the ice edges through time; (b) the changing glacier in a perspective view; (c) differences between 1968 and 2015 in detail

In a GIS environment with extensive surface modelling functionality, more absolute values can be calculated. Other map products derived from digital elevation models will be elaborated in Chapter 5.

FURTHER READING

Dale, P. F., and J. McLaughlin. 2000. *Land Administration*. Oxford: Oxford University Press.

Kimerling, J. A., A. R. Buckley, P. C. Muehrcke, and J. O. Muehrcke. 2016. *Map Use: Reading, Analysis, Interpretation*. 8th ed. Redlands: ESRI Press.

Laurini, R., and D. Thompson. 1992. *Fundamentals of Spatial Information Systems*. Vol. 37, *APIC Series*. London: Academic Press.

Samet, H. 1989. *The Design and Analysis of Spatial Data Structures*. Reading, MA: Addison-Wesley.

Tomlin, C. D. 1990. *Geographic Information Systems and Cartographic Modelling*. Englewood Cliffs, NJ: Prentice Hall.

Williamson, I., S. Enemark, and J. Wallace. 2010. *Land Administration for Sustainable Development*. Redlands, CA: ESRI Press.

5

Map Design and Production

5.1 INTRODUCTION

In previous chapters, we have seen that maps are geospatial images that can influence people's conception of space. Maps have this influence partly because of convention and partly because of the general characteristics of the graphic cues used, either on paper or on the monitor screen. Convention especially plays a role in topographic mapping: most of the symbols used on topographic maps (see Chapter 6) have come down to us in a form conditioned by the 18th-century examples, and we have stuck to them ever since. Among these conventions are the rendering of water by a blue colour, forests by a dark green and built-up areas by a red, grey or pink colour. Association may have been at the root of this usage, but may not be valid any more, and so it has changed into convention. The convention of using specific symbols on topographic maps originated in the example provided by French topographic mapping practice in the 18th century. This convention has been strengthened by the fact that in the 19th century, all topographic maps were produced with the same objective, i.e. infantry warfare.

The result, for topographic maps, is a large collection of symbols – for buildings, infrastructure, terrain aspects, hydrography and administration – that has been more or less standardized, and that

works because it has been standardized. It works moreover because people are used to this kind of symbology on these kinds of maps; it can be learned by those that use topographic maps (see also the legend in Figure 6.45).

There is an ever-increasing proportion of maps, however, that have nothing to do with descriptions of the terrain and its fixed assets, and instead have other objectives: the thematic maps defined in Section 3.2. Here, because of the ever-changing themes, and the ever-changing aspects of reality that are visualized, one is not governed by convention but is able to improve information transfer by using the innate characteristics of the variation in graphic characteristics (e.g. shape, colour, size, texture) of the symbols we use. When we study map symbols as such, i.e. not as a representation of the Earth's surface but as a set of dots, dashes and patches, we find that it is this variation in graphical aspects, which conveys meaning to the map reader, like a sense of varying magnitude or differences in nature. It was the French geocartographer Bertin who placed the various sensations created by the variations in graphical aspects in a logical structure (Bertin, 1967, 2011).

Indeed, when one looks at Figure 5.1a, one will perceive a dark circle against a light background, but this will not tell us much. Even when a legend is added to indicate that the circle represents

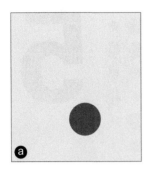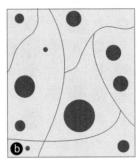

FIGURE 5.1 A map: (a) without information because of the lack of context or reference; (b) with information due to its geographical and thematic context

5.2 SYMBOLS TO PORTRAY DATA RELATED TO POINTS, LINES, AREAS AND VOLUMES

The data that have to be visualized will always refer to objects or phenomena in reality. These can be heights measured at specific points, traffic intensities measured along a route network, numbers of inhabitants living in an area or the volume of a hill in thousands of cubic metres. In Section 7.2, we refer to specific aspects of the data; here, we will show the graphical means we have at our disposal to represent them.

In cartography, we use dots, dashes and patches to represent the location and attribute data of point, line, area and volume objects as in Figure 5.2. (One could mention here that the definition of point, line and area objects, i.e. objects that refer to point, line and area locations, is a matter of scale: a line that represents a river would have to be exchanged for an area if the scale of the map would increase. The built-up area of a settlement would be rendered by a dot if the scale of the representation were to decrease enough.) It seems to be obvious that point data are represented by dots and that area data are represented by patches, but there is more to it than that.

Figure 5.3 provides an example of various kinds of point data: equal-sized dots, each denoting the same value (e.g. ten inhabitants); dots that vary in size and thus represent different quantities for specific point locations (as in Figure 5.2); and – and this is represented by the addition of boundaries – proportionally sized dots that can also refer to enumeration areas. In the latter case, the dots could be considered area symbols even though each of them is centred on a point location (e.g. the respective area's gravity point). Another application of dots for rendering areal data is in a regular grid mode. The value valid for an area can be assigned

40 000 employees in an automotive factory, this information seems to be a waste of space, as it could have been expressed better in alphanumeric form. As there is no variation, there is no frame of reference, no context. But as soon as this single symbol is put into a geographical and data context, as in Figure 5.1b, this variation in graphical cues will immediately render geospatial information more effectively than an alphanumeric description could. Differences in size are obvious and will be conceived automatically as differences in number (of employees in automotive factories). So, a hierarchy will be perceived based on these differences in number. In addition to the hierarchy, a pattern will be discerned, influenced by the relative distances between the various symbols.

So, the differences in symbol size are an important characteristic, conveying to the map reader the sensation of differences in number – in our example here – to convey some idea of the relative importance and distribution of the automotive industry in an area. It is on these automatic reactions to variations in graphical cues that the graphical grammar is based (see Section 5.3).

So as the symbology of topographic (or 'inventory')-type maps covered in Chapter 6 is based more on convention than on this graphical grammar, and as thematic, communication-oriented maps (like those discussed in Chapter 7) are more based on this grammar, it is here, in Chapter 5, that this grammar will be discussed. Following the descriptions of data gathering techniques (Chapter 2) and map functions (Section 3.4), here the characteristics of the graphical signs will be explained. These perceptual characteristics of the graphical signs will have to be matched to the data characteristics analysed in Section 7.2 and the communication objectives (Section 7.4) in order to satisfactorily portray the information requested.

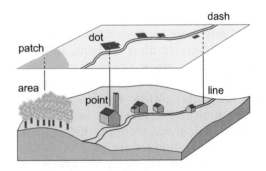

FIGURE 5.2 Relation of dots, dashes and patches to the point, line, area and volume objects which they can represent

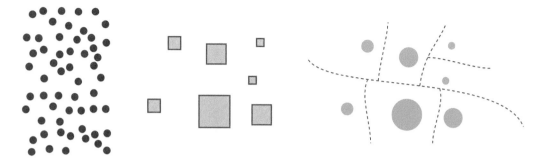

FIGURE 5.3 Various types of dots that are used as point or area symbols

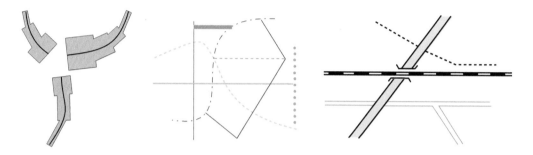

FIGURE 5.4 Various types of dashes that are used to symbolize linear objects

to the nodes in a regular grid superimposed over the area (see Section 7.5.3, Figure 7.25).

Figure 5.4 presents some dashes used to express various types of linear data: boundaries, roads and railways, flow lines proportional to the numbers of passengers transported, etc. Lines can also be used to represent areal data, by using them as shading, but, as is the case for point symbols, they must be combined in such a way that they are perceived as patterns and not as individual points or lines. Lines can also be used to indicate volumes (as in Figure 4.13c or 7.16d and e). In the same way, a string of dots representing a linear feature could be referred to as a 'line symbol'.

Figure 5.5 illustrates a number of patches used for representing areal data: patches that suggest qualitative or quantitative differences between the various areas concerned. As said above, it is their repetition that leads us to perceive the dots and dashes as area symbols. In all the cases in Figure 5.5, within the boundaries of each area, the patterns are homogeneous. If patterns were not homogeneous, they could be used to indicate volumes: hill shading would be a good example (see Figure 6.30).

FIGURE 5.5 Patches or patterns used to symbolize area objects

5.3 GRAPHIC VARIABLES

As in the example in Figure 5.1, it is the difference in symbol size which map readers perceive as a difference in numbers. In order to systematize the perceptual characteristics available, we will list them here:

Difference in numbers and ratios;
Difference in distance;
Difference in order;
Difference in quality.

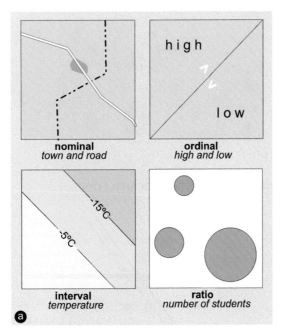

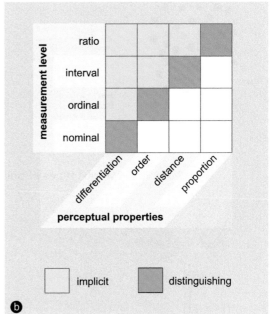

FIGURE 5.6 Portrayal of the measurement levels discerned (a) and the perceptual characteristics required in graphical variables to render them (b)

When confronted with differences in (grey) value or lightness of tones, one will experience a sensation of perceiving differences in distance. In the legend of Figure 5.6a, the population density is rendered by grey tones that have an increase which is perceived as regular: the distances between the classes are similar. That is why, for most observers, the population density values rendered would show a regular increase. This same difference in population density value (i.e. the distance between successive values) can also be perceived from point symbols different in size that are applied as area symbols in grid-type maps. These would have the added characteristic of also allowing ratios to be perceived (this density is so-and-so many times higher than that density (Figure 5.6b) (see also Figure 5.7).

As Figure 5.8 shows, differences in order will be perceived from differences in symbol size, from differences in value or lightness and from differences in grain (Figure 5.6c). Nominal or qualitative differences will be perceived from differences in colour hue, shape or orientation, but from these three variables, colour differences give the best performance.

In order for the symbol differences to be perceived as qualitative differences only, they must be perceived as having similar values. If one colour were to be perceived as much darker than another, then ordered differences would be experienced as well, the darker colour denoting areas that would be both

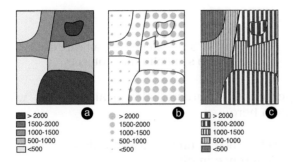

FIGURE 5.7 Differences in population density rendered by differences in (a) value, (b) size and (c) grain

	nominal	ordinal	interval	ratio
dimensions of the plane	+	+	+	+
size	−	+	+	+
(grey) value	−	+	+	
grain/texture	+	+	+	
colour hue	+			
orientation	+			
shape	+			

+ yes
− no

FIGURE 5.8 Relation of graphical variables to perception characteristics (based on Bertin's Semiology of Graphics, 1967, 2011)

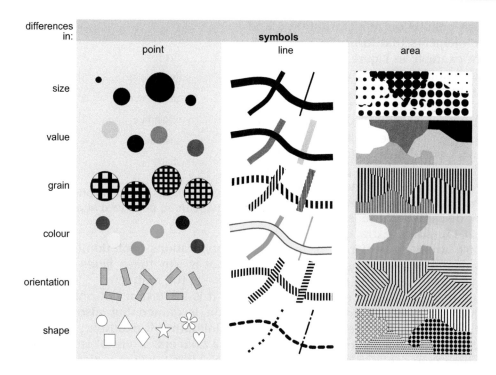

FIGURE 5.9   Basic graphic variables

different and more important than the lighter areas. In practice, darker colours can only be used to represent qualitative information for small areas; otherwise, they would dominate the image too much.

While discussing these perceptual characteristics of the graphic cues, we encounter various basic differences in the graphic character of the symbols we discern. All the differences imaginable between symbols can be summarized as being cases of six graphical variables. Bertin (1983, 2011) discerns, as basic graphic variables (Figure 5.9):

+ Differences in (symbol) size;
+ Differences in lightness or (colour) value (of the symbols);
+ Differences in (symbol) grain or texture (mainly for black and white maps; not much used anymore);
+ Difference in colour hue (of the symbols);
+ Difference in orientation (of the symbols);
+ Difference in (symbol) shape.

Figure 5.1 shows one example of difference in symbol size – but differences in line width (Figure 5.6c) or in area symbols like proportional dots in grid patterns (Figure 5.6b) also qualify (the graphic variable 'differences in size' never refers to the surface of the areas the symbols refer to!). Differences in (grey) value or lightness are shown in Figure 5.6(a) and also in Figure 5.10.

With differences in grain or texture, Bertin referred to differences that emerge when a specific pattern is being enlarged or reduced. The ratio between the areas that are black and white, respectively, will remain the same during this (photographic) process, but at the same time the coarser the pattern, the higher it will be perceived in the resulting hierarchy. Figure 5.6c shows a map using differences in texture in order to generate an impression of order among the categories discerned. In practice, this difference in grain isn't much used.

Differences in colour hue only work in providing a suggestion of qualitative differences when they are perceived as having similar lightness. Totally saturated colours (i.e. when the whole area is only covered by ink with one specific wavelength, so is not mixed with either white, black or any other colour hue) have different lightness values. Figure 5.11 shows this.

FIGURE 5.10   Differences in value or lightness

FIGURE 5.11   Differences in colour

Differences in orientation refer to patterns and not to the line elements that form the base map. The upper-right example in Figure 5.5 shows such differences in orientation. These can refer either to line patterns or to dot patterns.

Differences in shape can refer to differences in the dots, in the lines or in the patterns used for area symbols. Again, shape differences – as a graphic variable – would never refer to the shapes of the areas the various colours, patterns or symbols refer to – they only refer to the symbols themselves, as shown in the map legend.

After Bertin discerned these six graphic variables (1983, 2011), some extra ones have been proposed. These would be differences in colour saturation, arrangement and focus. Differences in arrangement refer to the regularity or non-regularity of the distribution of symbols. Focus refers to the clarity with which the symbols are visible, and so to their definition on the plane. We will not include arrangement and focus among the basic graphic variables, because they can also be thought of as aspects of definition or pattern of real-time objects. Colour saturation (also called 'chroma') can be defined as the percentage of the reflection of light from an object composed of colour of a specific wavelength. The larger the reflection percentage of the light with this wavelength, the more saturated or brilliant the specific colour will appear; therefore, differences in colour saturation would also be perceived as differences in order.

But what do we do with these basic graphic variables? The importance of discerning them and the perceptual characteristics of the differences in each of them is that they help map designers in selecting those variables that provide a sensation which matches the characteristics of the data or the communication objectives. Figure 5.8 relates graphic variables and perception characteristics to each other, and it is the key illustration of this chapter. As can be seen, it also refers to the dimensions of the sheet of paper or monitor screen on which our maps are drawn.

These planar dimensions of the map also have perceptual characteristics: as one location is not equal to another, space differentiates. Because of contiguity, individual point objects can be grouped. If on the way from A to C one has to pass B, there is a distinct order A–B–C, which cannot be changed. Distances or angles measured on the ground have numerical connotations.

It would perhaps be expected that variations in size, value and texture would also be able to denote nominal differences (Figure 5.8). They can, but at the same time they have hierarchical

	dots	dashes	patches
size	4	4	5
(grey) value	3	4	5
grain/texture	2	4	5
colour hue	7	7	8
orientation	4	2	-
shape	-	-	-

FIGURE 5.12 Visual isolation: the number of categories that can be perceived at a glance (Bertin, 1967, 2011)

connotations that dominate overall impressions. That is why these fields have been indicated as – no – in the matrix.

It is important not only that the correct impression is gained but also that it is gained with a minimum of exertion. Here, we may introduce the concept of visual isolation (which Bertin calls 'sélection'), which indicates whether or not all the relationships that can be perceived between the various categories discerned on the map can be perceived at a glance. Not all graphical variables work equally well in this respect, depending partly on the number of categories one wants to be differentiated on the map.

If it is nominal differences one is interested in, with areal objects (patches) different in colour, eight classes is the maximum that can be differentiated; if more than eight classes are selected, it will no longer be possible to discern the distribution of each of them.

So, if five classes of linear objects are to be distinguished between at a glance, then size, colour hue, value and texture are to be selected according to Figure 5.12. If the data to be rendered were ordered, colour hue would not qualify any more. If the pattern were dense, there would be no place for lateral extensions, so size could be discounted. Whether texture or value is used is left open to one's personal choice: they would be equally effective.

5.3.1 Visual Hierarchy

The selection of the most suitable graphic variables and processing them according to the proper mapping method is still not enough from a map design point of view. Legibility considerations would also play a role here; legibility is also determined by contrast, graphical density and angular differentiation.

The introduction of contrast is based on the assumption that the map data will consist of a number of categories that will each have a different role to play in the geospatial message. The data analysis process (Section 7.2) will result in the identification of more and less important data categories. The examples in Figure 5.13 both show the number of employees in the service industries in the Netherlands. Figure 5.13a does so, but poorly. Although the map does portray the numbers of employees per province, it is the sea which is most conspicuous, followed by the surrounding countries. Both the Netherlands itself and the proportional circles score equally low in their conspicuousness. The map in Figure 5.13b shows how the information hierarchy (the sequence from most to least important aspect of the data to be shown) should be portrayed. It is the employees that should stand out most, followed by the province they work in; the surrounding areas should come last. One need not even differentiate between the sea and the neighbouring states, as they are equally unimportant to the map theme (unless we expect the proximity of either of them to influence the distribution of the employees in this sector).

5.3.2 The Use of Colour

There is more to colour in maps than its suitability for distinguishing nominal categories. Among the aspects of colour that are differentiated between are colour hue (the dominant wavelength), colour saturation (the proportion of the light reflected, which consists of this particular wavelength; this can be diminished by adding white or black or other colours) and the (grey) value or lightness (the impression it would give when shown on a black-and-white monitor). The number of different (grey) values that can be discerned within one colour depends on its hue: for yellow, only three steps can be discerned, while for red and blue, six or seven can be distinguished (see also Section 5.8).

Colour perception has psychological aspects, physiological aspects, connotative/subjective aspects and conventional aspects. Among the physiological aspects, it has been noted that on small areas, it is difficult to perceive colours and that between some colours, more contrast can be perceived than between others (so this combination could be used in order to improve acuity).

Saturation differences can be effectuated in practice by adding black screens to the colour. Figure 5.15 compares the effect of value differences and saturation differences on a colour scale. The figure in Figure 5.14 shows a colour scale, which was lengthened by adding saturation differences to it. Colour differences are also used in situations where deviations from a central situation are indicated. The simplest case would be a binary map: showing those areas or points below or above a central value or threshold value. As soon as one would want to show gradations above and below this threshold value, several solutions leading to diverging maps would be possible (Figures 5.16 and 5.17, see also Brewer, 1994). As can be seen

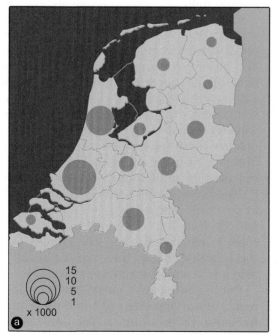

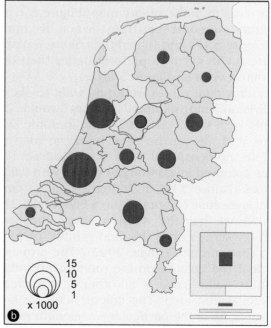

FIGURE 5.13 Graphical or visual hierarchy: (a) poor; (b) good

FIGURE 5.14   Colour-scale bar lengthened by adding saturation

value differences

saturation differences

FIGURE 5.15   Comparison of value and saturation

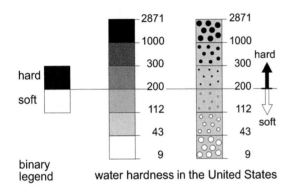

FIGURE 5.16   Options for the legend of a black-and-white map of water hardness in the United States, with left a binary map and in the middle a sequential scale using variations in (grey) value and at right a diverging or bivariate scale, using variations in grain

from the resulting maps, it is easiest to find the area one lives in in the binary map (Figure 5.17a); general trends are best seen from the non-diverging scales (Figure 5.17d and f), and the coloured maps (as Figure 5.17d) seem to perform better than the non-coloured ones (as Figure 5.17f).

If it is individual values we would want to show, a subdivision into different colours would be best. This is in fact a default setting in some GIS programs, but, as can be seen from the map in Figure 5.18, no geospatial trend is visible from the map thus coloured in, contrary to a map with value differences (Figure 5.19) or, to some degree, conventional layer zone colours (Figures 5.20 and 5.21).

How do we get the same colours that are selected on screen onto the final printed or plotted map (Brown and Feringa, 2003)? The problem here is that the colours on the monitor are additive colours – like those of all other light sources (sun, neon lights, etc.). The colour hues we see are perceptions of the particular wavelength radiation from these sources. In Figure 5.31, one can see that the primary colours emitted from these

sources are red, green and blue. When there is an overlap between these primary colours' wavelengths, the addition of them leads to the perception of secondary additive colours: yellow, magenta and cyan. And when all three primary colours are added, this will lead to a sensation of white. The more wavelengths added together, the brighter the image will be.

In contrast, printing or plotting inks, when applied to paper, act like filters. Therefore, they are called 'subtractive colours'. A paper sheet printed cyan will absorb the red wavelengths and only reflect blue and green, which together (Figure 5.31, subtractive colour scheme) will give us the impression of cyan. One can see that it might be problematical to convert the additive colours from the monitor into subtractive colours on a printed or plotted map.

Colours on CRT (cathode ray tube) monitor screens used to be the result of red, blue or green phosphor dots being activated by an electron beam. The more electrons emitted, the more intense the light of the phosphor dots would be. When activated by the respective electron beam, a green dot will emit a green radiation; when not activated, it will not radiate any colour. When one perceives colours on the monitor screen, it is a result of the differences in intensity of light emitted by green, red or blue phosphor dots. Liquid crystal display (LCD) screens work with continuous backlight and filters consisting of red, green and blue pixels that can be switched on or off by liquid crystals that react to electric current. The colours perceived on the monitor screen now are the result of impression caused by the activated pixels.

Printed onto paper, variations in colour will be the result of varying sizes of the dots printed in cyan, magenta, yellow or black. The dots can be superimposed or they can be printed next to each other, so that dots in various colours are intermixed. In that

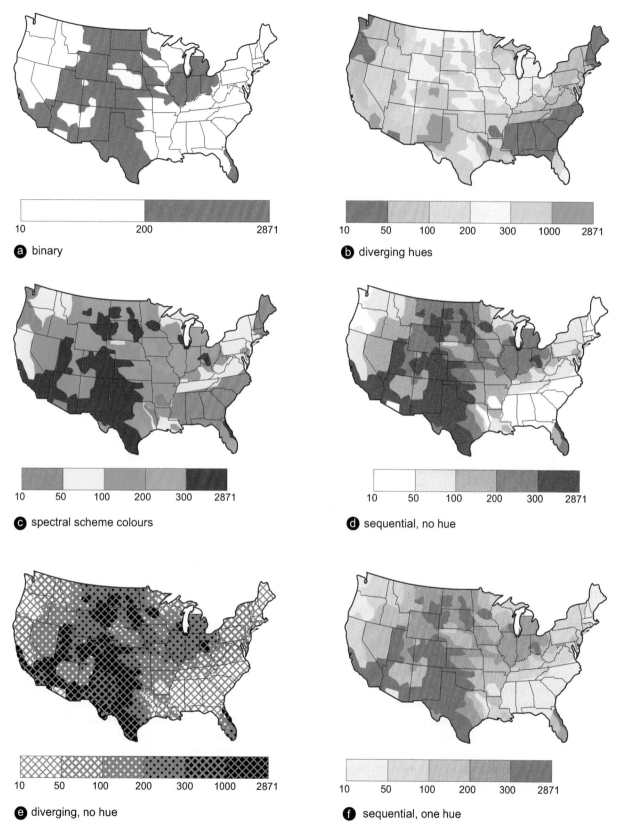

FIGURE 5.17 Water hardness in the United States: gradation over and under a threshold value versus non-threshold choropleths. Binary (a), diverging (b), spectral (c) and sequential (f) colour schemes, and no hue sequential (d) and diverging (e) schemes

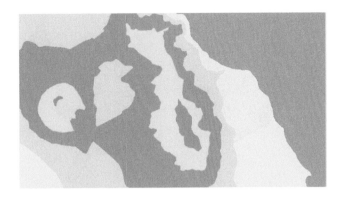

FIGURE 5.18 Layer zones with different colours for optimal retrieval of zones

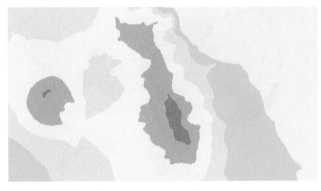

FIGURE 5.20 Layer zones with different conventional colours

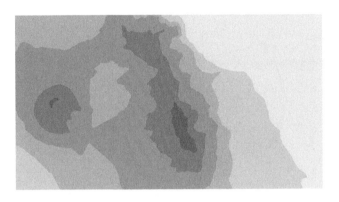

FIGURE 5.19 Layer zones with different tints for optimal recognition of trends

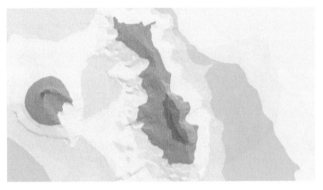

FIGURE 5.21 Layer zones draped over shaded relief model

case, we call it 'dithering', and this will result in the eye perceiving shapes in an average colour, which is less acute than when the dots are overprinted.

In order to really get a good fit between screen and plot colours, it is wise to produce colour charts (Brown, 1982). These are sheets with little squares of different colours, resulting from different intensities from yellow, magenta and cyan, with which the screen colours (which are red-, blue- or green-based) can be matched. The purpose of these colour charts is to judge in advance what the colours on the screen will look like when printed.

Some colours (red, yellow) seem to come forward from the paper they have been printed on; others (green and blue) seem to retreat. This rather weak phenomenon was once considered so important that the layer zones in height representation were coloured according to this phenomenon: blue-green for the area below sea level, green and yellow for lowland and hills, and brown colours for mountains, with red for areas over 5000 m. The effect was meant to be that those areas that appeared to be nearest to the viewer were in fact those farthest away from sea level. This effect, however, was impossible

to measure then – but by the time this was realized it had already been applied in school atlases and had become so popular that it turned into a convention. *The Times Comprehensive Atlas of the World* (Bartholomew/Harper-Collins 2018) provides a prime example of this convention.

5.4 TYPOGRAPHY: CONCEPTUAL AND DESIGN ASPECTS OF TEXT ON THE MAP

By the phrase 'text on the map', we mean the text within the map's frame, and not the additional information (title, legend, etc.) in the map's margin. Text on the map itself has the primary function of providing geospatial addresses – by naming the various map objects (geographical names or toponyms are used for this purpose; Kadmon, 2000). A secondary function would be to indicate the nature of objects. On topographic maps, terms such as 'factory', 'cemetery' and 'airfield', are printed in a special typeface, indicating these are generic terms and not toponyms or geographical names.

When compared with general typography or with texts in books, texts in maps have some special characteristics. Map texts consist of individual words instead of sentences; the words are unfamiliar instead of familiar, and there might be larger spacings in between the letters than is customary in book texts. In contrast to book texts, names on maps do not have to be horizontal, and they certainly are not neatly placed in lines: there is a jumble of different styles and sizes; the words refer to symbols instead of to each other as is the case in book texts and – and this is the worst aspect – the texts on maps superimpose lines and patterns. For all these reasons, text on maps has some extra requirements.

They should be easily identifiable and legible even if larger interspaces apply. It should be possible for the lettering styles selected to be differentiated through differences in boldness and size. If these requirements can be met, the next requirements for the selection of letter types are as follows:

They should be able to convey hierarchies (differentiating between more and less important objects or object categories);
They should be able to show nominal differences (between different data categories);
It should be possible to use them for relating to point, line and area objects.

Let us look how aspects of lettering can be used to meet these requirements. A hierarchy can be achieved in a number of ways:

Variation in boldness;
Variation in size;
Variation in spacing;
Variation in colour value;
Variation between upper case and lower case.

Nominal differences can be created by:

Variation in colour;
Variation in style (shape);
Variation between roman script and italics.

Figure 5.22 illustrates these differences, and Figure 5.23 shows an application of these differences. Frequently, in order to result in a hierarchy, a combination of the above-mentioned variations is used, resulting in, for instance, a series that goes from lower-case italics, 3 mm high, to bold roman upper-case letters, 10 mm high, differentiating between settlements on the basis of their numbers of inhabitants.

These nominal and hierarchical differences are important because they help one in finding specific names on the map. In the legend, one will always be able to find out how specific object category names are rendered on the map. If one finds that a name belonging to a category (e.g. a water name or hydronym, or a place name or the name of a physical object like a mountain) is displayed using a specific script type, one will know what shape of name to search for on the map, so the time needed to find this name on the map will be substantially reduced. This is because only names given in this specific script type will be targeted visually.

For printed maps, a requirement for map text would be that it is visible without having to use a magnifying glass; it should not be too thick (and risk obstructing map detail) or too thin (and risk the danger of being lost in the map detail). There should be good differentiation between the letters e and c, between o and u, and between u and v, between 3 and 5 and 8, and between 1 and 7. Lettering should be resistant against zooming in or out to a reasonable degree.

Different objects on the map have different requirements: point objects like cities are preferably named by text that is placed slightly above or below the line the object symbol is on, and preferably to its right. Linear objects like rivers are preferably named by text that is parallel to and close to the lines, and even follows their bends (this is something difficult to achieve with current mapping

		BERN	GENÈVE	LUZERN	BEX	SION	SCHWEIZ
	(high)						
difference in hierarchy		spacing	case	size	boldness	width	grey value
	(low)						
		SPIESS	Gryvon	VILARS	GSTAAD	SION	SCHWEIZ

		Argentine		MURTENSEE		VALAIS
difference in quality		colour		style		roman/italic
		Lac Léman		LAC DE MORAT		RHÔNE

FIGURE 5.22 Variation of map scripts in order to show hierarchical and/or nominal differences

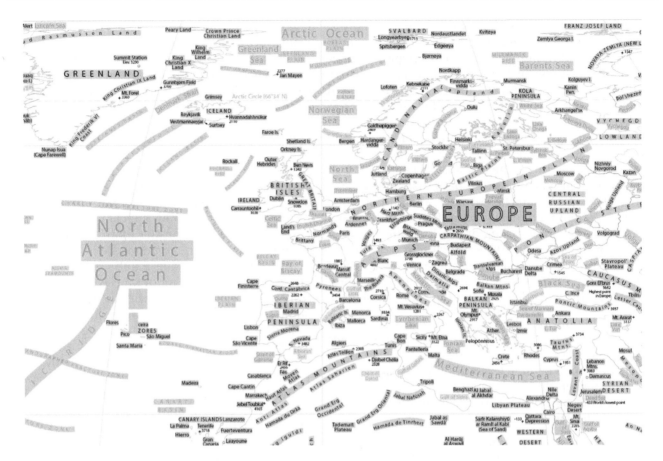

FIGURE 5.23 Detail of the type layer for the Equal-Earth physical world map. Tom Patterson, Equal-Earth.com

software). In order to relate to area objects, one would try to show the extent of the object by covering its largest extension with its name, which means both interspacing and, when this largest extension is not horizontal, tilted names. It is with these techniques that an optimal relation between text and map will emerge (see Figure 5.24).

Finally, in order to promote rapid identification, a short note about map title and headline. From the data analysis procedure covered in Section 7.2, the invariant aspect (the information aspect common to all map data elements) will be deduced. This common aspect can be expressed as the title of the map. It is customary to represent the most important aspects of the title (those referring to area and theme) as prominent keywords or headline (called 'vedette' in French, e.g. 'Land cover in South Sumatra in 2010'), and this will be displayed boldly on the map. The complete title can be rendered less conspicuously underneath (e.g. 'Land cover as interpreted from SPOT imagery taken in the period 2010–2012 of Sumatra south of the 4th meridian'). This will help in speedy identification.

5.5 REQUIREMENTS FOR THE CARTOGRAPHIC COMPONENT OF GIS PACKAGES

This section covers the transformation of the digital cartographic model (DCM) discussed in Figures 1.4 and 2.10 into the map, be it permanent or virtual (Figure 5.25). This used to be a time-consuming aspect of cartography but has now been reduced to mere button pushing when output is required.

The functionality of software available for the production of maps varies strongly. The larger GIS packages have very extensive cartographic modules, with options for map design and production, while others such as Carto and Tableau have limited variation to visualize GIS data. However, with some scripting knowledge tools such as Mapbox and Kepler.gl, let you create advanced interactive cartographic visualizations, as well as by using libraries like OpenLayers. With even more scripting skills, the options using the data-driven visualization library D3.js are nearly unlimited. If an organization has to select the (GIS) software to produce maps, some specific aspects regarding

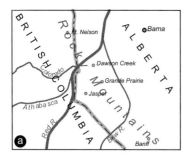 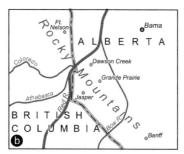 

FIGURE 5.24 Bad (a), good (b) and even better (c) text positioning: place names should be positioned to the right slightly above or below the symbol level, river names above and parallel to their course, and area names positioned so as to maximally cover the named area. Unless otherwise necessary, all names should be horizontal. Wherever there was a danger of lines crossing (and obliterating) names in (b), the former have been discontinued to leave open spaces for the letters. In (c), rivers and river names are in blue in order to more easily identify them

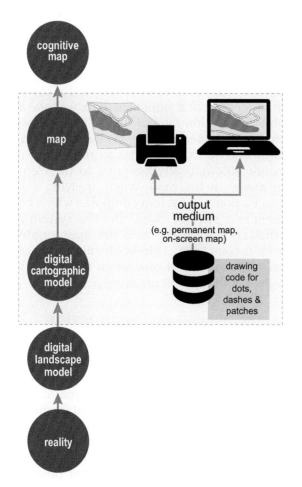

FIGURE 5.25 From DCM to permanent or virtual maps

cartographic functionality have to be considered. These reflect general use, design and output. The packages should be able to handle topographic and thematic data. It would be useful if the topographic data could be split over different layers. Depending on the map design, the relevant layers can be switched on or off (examples are layers with roads, boundaries and administrative units). Packages should have the functionality to produce thematic map types to deal with qualitative and quantitative data (e.g. choropleth and chorochromatic maps). A link with a database is preferred because it allows one to use the same map design for different (temporal) data without too much effort (see Chapters 6 and 7).

The software should have (interactive) facilities to create a legend, north arrow, scale bar and map title. Often the legend is created by default, but users should have the possibility of choosing their own symbology and colours. For incorporating a scale bar and a north arrow, many packages have a symbol library available from which the users can use the most appropriate representations. For titles and other texts, many different fonts are often available. Some packages are not able to display any marginal information at all when dealing with web maps, because screens are relatively small. In these situations, the user can sometimes access the marginal information through pop-up menus. In other words, the necessary information would be available on demand.

5.5.1 Data Manipulation

Options to change the coordinate system, to generalize topography or to classify attribute data are needed. For most GIS software packages, this is no real problem since these facilities are seen as generic GIS functions. However, generalization is often limited to a single-line generalization algorithm, and the possibility to classify attribute data will depend on the database linked to the GIS package. In those packages devoted to mapping only, data manipulation facilities are often limited, and the user can, for instance, choose between three data classification methods only (see also Section 7.3).

5.5.2 Output

Facilities to send maps to printers, to pdf files, to publish on the Web or to mobile phones should be available. Pdf files, i.e. Adobe portable document files, are a highly suitable way to distribute documents of any kind over the Web or as stand-alone files as they can be viewed with a free viewer. It should be possible to visualize the maps designed on any output medium. For interactive web maps, the visualization adheres to responsive web design.

When large print runs are needed, offset presses are called for. Since printing presses can print only one colour at a time for each printing plate, in order to reproduce all possible colours, one needs at least four printing plates (for the primary colours and black, see Section 5.3). The preparatory process that leads up to these four printing plates is called 'colour separation' (see also Figure 5.30). When it comes to automatic colour separation, only the larger GIS packages and the generic desktop publishing packages can handle this. However, today digital presses exist that do not require printing plates. Some of the larger GIS packages have dedicated modules for cartographic production ... When the focus is on online interactive map publication, the webpages will be encoded in the HTML5 standard, often combined with standards for styling and layout (CSS3), using scalable vector graphics (SVG) for resolution-independent graphics. This is often in combination with WebGL, the JavaScript API, for rendering interactive 2D and 3D graphics.

5.5.3 Graphical User Interface

A big problem with most GIS-related software is their user interface. If users agree upon one issue, it is the lack of user-friendliness of GIS-related software. If one realizes the enormous number of commands and options available, this is not strange. A graphical user interface (GUI) is regarded as a minimum requirement, especially when interactive map design has to be executed.

Some general questions to consider when opting for a mapping package are related to the following issues:

Quality: How is the functionality, as described by the vendor, realized? Is it linked to cartographic theory? What kind of output options (e.g. map types) are available?

Usefulness: Does the package fit into your organization's computer environment (MS Windows, Mac-OS or Linux)? Is it compatible with other available (GIS) software packages (to allow for data exchange, for example)? Most important is that the software can handle the standards of the Open Geospatial Consortium (OGC) and can work with the different data (exchange) formats offered.

User-friendliness: How easily will a new user be able to work with it? What level of training is required? What are the manuals like? How good is the support by the vendor? How are errors reported and processed?

Cost: How much has to be spent (initial cost and maintenance)?

Tomorrow's users of GISs will require easily produced on-screen maps as a direct and interactive interface to their geographical data, which will allow them to search for spatial patterns. In this process, the on-screen map plays a key role as an interface to the data, and it becomes central to spatial data handling and analysis activities.

In order to serve as an example for the production procedure of a statistical map, we take a population distribution map of the United States for the year 2020. Based on the available statistics per county and the county codes stored in the GIS, values are automatically assigned to their enumeration areas on the map. When applying a specific classification, the system computes which counties belong to a specific class, and the system applies to that county on the map the appropriate shading or colour tone. This is all relatively easy and straightforward. It is only when we want to adapt the design that problems emerge.

5.6 MAP DESIGN AND PRODUCTION

On-screen interactive design has been mentioned in the previous section. The interactive design and production functionality of most GIS software packages has matured over the years. However, some problems remain like text placement, even though label placement routines are available to assist. Often the final touch is still done using software like Adobe Illustrator, which would allow the user to drag the text to the correct position with the mouse, and even scale it interactively. Also, the application of specific design and construction functions available in graphic packages gives the graphics a fresh and sophisticated look that is missing or very difficult to realize in most maps created with GIS software, even though these GISs do have extensive cartographic functionality. Figure 5.26 shows the difference between a default GIS map (at the left) and the sophisticated version produced from it, enhanced by a

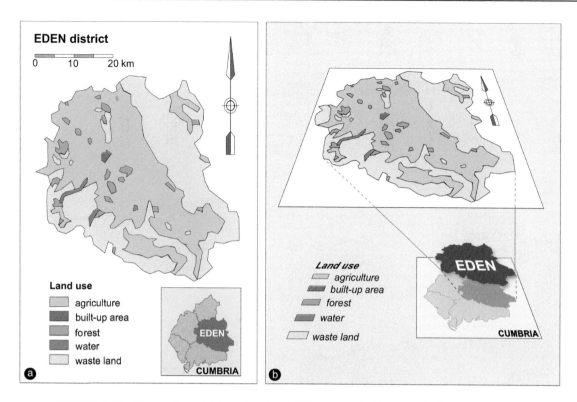

FIGURE 5.26 Examples of (a) a standard GIS map and (b) a sophisticated DTP map

graphical package. These packages, on the other hand, are only about graphics and elementary options, like input of geographical boundaries, are often problematical, although plug-ins (like MAPublisher) give a graphical package (like Adobe Illustrator) some GIS functionality. Most graphical packages offer libraries with maps, just as they offer libraries with flags, animals, cars, airplanes and other clipart symbols, and have many text or symbol fonts available as well (see Figure 5.27 for some examples). Natural Earth (https://www.naturalearthdata.com) provides free vector and raster map data at 1:10, 1:50 and 1:110 m scales. Natural Earth Vector files are provided in ESRI shapefile format, while the Raster files in TIFF format. The geographical coordinate system (projection) is used and can be adjusted for the vector files to other projections.

Today, the ability exists within graphic working environments to paste and copy graphics from one application into another. When dealing with a document including text, graphs, tables and maps, clicking on a table will make spreadsheet functionality available, clicking a map will activate GIS or graphics software, and all changes can be saved in the original document. This is the strength of the current desktop environments in which geographical data can be as easily dealt with as today's texts and tables.

The latter part of this section is organized according to the scheme in Figure 5.28. It splits the map production process into five distinct phases. The text will concentrate on the digital approach. The scheme does not pretend to be the only possible or complete approach, but it presents a valid generic method of map production. Other cartographic textbooks, such as *Elements of Cartography* by Robinson et al. (1995) and ICA's *Basic Cartography* series (Anson and Ormeling, 1995–2002), go into much more detail, regarding the conventional approach. The phases distinguished in the scheme are setting the map objectives and map specifications, collecting the data, creating the map image and its production, and distributing the final product.

1. The first phase is setting the map objectives: what area and map theme should it visualize, and for which year, and what should the map convey to the reader? The map type, aggregation level and resolution have to be defined, as well as the audience the map is intended for. In a traditional mapping environment, a sketch map would have been produced, to help this decision-making process. Map design specifications have to be set as well: will the map be in colour or black and white, and which font is to be used for lettering the map?

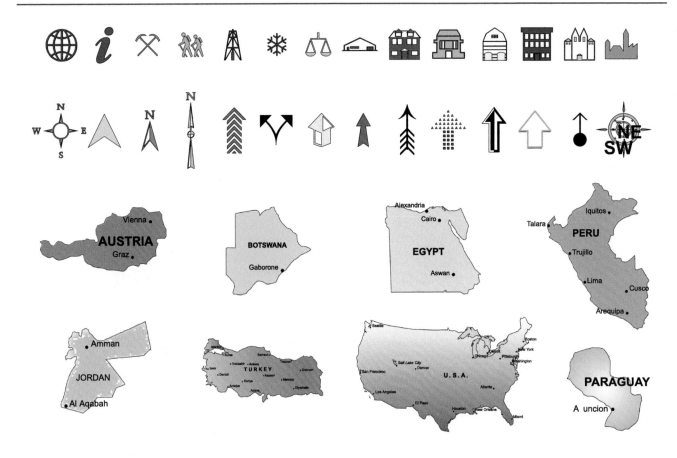

ABCDEFGHIJKLMNOPQRSTUVWXYZ   1234567890 abcdefghijklmnopqrstuvwxyz

ABCDEFGHIJKLMNOPQRSTUVWXYZ   1234567890 abcdefghijklmnopqrstuvwxyz

ABCDEFGHIJKLMNOPQRSTUVWXYZ   1234567890 abcdefghijklmnopqrstuvwxyz

ABCDEFGHIJKLMNOPQRSTUVWXYZ   1234567890 abcdefghijklmnopqrstuvwxyz

ABCDEFGHIJKLMNOPQRSTUVWXYZ   *1234567890 abcdefghijklmnopqrstuvwxyz*

ABCDEFGHIJKLMNOPQRSTUVWXYZ   1234567890 abcdefghijklmnopqrstuvwxyz

FIGURE 5.27 Desktop publishing packages and available fonts and clipart for cartographic purposes

2. In the second phase, the data will have to be selected from one or more (edited) digital landscape models and statistical files available in a country's spatial data infrastructure. An example from the Netherlands is PDOK, where both thematic files and background maps can be downloaded. The selection will include topographic base data and thematic content representing the results of spatial analysis. This selection will have to be processed: the topographic base data might be needed in a different projection, only a specific part of the mapped area might be needed, and it might have to be generalized and updated. The thematic attribute data will have to be processed as well, in order to have it answer the requirements set in phase 1. Ratios might have to be computed, or densities, or trends, might have to be assessed. And the topographic and attribute data would have to be linked as well.

According to the map design specifications set in phase 1, the result of the data collected and combined in phase 2 will then take form in phase 3.

3. This phase consists of the definition of the DCM, which will contain all graphical attributes (such as line colour, shading patterns and text fonts) of the geographical objects to be represented in the map. A similar process takes place in the conventional environment

production phase	result

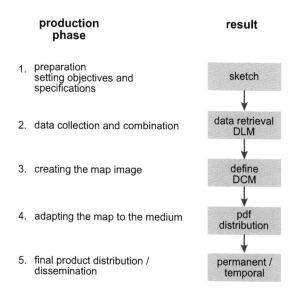

1. preparation
   setting objectives and
   specifications — sketch

2. data collection and combination — data retrieval DLM

3. creating the map image — define DCM

4. adapting the map to the medium — pdf distribution

5. final product distribution / dissemination — permanent / temporal

FIGURE 5.28 Map production scheme: five production phases can be distinguished, and for each of these, digital and analogue options are given. It is possible to switch from digital to analogue, and vice versa, at almost any phase

where the sketch map is populated with data and converted to a map that includes the information similar to a DCM. The layout of the maps represented by DCMs is inspired by the contents of Sections 6.2 and 6.3 and Chapter 7. Produced digitally, the result of the DCM will be visible on-screen. This medium has its limitations in size, resolution or colour as there is no one-to-one relationship between the image as displayed on screen and printed on paper. The map image displayed can be used to check if it matches the specifications. Phases 1, 2 and 3 might be part of an interactive design process. During this process, one deals with questions such as: 'Do we have to include additional data or should we delete some to keep the map legible and informative?' and 'Is the line thickness correct?' 'Are the place names positioned correctly?' If changes have to be made, one can return to phase 1 or 2 as the lines in the scheme show. During this phase, one cannot determine whether the colours of the symbols match harmoniously or contrast sufficiently. For that purpose, some calibration work would be needed beforehand.

4. The on-screen (web) map could even be the final product. Maps produced in a GIS environment can be viewed in that GIS application or exported (as a png or jpg, for instance, or as a pdf). In the first case,

they would have additional capabilities: not only could their production be effectuated interactively but also, more importantly, it is possible to keep a link with the GIS database. The map can be queried and so more in-depth or additional information can be obtained by the viewer from the map. For instance, the real value of an individual geographical unit in a classified choropleth can be accessed. It is possible to select a group of units and ask for statistics.

If one intends to keep the map image in digital format only, one might store it as a bitmap, pdf file or in any other format. It can be made available on the World Wide Web (WWW). Other options would be to incorporate the map in an electronic atlas or in an animation or multimedia environment. If the digitally produced map is to be reproduced on paper, conversion software and hardware have to be used that transform the DCM into a tangible format.

From Figure 5.29, it can been seen that there are multiple routes to create maps, either on paper or on screen. And it should be realized that this scheme presents one of the many other options possible. Starting point is an assumed data store that provides the necessary data to create the maps. Its data can be the result of spatial analysis processes in a GIS like ArcGIS, or made available via the data infrastructure (geofiles), be available via direct real-time feed through sensors (SWE – Sensor Web Enablement) or collected by volunteers (citizen science or VGI (volunteered geographical information). These sensors could, for instance, be water gauges or meteorological stations.

With a GIS, one can create maps, but often one requires just a better design quality. If one would opt for a paper map, it is possible to retrieve GIS files in shapefile or PDF format in graphics software and improve the design. Use of a special GIS plug-in for graphics, like Avenza's MAPublisher editing in Adobe Illustrator, results in a two-way process and changes in the graphics files will also occur in the original GIS database. Alternatively, maps produced in a GIS are exported as pdf, tiff or jpg files and imported in graphics software such as Adobe's Illustrator or Photoshop. Manipulation of the graphic image is possible but there is no link to the original database. The graphic software can create a final pdf file, which can be sent for hard copy output

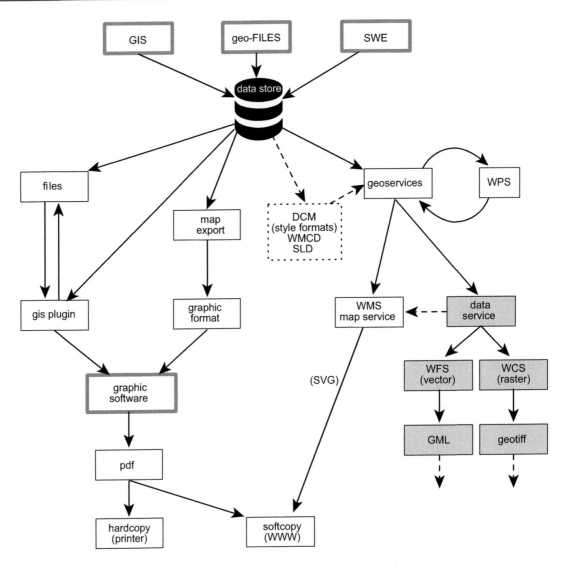

FIGURE 5.29 Options to display permanent and virtual maps

through a plotter or printer, or be viewed on screen as an (interactive) softcopy.

Another option to create on-screen maps in, for instance, a web browser is through the use of a geoservice. In its simplest approach, based on the client–server architecture, the user sends a request to a web server which will respond by returning a file that contains the image of a map. The maps presented like this are raster-based, static pictures (JPG, PNG, without any interactivity). For interactive maps, the client–server architecture gets more complex and has to allow interoperability of heterogeneous components based on World Wide Web Consortium (W3C) and dedicated standards of the OGC. They defined standards for Web Map Service (WMS – for portrayal of maps), Web Coverage Service

(WCS – an interface to raster data/geotiff) and Web Feature Service (WFS – interface to vector data (Geography Markup Language (GML)/ GeoJSON (JavaScript Object Notation)).

Regarding the production of the hard copy mentioned in Figure 5.29, many different routes are possible that result in displays on paper. The choice will depend on the type of output needed, but in most cases Adobe Illustrator will be the intermediary to define the task of the printers or plotters. Software used to design maps will normally translate its drawing commands into a special language. The most common of these languages is PDF, developed by Adobe. The code contained results in a resolution-independent textual description (normally in ASCII (American Standard Code for Information Interchange)) of the map. A polygon, for instance, is described by a list of coordinates and some codes that can provide information on the polygon's interior, line type and thickness. Theoretically, it should be possible, when using PDF, to send the same map to devices with different output qualities, as tags with different production commands can be added to it. Those devices capable of processing PDF will transform the code via a raster image processor to a bitmap, which can be printed on paper. The coded image can be viewed on-screen before it is produced on paper to check the result. Among the (raster) output devices are laser printers, electrostatic plotters and ink-jet printers. Laser printers work by an electrophotographic process, either in black and white or in colour. Their resolution varies from 300 dots per inch (dpi) to 1200 dpi, on paper up to A3 size. Electrostatic plotters are very fast machines that can handle paper up to A0 size, and can handle full colour. Ink-jet printers can also handle colour and different sizes of paper.

With current technology, the hard copy devices mentioned above are less suitable to produce many copies of the same map at reasonable prices. For this type of duplication,

for instance to print topographic maps, one still relies on a printing press. Here, the digital method currently goes up to digital offset printing on demand. Since a printing press can only print one colour at a time, colour separation is necessary. Normally, one separates the colours of a map into four colours (see Figure 5.30). A combination of the basic colours, cyan, magenta and yellow, with black added for crisper definition (CMYK), allows one to print any colour (see Figure 5.31).

Many graphical packages have the possibility for digital colour separation as shown in Figure 5.30. Here, the software creates four files, one for each of the basic colours. For each of the symbols used in the map, the colours are known, as well as the percentages of cyan, magenta, yellow or black used. If the forest in the map is represented by a green polygon, and the shade of green is composed of 40% yellow and 30% cyan, this polygon is sent to both the yellow and cyan files with a screen density of 40% and 30%, respectively.

Conventional colour separation is much more complex (see the four negatives in Figure 5.30), one for each of the four basic colours (CMYK). On each of these negatives,

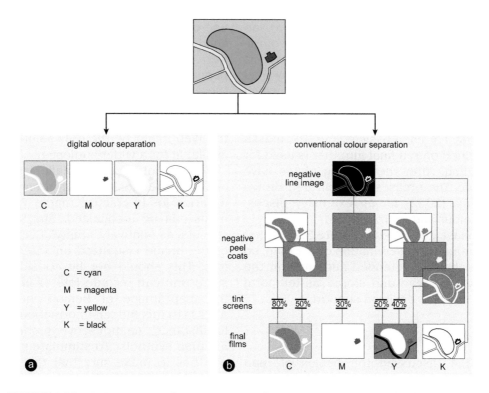

FIGURE 5.30 Colour separation. For conventional colour separation, with peel coats masks can be created that, when exposed through a tint screen, result in halftone images. The final films will be printed in cyan, magenta, yellow and black, respectively

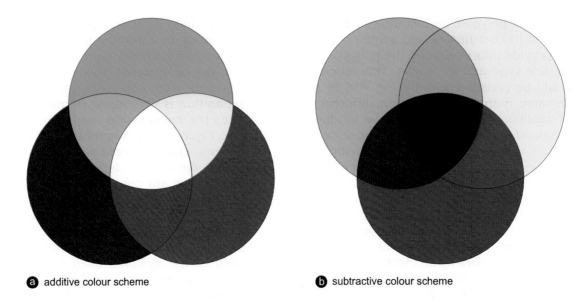

a additive colour scheme   b subtractive colour scheme

FIGURE 5.31 Primary colours. In an additive colour scheme red, green and blue colour light sources are combined to generate other colours. Mixed in equal intensities they result in white. In a subtractive colour scheme the primary colours cyan, magenta and yellow are used to generate other colours. Mixed in equal intensity they result in black. This is used for paint and print.

only those lines remain which indeed must be printed in that particular colour. The blue lines in the original maps, representing the shorelines of a lake, are removed from the yellow, magenta and black negative. The green forest border, however, will remain on both the cyan and yellow negatives since green is a combination of these two colours. The next step is to create for each basic colour as many peel coats as necessary to create masks, which are used to incorporate, for instance, a 30% screen (to produce lighter tints of the printing colours). For each basic colour, the masks are combined onto a final film that is used for plate making. After the printing plates have been made, the presses can roll. See Brown and Feringa (2003) for an extensive discussion on colour and reproduction.

5. The last phase described in Figure 5.28 is that of distribution or dissemination of the map, so that it reaches the intended audience, in the hope it will be read and stored as temporal map and used for decision support.

5.7 WEB MAP DESIGN

Sections 5.2–5.4 discussed the basics of map design. These basics do not change because of the web environment, although it does offer interesting extra opportunities, the most important of which is accessibility. The Web as a medium to display maps also has some disadvantages. When creating a web map, one has to consider the physical design, in both file and display size. The first is important because people are not eager to wait for long downloads, and the second because the use of scroll bars to pan the map is also discouraged. Due to these characteristics, the design of web maps needs extra attention. Just scanning paper maps or using default GIS maps and putting them on the Web is not a good procedure, although in some situations there might be no alternative. Scanning historical maps and publishing them on the Web might be the only solution to make them available for a wider audience.

For symbol selection, Leaflet is referred to here; it is one of the most popular open-source JavaScript libraries for interactive maps. Well-designed web maps can be recognized, due to the above constraints, as relatively 'empty'. Every part or element of the image visualized on screen should be legible. This should not be considered problematic since one can include lots of information behind the map image or behind individual symbols. Access to this hidden information can be obtained, for instance, via mouseover techniques or by clicking map symbols. To stimulate such an approach, one has to make sure that the symbols have an appearance that invites clicking them. In case of mouseover techniques, the appearance of a symbol will change when covered by the cursor, or textual information will appear on the map. Clicking the symbol might open new windows or activate other web pages.

Additional information could also become visible by a so-called texture filter. This 'tool' could have the shape of a magnifying glass that is moved over the map. The result can be multiple (see Figure 5.32). The content of the looking glass could, for instance, be an enlargement of the map, either revealing more details of the area covered or not. On the other hand, it could also result in less detail being shown, allowing the viewer to concentrate on particular map data, or visualizing the data in a particular way. Another option could be that the map area under the lens changes into a satellite image or thematic map.

The possibilities offered by the WWW have extended the traditional cartographic variables as proposed by Bertin (1967, 2011) and explained in Section 5.3. Web design software enables the application of new variables, such as blur, focus and transparency, while shadow and shading play a prominent role as well. Blur gives symbols a fuzzy appearance and can, for instance, be applied to visualize uncertainty, while focus will introduce blinking symbols to attract attention, as through the filters in Figure 5.32. Both transparency and shading/shadow can be used to simulate a three-dimensional look. Transparency can be seen as a kind of fogginess, by which part of the map content is obscured or faded in favour of other information.

It can, e.g., be applied to subdue the background in a map in order to enhance the main theme in the foreground (for instance, a drape of geology over terrain features). In a three-dimensional 'land-scape' environment, it can also be used as a depth cue. The use of shadow and shading increases the sense of depth. Shading is commonly used to increase the contrast between 'figure' and 'ground' or, as in relief maps, to create a three-dimensional terrain impression. Shadow, also known as 'cast or drop shadow', can be applied to give the symbols a three-dimensional look. In web maps, this three-dimensional feel of the symbols invites the user to click on them in order to activate a hyperlink or mouseover effects. The visual effect of shadow is casting a shadow of the symbol on to the background. Figure 5.33 gives some examples. The new variables are applied a lot in many web maps but they still have to be further researched effectively as, despite all technological progress, it remains important to find out whether the new representations and interfaces actually work.

Developments around the Internet have created the possibility to disseminate three-dimensional virtual environments via the WWW using WebGL. The use of this programming language allows one to display three-dimensional maps that can be looked at from any viewpoint. With virtual reality tools, one can immerse oneself in the map environment.

FIGURE 5.32   Examples of a texture filter in use on the Web

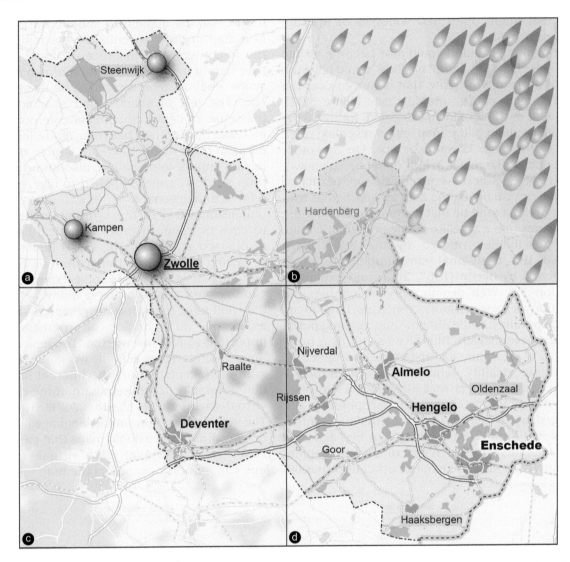

FIGURE 5.33 Additional graphic variables: (a) shadow/shading; (b) blur; (c) transparency; (d) blinking (focus)

At the beginning of this chapter, it was mentioned that the provider of data or maps does not always have control over the final appearance of the map. This remark was made in the context of a user who, for instance, can decide which layers to switch on or off and define the map content. However, especially in relation to colour, the appearance of maps that are made visible on the user's monitor will depend on the particular monitor configuration and settings. For this reason, web map designers often adopt a cautious approach and assume the minimum configuration and lowest settings.

5.8 WEB MAPS AND MULTIMEDIA

The WWW is an ideal platform to combine different multimedia elements with maps. Multimedia is defined as interactive integration of sound,

animations, text and (video) images. When realized in a geodata environment, it allows for a map to link through to all kinds of other information items of a geographical nature. These could be text documents describing a parcel, photographs of objects that do exist in the GIS database or videos of the landscape of the current study area. The multimedia definition used here is the one by Laurini and Thompson (1992): 'a variety of analogue and digital forms of data that come together via common channels of communication'. For a typical cartographic context of multimedia, see also *Multimedia Cartography* (Cartwright et al., 2007). Figure 5.34, taken from the Encarta Interactive World Atlas online, shows a good example: querying Kilimanjaro leads to the map of north-eastern Tanzania; further clicking on the map gives access to texts about the mountain,

FIGURE 5.34   Collage compiled from an electronic atlas (Microsoft Encarta Interactive World Atlas 2001)

its place in the literature (Hemmingway), terrain photographs and satellite imagery that can be tilted; further web links about the mountain can be accessed as well.

Current (2020) GIS packages have only limited capacity to handle (except for a bleep to signal an error) video and animations. Text can be generated from the database, and often scanned images of photographs of geographical objects, paper maps and text documents can be displayed. Currently, the GIS packages are nearly fully integrated in the desktop environment, as explained in the previous section. The user is guided by a generic GUI that allows the display of maps and provides access to the data behind these maps. Pointing at an object on the map would immediately highlight the corresponding record in the database or diagram. In the same way, they could open up towards multimedia and to the general desktop environment, as the map would allow for direct links with spreadsheets, video and animation. But if we want to incorporate this technology into our geospatial data handling systems, questions have to be answered first. Among these are how to structure multimedia information (in order to allow for profitable navigation strategies), how multimedia can be used to enhance spatial analysis, what the role of the map will be and what will be involved in interface design. Today this has all culminated in the story map (see Section 9.5). The following sections present the relation between the map and the individual multimedia components (sound, text, (video) images and animation).

5.8.1 Sound

Maps can function as indexes to sound libraries. In some electronic atlases, pointing at a country on a world map would initiate the national anthem of the particular country being played, or some general phrases being spoken in the country's language. In this category, one can also find the application of sound as background music to enhance a mapped phenomenon, such as industry, infrastructure or history. Experiments with maps in relation to sound are known from topics such as noise nuisance and map accuracy (Fisher, 1994; Krygier, 1994). In both cases, the volume of noise is controlled by pointing at a location on the map. Moving the pointer to a less accurately mapped region would increase the noise level. Both are examples related to analysis. The same approach could be used to explore a country's language. Moving the mouse to a particular region would start a short sentence in that region's dialect. Krygier (1994) and Schiewe (2014) have experimented with sound as an additional variable to graphical variables such as colour and size. In virtual reality environments, three-dimensional sound is used for orientation purposes.

5.8.2 Text

GIS is probably the best representation of the link between a map and text (the GIS database). Imagine a map showing a country's population density in which all provinces are coloured according to one of the four different classes discerned. To amplify the presentation, the user can point to a province, which will result in the display of its name and its actual individual population density

value. Electronic atlases often have all kinds of encyclopaedic information linked to the map as a whole or to individual map elements. It is possible to analyse or explore this information. Country statistics can be compared. Clicking Lisbon on the map of Portugal would reveal a list with the most important tourist sights, or even activate other multimedia components like starting a video tour of the city.

5.8.3 Images

Maps are models of reality. Linking video or photographs to the map will offer the user a different view of reality. Topographic maps present the landscape. Next to this map, a non-interpreted satellite image or aerial photograph can help users in their understanding of the landscape. While analysing a geological map, it can be enhanced by showing landscape views (video or photographs) of characteristic spots in the area. A real estate agent could use the map as an index to explore all the properties he or she has for sale. Pointing at a specific house would show a photograph of the house and the construction drawings, and a video would start showing the house's interior.

5.8.4 Video/Animations

Maps often represent complex processes. Animations can be very expressive in explaining these processes (see also Chapter 8). To present, for instance, the structure of a city, they can be used to show subsequent map layers which explain the logic of this structure (first relief, followed by hydrography, infrastructure, land use, etc.). Animations are also an excellent means to introduce geospatial data's temporal component: the evolution of a river delta, the history of the Netherlands coastline or the weather conditions of last week. Videos with real-world imagery can be blended in too.

From a technical point of view, there are almost no barriers left. The user is confronted with a screen with multiple windows displaying text, maps and even video images supported by sound. The important issue is to keep in control of all the options, and to allow the users to manage all the information that will reach them. It is most important, therefore, to state the objectives (tasks) or the purpose of use (exploration, analysis or presentation) in advance, so that there will be a yardstick by which the result can be measured.

The (web) map can be seen as an explanatory integrating interface to the real world.

FURTHER READING

Anson, R. W., and F. J. Ormeling. 1995. *Basic Cartography.* Vol. 1–3. London: Elseviers - Butterworth.

Anson, R. W., and F. J. Ormeling, eds. 2002. *Basic Cartography for Students and Technicians.* 2nd ed. Vol. 2. Oxford: Butterworth-Heinemann.

Brown, A., and W. Feringa. 2003. *Colour Basics for GIS Users.* Upper Saddle River, NJ: Prentice Hall.

Cartwright, W., M. Peterson, and G Gartner, eds. 1999. *Multimedia Cartography.* Berlin: Springer.

Field, K. 2018. *Cartography.* Redlands: ESRI Press.

Keates, J. S. 1993. *Cartographic Design and Production.* 2nd ed. New York: John & Wiley Sons.

Kraak, M. J., and A. Brown, eds. 2000. *Web Cartography - Developments and Prospects.* London: Taylor & Francis.

Monmonier, M. 2018. *How to Lie with Maps.* 3rd ed. Chicago, IL: Chicago University Press.

Muehlenhaus, I. 2013. *Web Cartography: Map Design for Interactive and Mobile Devices.* Boca Raton, FL: CRC Press.

Peterson, M. P. 2014. *Mapping in the Cloud.* New York: Guildford Press.

Topography

Chapters 6–8 will each deal with specific cartographic characteristics of geospatial data as discussed in Chapter 1 (see Figure 1.3). In this chapter, it is all about location and the 'Where?' question (Figure 6.1). Traditionally, the Earth's topography is graphically rendered by topographic maps. Today, these topographic maps are derived from data sets known variously as 'topographic framework data',

'geospatial core' or 'foundation data'. These encompass geodetic control data (based on a geospatial reference system), data related to built-up areas, hydrography and infrastructure, the digital elevation model (DEM), administrative boundaries and postal codes (essential to link socio-economic data to physical data) and geographical names. This chapter deals with georeferencing (how to locate objects), map projections and transformations (how to convert locations from the 3D Earth to the flat plane), generalization (how to aggregate locations for scale reduction), relief (how to deal with location in 3D terrain) and geographical names (for expressing location in daily contacts). This chapter ends with examples from several data-providing organizations.

6.1 GEOREFERENCING

The locational component gives geospatial data their unique character. It distinguishes these data from all other data. Another typical characteristic is that the geographical objects that geospatial data represent occur in complex and irregular patterns. This makes it difficult to describe these objects. Try, for instance, to give a description of the basin of the River Zambezi. It should include the shape of the river, its width and its relation with its tributaries, as well as phenomena such as swamps, islands and forests. Also included should be relationships

FIGURE 6.1 Working with geospatial data from the perspective of the data's locational component

with features such as bridges, dams, other infra- structural artefacts and administrative boundaries.

Depending on the scale on which this description of the digital landscape model (DLM) is based, it can be very complex and extensive, but it will probably still be incomplete. Selections have to be made, and terminology used such as 'close to' and 'left of' will not increase the accuracy of the description. What methods are available to indicate location? Each georeferencing method has its own advantages and disadvantages, as Figure 6.2 shows. Names of geographical features (toponyms) can be used (Figure 6.2a). These distinguish one feature from another. However, their use is no guarantee that a unique description will result. 'Springfield' could be located in Massachusetts, Illinois, Ohio or Oregon

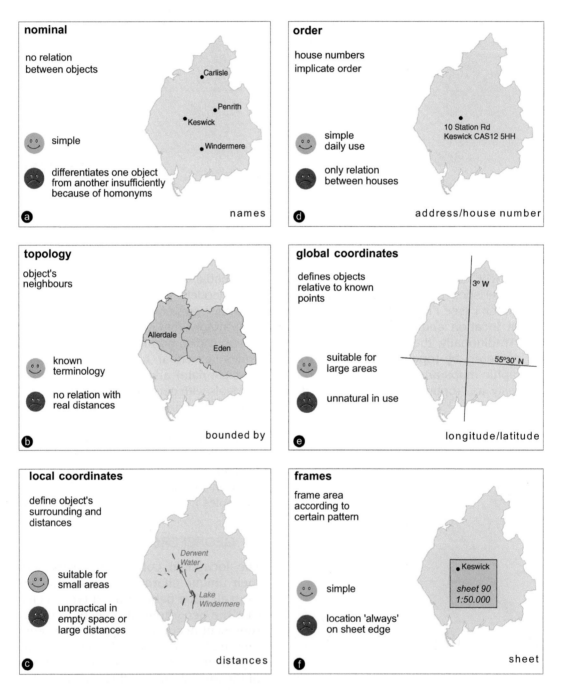

FIGURE 6.2 Maps and location: georeferencing methods for localizing geographical objects: (a) nominal; (b) order; (c) topology; (d) local coordinates; (e) global coordinates; (f) frames. Homonyms (in a) are different geographical objects that have the same name

in the United States, or in New Zealand or even in South Africa. A practical approach seems to be the use of addresses and postcodes (or ZIP codes). This has an implicit order: 10 Station Road will be close or next to 8 Station Road (Figure 6.2b). However, it only refers to buildings and does not include natural features such as rivers, lakes or mountain ranges. The use of topology is another option. The terminology is familiar: Allerdale bounds Eden in the west (Figure 6.2c). However, it does not include any information on real distances (Figure 6.2e). The use of a coordinate system is also quite common. It could be a global coordinate system, where each location is defined by longitude and latitude (Figure 6.2d), or it could be a position in a national coordinate system. National coordinates provide a unique indication of location, but are less natural in use than a system of parallels and meridians. Local coordinate systems for small areas are in use as well. A typical cartographic reference system is the reference to map sheets. This method is simple to use, but in practice, locations needed are often on two or more sheets. Figure 6.2f shows that Keswick can be found on sheet 90 of the Ordnance Survey's (OS) 1:50 000 map series.

GISs always use coordinate systems, often in combination with topology. As explained in Chapter 1, GIS is about data integration and spatial analysis. In order to be able to combine two or more data sets to execute a spatial analysis or a cartographic compilation, it should be possible to reference both sets in a common coordinate system. In practice, this is often quite difficult. Different coordinate systems are used (if any), as well as different map scales. This chapter elaborates on the methods and techniques available to deal with these problems. They relate to converting one system to another, the simplification of maps and the conversion from three-dimensional reality (the globe and local relief) to the two-dimensional plane. Without solving these problems, or knowing their exact nature, spatial analysis or cartographic compilation will not give a valuable result, and consequently, the maps produced cannot be trusted. Since the Earth is almost a perfect sphere, the elementary globe referencing system is based on spherical coordinates. This system, called the 'geographical coordinate system', defines a location by latitude and longitude. Figure 6.3a shows this system, where latitude and longitude are measured in degrees, minutes and seconds. The origin of this spherical system is the intersection of the Equator and the Greenwich Prime Meridian. For latitudes, the Equator is defined as 0, the North Pole as +90 and the South Pole as −90. The intersections of

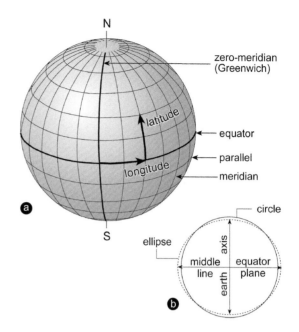

FIGURE 6.3 (a) Geographical coordinate system; (b) flattening of the Earth

all planes of a certain latitude and the globe are called 'parallels'. All half-circles from the North Pole to the South Pole are called 'meridians'. East of the Greenwich Meridian, which is defined as 0 longitude, they increase up to +180, and west to −180. Both parallels and meridians make up the global gratiular network. From Figure 6.3b, it can also be seen that the Earth is not a perfect circle. It is flattened at both poles, resulting in the three-dimensional shape of the Earth being an ellipsoid rather than a sphere. The ellipsoid is used for calculation purposes to convert the three-dimensional Earth to the flat paper (see next section).

To define a location, its latitude (ϕ) and longitude (λ) are measured from the Earth's centre to the location on the Earth's surface. From Figure 6.4, it can be seen that a point's geographical latitude (ϕ) is defined by the angle, in a meridian plane, between the Equator and the line from the Earth's centre to this point on the globe. Geographical longitude (λ) is defined by the angle, in the Equator plane, between the Greenwich Meridian and the meridian of the particular point. Accordingly, the location of Helsinki can be defined as 60° North of the Equator and 25° East of Greenwich. To be more precise, this would be 60°10′N and 24°58′E. The relation between the geographical coordinates and distances on the Earth is determined by the place on Earth. Along the Equator, and along all meridians, 1° is 111.11 km in distance, assuming that the circumference of the Earth is 40 000 km and has a radius of 6370 km. At 45°N or S, a parallel has a

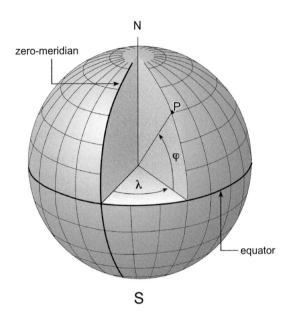

FIGURE 6.4 Latitude and longitude

circumference of 28 301 km, resulting in degrees that are 78.6 km in length. At both poles, this length is zero (Figure 6.4).

Internationally, WGS84 (World Geodetic System 1984) is used as a standard for calculations of position, distances, etc. and is used in fields varying from surveying to aeronautics and GPS (Global Positioning System). The standard describes a set of parameters. Among them are data related to the shape of the ellipsoid (for instance, the ellipsoid's major axis is 6378.137 km). Additionally, it includes data on the gravity model of the Earth.

Most national mapping organizations have introduced a national coordinate system for unique and consistent reference and calculation purposes. Figure 6.5 shows the British system. It is called the 'National Grid system', and it adheres to a Cartesian rectangular coordinate system. The map in Figure 6.5b shows how Britain is covered by 100 km². These squares are a refinement of a 500 km² grid (Figure 6.5a). In relation to the geographical coordinate system, the true origin is found at the crossing of meridian 2°W and parallel 49°N. The relation between both systems is expressed in the figure by the bold grey graticule lines of the transverse Mercator projection and the black National Grid lines. However, to avoid confusion with negative and positive coordinates when defining a location in the National Grid system, a false origin has been introduced. The true origin has been shifted 100 km north and 400 km west, always resulting in positive coordinates in either direction. To make the system even easier to use, a location is denoted by letters as well. The 500 km² provides the first letter (for Keswick in Cumbria, an N), and the second letter is derived from one of the twenty-five 100 km² that make up the 500 km². For Keswick, this is a Y. In Figure 6.4b, other letters have been added to the 500 km²N as an example. In this system, the 100 km² can be further refined, as Figure 6.5c shows. The location of Keswick can thus be pinpointed by NY268232. This possibility of refinement guarantees that all map scales produced by the OS are covered by the same National Grid.

Figure 6.6 demonstrates the use of a local coordinate system. To position objects found during an archaeological excavation along the River Nile in Egypt, a local Cartesian grid has been created to cover the site. To do so, a base measurement was effected along a horizontal railway track between Elkab and Nag'Hilal. Then, a triangulation network was set up, based on this baseline. In the actual excavation area, a rectangular grid was established, with as positions of its nodes determined in this triangulation net. This rectangular grid was densified further locally according to requirements.

6.2 MAP PROJECTIONS

Once upon a time, cartographic education mainly consisted of training in plotting different projections. Since the introduction of the computer, this has no longer been the case, one of the first major contributions of the computer being the ability to plot any area with known coordinates according to any projection. As important as plotting an area according to one of these projections is the ability to spot the nature of projections maps are plotted in, as this knowledge is a prerequisite in dealing with geographical information in GISs. All the spatial operations possible in a GIS are only relevant if the files combined are all based on data in the same reference frames or projections.

Mathematical formulae are used to transform spherical geographical coordinates to the two dimensions of a plane. This transformation process is referred to as 'map projection'. The transformation from the three-dimensional ellipsoid to the two-dimensional plane is not possible without some form of distortion. The distortion affects shapes, distances and directions. Each of the many formulae available (Canters and Decleir, 1989; Kessler & Battersby, 2019, Lapaine & Usery, 2017, Maling, 1992; Snyder, 1987; Snyder and Stewart, 1997; Snyder and Voxland, 1989) will result in different distortions. This determines whether each map projection will be suitable or unsuitable for a certain purpose.

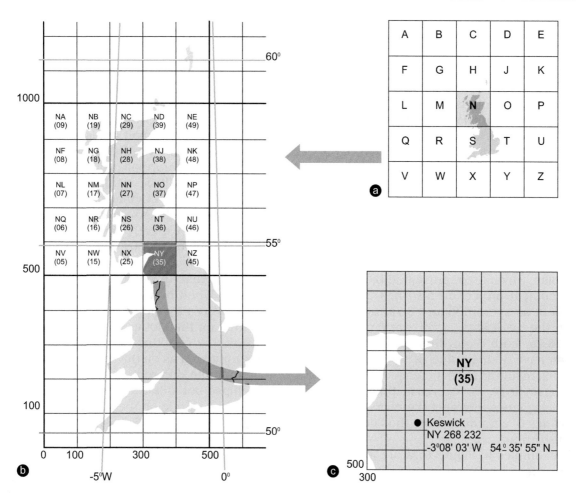

FIGURE 6.5 National coordinate systems: the British National Grid. (a), (b), and (c) zoom in to a more detailed coordinate notation

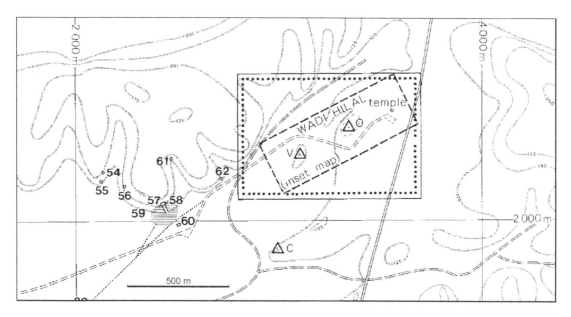

FIGURE 6.6 Local coordinate systems: the system shown has been established for an archaeological excavation along the Nile in Egypt (Depuydt, 1989)

Map projections can be categorized on the basis of the shape of the projection plane, as seen in Figure 6.7. From this perspective, conical, cylindrical and azimuthal projections are distinguished. The point or line where the projection plane touches the ellipsoid is called the 'point or line of tangency'. No distortion is found at this point or along this line. Azimuthal projections have a single point of zero distortion. In the normal aspect, meridians are straight lines, and parallels are concentric circles centring on the pole. Cylindrical projections have a single line, called the 'standard line', of no distortion. In the normal aspect, this line touches upon the Equator. Both parallels and meridians are straight lines perpendicular to each other. Distortion increases dramatically towards both poles, which are represented by lines.

The conical projection also has a line of zero distortion, as seen in Figure 6.7a. In its standard aspect, the meridians, again, are circular arcs. In Figure 6.7a, both the cylindrical and conical projections have a single line of tangency (the standard line), resulting in an increased distortion away from these lines. To decrease this effect, some projection planes intersect the ellipsoid, resulting in two lines of zero distortion, thus decreasing the total distortion.

The examples in Figure 6.7a are all in direct or normal aspect, i.e. the projection form that provides the simplest graticule and calculations. For an azimuthal projection, this is the one which touches at the pole; for conical or cylindrical projections, this is the one in which the axis of the cone or cylinder coincides with that of the ellipsoid. Next to the normal aspect, projections can have a transverse or oblique aspect (Figure 6.7b). The transverse aspect is characterized by a distortion pattern that, with respect to the normal, is rotated by 90°. The oblique aspect includes all possible cases between the normal aspect and the transverse aspect.

Each projection has its specific patterns of distortion, with the least distortion at the tangential point or line. In order to have as little distortion as possible, it is feasible to change the map projection aspects, and by doing so move these tangential points or lines, so that the area of least distortion overlaps the region to be mapped.

The nature of the distortion pattern is another parameter with which to classify map projections. It is closely linked to the question of which map projection is used. This question will be addressed later in this section. In the transformation from a three-dimensional sphere to two-dimensional plane, only a few characteristics can be preserved. These refer to object shapes, area sizes or distances between objects. On the basis of the characteristics they retain, map projections are classified as conformal, equal area or equidistant or as being none of these. If the nature of the mapped phenomena requires that the shapes of the features should be preserved, a conformal map projection has to be chosen. However, preserving shape can only be done for small areas, and it is impossible to maintain it all over the mapped Earth. The application of a conformal projection guarantees that local shapes will be retained: circles drawn on a sphere will have circular shapes on the plane as well. However, the size of the circles will be different and may be greatly distorted, depending on the location on the sphere with respect to the point or line of tangency. If one wants to preserve the size of areas, an equal-area projection has to be chosen. When moving a circle over the globe, the area covered by the circle will be the same at any location on the globe. Applying an equal-area projection, the resulting area on the plane will cover the same area as on the sphere. However, its shape can be quite different. A circle can be distorted to an ellipse but still cover the same area as the corresponding circle on the globe. This can be seen in Figure 6.10b. Figure 6.8a shows the globe from two positions, highlighting Greenland and Saudi Arabia, which have about the same size. Figure 6.8b shows the effect on both areas of the application of an equal-area map projection, and Figure 6.8c shows the effect on both areas of the application of a conformal projection.

Equidistant map projections preserve distances between certain points. This can result in maps in which the distances from one or two points to all other points or certain lines are correct. Projections do exist that do not adhere to any of the three distortion characteristics described, and neither are they related to the cylinder, cone or azimuthal projection approach. To reduce distortion for a specific application, they are often modified versions of other projections, while some may be totally different. An example is the gnomonic projection, which extremely distorts but provides a great service to navigation as it renders all shortest distances as straight lines.

Two examples will be elaborated in more detail: the Universal Transverse Mercator (UTM) projection and the Eckert IV projection. The UTM projection is derived from the transverse Mercator projection, which has characteristics similar to those of the Mercator projection. The original Mercator projection is a conformal cylinder projection where parallels and meridian intersect at 90° angles. The projection is useful for navigation purposes

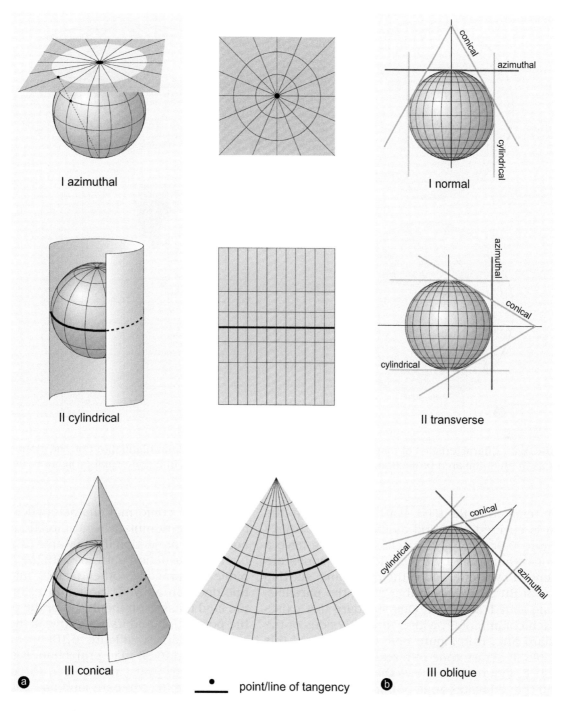

I azimuthal

I normal

II cylindrical

II transverse

III conical

• —— point/line of tangency

III oblique

FIGURE 6.7 Map projections: (a) projection planes; (b) projection aspects

(conformality) or for mapping regions near the Equator. As can be seen in Figure 6.9a(I), the areas near the poles are very much distorted. This projection is still inappropriately used for world maps in atlases or on wall charts. Because of its characteristics, it presents a misleading view of the world. One of the reasons it is found in most global web map services, such as OpenStreetMap (OSM) or Google Maps, is because the longitude and latitude are perpendicular to each other, which makes it easy to prepare and serve map tiles at all 20 scale levels found in those services. The transverse Mercator projection (III in Figure 6.9a) has the same characteristics as the original Mercator. In an adapted version, this projection is often used for surveying and mapping applications for relatively small areas. Because distortion near the standard line is limited, more than one standard line was introduced,

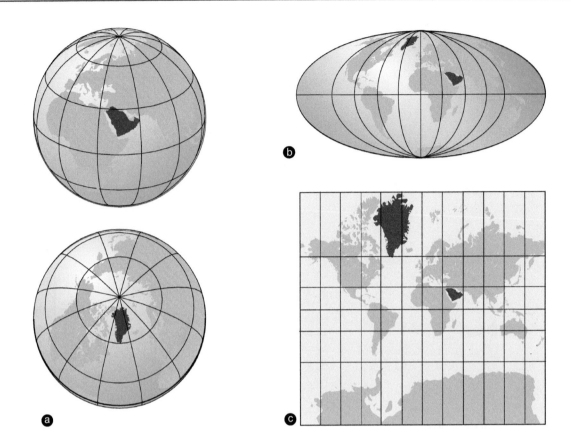

FIGURE 6.8 Characteristics of map projections: (a) Arabia and Greenland highlighted on the globe; (b) the same areas in an equal-area projection, here Mollweide; (c) the same area in a conformal projection, here Mercator

which resulted in the UTM. For this projection, the world is covered by a grid as can be seen from I in Figure 6.9c. The grid patches or zones have a size of 6° longitude by 8° latitude. The zones are referenced by a letter along the meridians, starting with C at 80°S, and a number along the parallels, starting with 1 at 180°W. One standard line exists for each column of 6°, with 3° on both sides of the standard line. II in Figure 6.9c demonstrates this principle. It shows zone 59E and the standard line of 174° E. Figure 6.9d shows the column between 6° and 0°and locates zone 30U in Western Europe. Each of these zones is further subdivided into several 100 km², referred to with letter combinations as can be seen in Figure 6.9d. The location of Keswick in Cumbria in UTM coordinates is 30U VF913522. To map the world with as little distortion as possible is quite difficult. Only a few map projections are suitable for world maps: while all projections have disadvantages, some have more disadvantages than others (like the Mercator projection discussed above). Choosing a map projection for a world map will depend on its intended use. The search will come up with, for instance,

a map with conformal characteristics if one wants to visualize dominant wind directions, or with an equal-area projection if one wants to visualize population distribution patterns with the dot map technique. In practice, other factors might influence this decision as well, for instance, the location of the land masses on the Earth, or the distribution of the population and its economic activities.

The Eckert IV (Figure 6.10) is an equal-area pseudocylindrical map projection. It preserves the size of areas and can be used to compare area, for instance, to compare land use patterns. Figure 6.10b show how the Tissot ellipses behave on the Eckert IV grid. Some projections are a compromise in overall distortion. The Natural Earth projection is such an example (Šavrič et al., 2016).

Which map projection to choose? The path to follow is similar to other decisions to be made in the mapping process. The first step is to define the purpose of the map. Does one want to compare areas, measure distances or display directions correctly? The nature of the distortion is the first guide in the selection process. A choice has to be made between retaining conformal, equal-area, equidistant or other

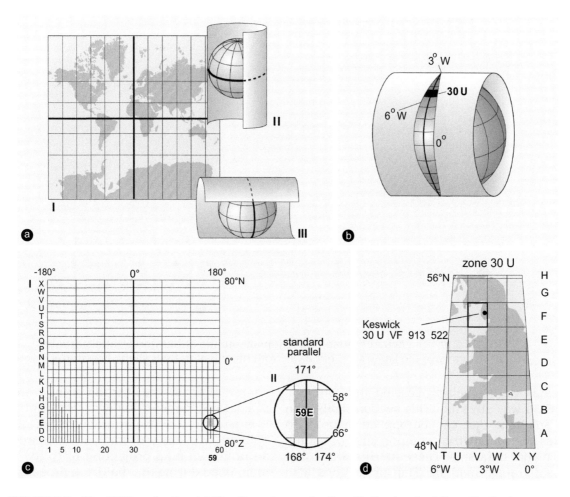

FIGURE 6.9 The UTM projection (a) The Mercator projection (I), its standard line (II), the transverse Mercator projection's standard line (III). (b) The standard line principle. (c) The UTM grid (I), zone 59E (II). (d) The location of zone 30U. Zone 30U's 100 km² and the location of Keswick

characteristics of the globe. Showing the flow pattern in the atmosphere or oceans requires a conformal projection: comparing vegetation coverage, or just the size of countries, requires an equal-area projection. This type of projection is also needed if one intends to indirectly compare areas in thematic maps, such as dot maps or proportional symbol maps. The second step is to look at the shape, size and location of the geographical area to be mapped. The projection plane and its aspect play a role here. The distortion pattern of the chosen projection should match the shape of an area as closely as possible. If one looks at Chile, South America, a projection with an N–S standard line would be very suitable since this country has a small E–W extension and a very large N–S extension. A map projection that could be used here is the transverse Mercator projection. Russia, on the other hand, has a large E–W extension, but also a relatively large N–S extension. It therefore needs one or two standard lines that run E–W. A conic projection would be a useful choice here. Azimuthal projections are very suitable for areas with an equal N–S and E–W extent. Finally, if relevant, the choice should be influenced by the manner in which the map extent fills the screen or paper. This is especially relevant for atlases. Concluding, one should realize that there are no good or bad projections but, rather, there are good or bad applications of map projections.

6.3 GEOMETRIC TRANSFORMATIONS

In Chapter 2, it was mentioned that it would be unlikely that all the data necessary for spatial analysis or map production will be available in the required format. In Section 6.2, the importance of being aware of the geometrical characteristics

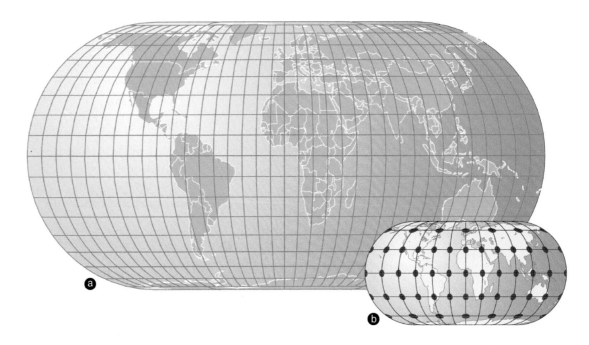

FIGURE 6.10 The Eckert IV map projection: (a) outline of the world; (b) distortion: all circles have the same size, but away from the equator, their shape is distorted

of the data prior to their processing and combination in GISs was stressed. This section deals with several solutions that can be applied in order to integrate the geometric components of different geospatial data sets. These are the affine and curvilinear transformations. Map projections also belong in this category when one has to convert one projection into another (Figure 6.11). An affine transformation converts all coordinates of a specific data set into the coordinates of another coordinate system. To apply the transformation, one needs at least three corresponding points in the old and new data sets. It is based on the formulae

$$x' = Ax + By + C$$

$$y' = Dx + Ey + F$$

where x' and y' are the coordinates in the new system. Parameters A, B, C, D, E and F are defined by comparing both points in the two data sets. The transformation is a combination of three basic operations: translation, rotation and scaling (Figure 6.12). The transformation allows for a different scaling along the x-axis with respect to the y-axis.

The curvilinear transformation applied in the process is also known as 'rubber sheeting'. It is used when two data sets do not geometrically match. However, one should only apply it when there are no alternatives, because it could be a

source of error. The starting point in this process is a data set geometrically correct for the purpose it is needed for. The characteristics of the other data set are less well known. This can be due to an unknown map projection or to the fact that a map, used during the digitizing process, has been affected by shrinkage or strains. The relative position of the lesser-known map is defined by clearly identifying corresponding points on both maps. These could be road crossings or characteristic points in a coastline (Figure 6.13). Choosing these points is done with both data sets visible on the screen. The corresponding points will be

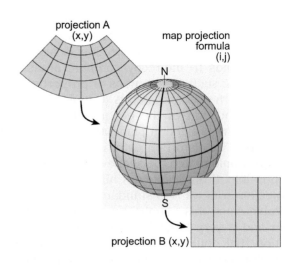

FIGURE 6.11 Conversion of map projections

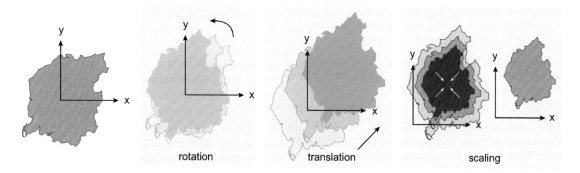

FIGURE 6.12 Geometric transformations: rotation, translation and scaling

linked by a vector. The transformation pulls and pushes the lesser-known data set until all vectors are reduced to zero. It is this process from which the name 'rubber sheeting' is derived. From the figure, it can clearly be seen that not all points are affected to the same degree, as is the case with the affine transformation. When the lesser-known data set is very much distorted, it is also possible to execute rubber sheeting on separate parts of the data set using different parameters. The quality of the result depends on the number of control points and their distribution, but even more on

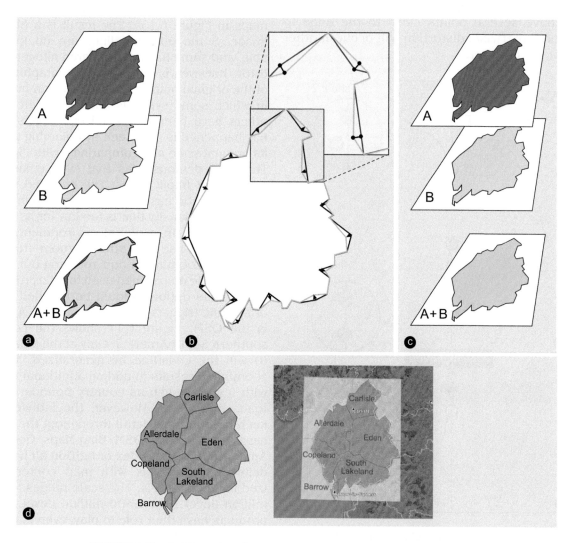

FIGURE 6.13 Rubber sheeting: (a) maps A and B do not match; (b) the transformation of map B based on map A; (c) maps A and B match

the accuracy of the data set that was assumed to be the better of the two. Google Earth offers this function and allows users to overlay their own scans on top of its satellite imagery. Figure 6.13d shows how the map from Figure 2.14a has been 'rubber-sheeted' on top of the imagery.

A special case of rubber sheeting is edge matching. Edge matching is applied to correctly connect features that are found at the edges of data sets. This problem is very prominent when data sets that have to be used in a spatial operation are produced by digitizing individual map sheets. Features that cross the map sheet border will almost never connect properly. There are several reasons for this: differences in shrinkage or strains, digitizing inaccuracies, errors in the original mapping process, etc. Figure 6.14 gives an example. It is important to remember that rubber sheeting can be a source of error. Edge matching should only be applied on those line features which have several points close to the edge in common; otherwise, distortion will occur in other parts of the data set.

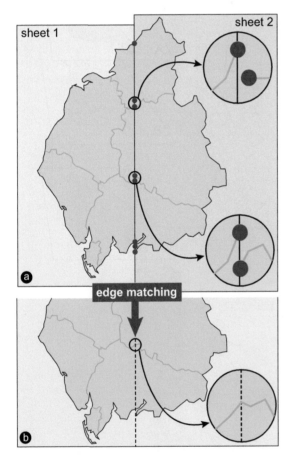

FIGURE 6.14 Edge matching: (a) sheets 1 and 2 do not match; (b) after local transformations, both sheets do match

6.4 GENERALIZATION

6.4.1 Background and Concepts

Each map within a certain scale range requires its own level of detail depending on its purpose. It is evident from Figure 3.3 that large-scale maps usually contain more detail than small-scale maps. However, even at the same scale, the level of detail might be different. A map for a reference atlas will hold much more detail than a map of the same area in a school atlas (see Figure 6.15). The process of reducing the amount of detail in a map in a meaningful way is called 'generalization'. The process of generalization is normally executed when the map scale has to be reduced. Map details in Figure 6.16 show why generalization is necessary. The map of Luxemburg in Figure 6.16a is drawn at a scale of 1:3 million. Figure 6.16b shows a photographic reduction of the same map at a scale of 1:12 million. It still contains the same amount of detail as the maps in Figure 6.16a. The result is a blurred map image. Some text and lines are no longer readable, and some have disappeared altogether. Figure 6.16c, however, is not just a photographic reduction of the original map at scale 1:3 million but a version in which some symbols have been left out, while others have been simplified or emphasized. What has happened is best seen by enlarging this map to its original scale and comparing it with Figure 6.16a. This also demonstrates that even enlargement is not without implications. The enlarged map detail in Figure 6.16d has an unrealistic emptiness and an information density that is too low for its scale.

Particularly in a digital environment, it is very tempting to use the offered zoom functionality. Unlimited zooming in and zooming out can result in unreliable maps when used to interpret their content. These options should be handled with care, or specific triggers should be built in. An example is the OSM. Figure 6.17 shows the coastline of southern South America. Only at the smallest scale (I) will the coastline be generalized. Sometimes zooming in results in adding additional map layers with features such as country boundaries, towns, roads and rivers. However, the individual layers keep their level of detail throughout the zoom process. Maps such as OSM, Bing Maps, Google Maps, Apple Maps, and Yandex or BeiDou all have 20 predefined scale levels with map content generalized for those levels. The scale ranges from 1:500 million (level 1) to 1:500 million (level 20). These products have their role to play, even if, from a cartographic point of view, they are not always correct. In a GIS environment, generalization tools should

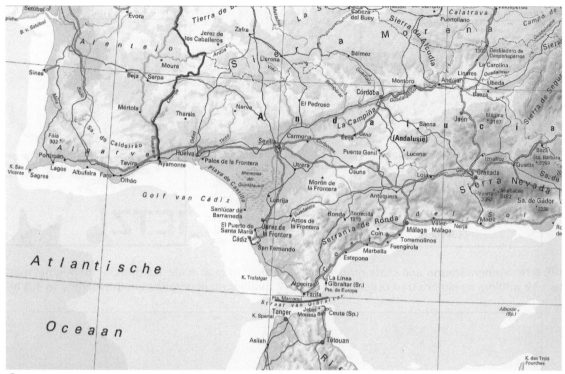

a From school atlas (Grote Bosatlas 2001, courtesy Wolters-Noordhoff)

b From reference atlas (Wolters Noordhoff Wereldatlas 2001, courtesy Wolters-Noordhoff)

FIGURE 6.15 Generalization related to atlas objectives: (a) from school atlas (Grote Bosatlas 2001, courtesy Noordhoff); (b) from reference atlas (Wolters-Noordhoff Wereldatlas 2001, courtesy Noordhoff)

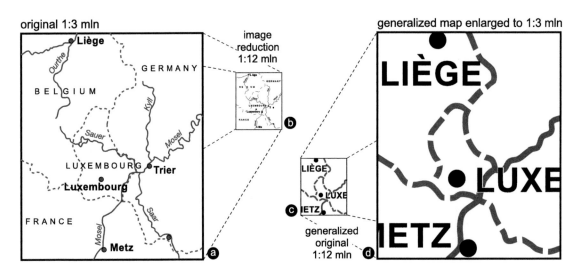

FIGURE 6.16 Generalization and scale reduction: (a) original map at scale 1:3 million; (b) original map reduced to 1:12 million; (c) generalized original map at 1:12 million; (d) generalized map enlarged to 1:3 million

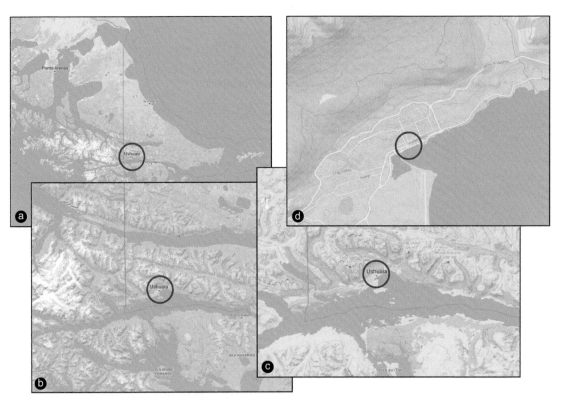

FIGURE 6.17 Generalization in OSM (https://cloud.maptiler.com/
maps/). (a), (b), (c), and (d) zoom in from overview to detail

be at hand. If a spatial analysis operation requires data from different sources, these data sets should be of the same level of detail; otherwise, it will be (in a logical sense, if not in a technical sense) impossible to use the data in a spatial operation. Results would be unpredictable. DLMs of GIS contain a selected description of reality as explained in Chapter 1. It is still likely that a further selection

may be needed. One reason for this could be that the digital cartographic model (DCM) requires a much smaller scale than the basic data in the landscape model will allow. Figure 6.18 shows how generalization fits in around the DLM and the DCM (Gruenreich, 1992).

In fact, to populate a DLM by selecting items from reality can be called 'object generalization'. A DLM

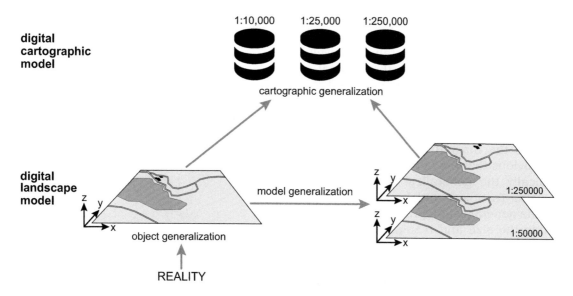

FIGURE 6.18 Generalization and DLMs and DCMs. In model generalization, elements in (multiple) layers are removed; in cartographic generalization, the resulting images are graphically simplified to solve potential graphic conflicts

can, according to specifications, be transformed into another DLM with a smaller scale. This process is called 'model generalization'. To create a DCM, cartographic generalization is applied, in order to be able to create legible maps without graphic conflicts. National Mapping Agencies who provide a series of products on different scales have to organize their generalization policy well. As Figure 6.19 shows, they have two generic options, the star solution and the ladder solution. In the first case, all derived products, whether DLMs or DCMs, are derived from a single – often large-scale – master DLM. The second case derives the smaller-scale version from the

scale level just above the new scale. In practice, the ladder approach is most frequently used, sometimes mixed with the star solution. Current multi-layer automated generalization options are close to derive maps at all other scales from a single master database, the ultimate requirement.

During generalization, one should keep several factors in mind in order to achieve proper results. Most important, as often in cartographic operations, is the purpose of the map and its audience. Generalization entails information loss, but one should try to preserve the essence of the contents of the original map. This implies maintaining

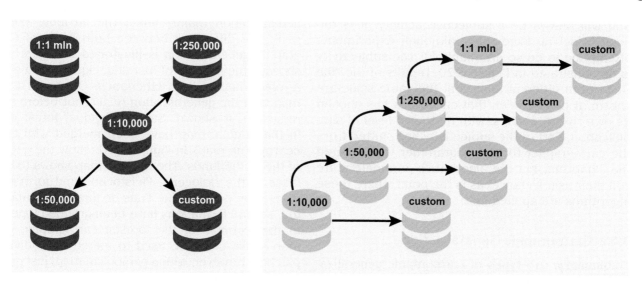

FIGURE 6.19 Star solution versus ladder solution in generalization from a master DLM. 'Custom' refers to a generalization scale on demand

geometric and attribute accuracy, as well as the aesthetic quality of the map. The visual hierarchy (see Section 5.3) should be maintained as well; for instance, prominent features in the original map should remain prominent in the generalized result. Depending on the audience, the results may be different, as in the example of the reference atlas and the school atlas discussed above. Another important factor is the magnitude of the scale reduction. It is obvious that the larger the reduction, the more radically the generalization will affect the original data. Technical and human factors also influence the generalization process. Technical factors include the size and resolution of a monitor screen. In a GIS environment, considering the computational elements is equally important. Which algorithm is most cost-effective and will result in maximum data reduction and minimum storage capacity? These factors interrelate with the human factors. The discriminating capacity of the human eye is limited. Finally, one should always consider the nature of the map contents. Does one deal with quantitative or qualitative information? An answer to this question defines the setting in which a cartographer or a software program can execute the generalization process. Qualitative map content requires a different approach from a quantitative content. The former will require a more extensive knowledge of the mapped features when compared to the latter. This split in the nature of the map contents results in two classes of generalization: graphic and conceptual generalization.

Even if everybody took into consideration the factors mentioned above, generalization results could differ from cartographer to cartographer and from algorithm to algorithm. Generalization is and will always be a subjective activity. It is difficult to set up fixed rules, although experiments attempting to do so are manifold. The subjectivity is demonstrated in Figure 6.20. Details of the Nile Delta from different atlases at the same scale are shown. It can be seen that each atlas has chosen its own river branches, with their own level of simplification. Part of the subjectivity has shifted from the cartographer to the programmer who created the algorithm, but the algorithm's parameters are still their user's choice, as is the point where these algorithms are applied first.

6.4.2 Cartographic Generalization

Traditionally, two types of cartographic generalization are distinguished: graphic and conceptual generalization. The difference between them is related to the methods involved in the generalization process.

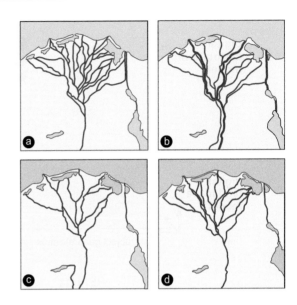

FIGURE 6.20 Samples of subjectivity: The Nile Delta at 1:5 million from: (a) the Atlas Jeune Afrique; (b) the Alexander Weltatlas; (c) the Atlas of Africa; (d) the Times World Atlas (after Pillewizer and Töpfer, 1964)

Graphic generalization is characterized by simplification, enlargement, displacement, merging and selection. None of these processes affects the symbology. Dots stay dots, dashes remain dashes, and patches stay patches. Conceptual generalization is also characterized by the processes of merging and selection, and in addition, it comprises symbolization and enhancement. As a result of these actions, the symbology in the map may change. Another difference is that the processes linked to graphic generalization mostly deal with the geometric component of geospatial data, while those processes linked to conceptual generalization mainly affect the attribute component. The difference between both types of cartographic generalization is illustrated in Figure 6.21, showing the process of merging. Figure 6.21a shows a forest map of the Netherlands. On top of it, the map with the generalization result, but before scale reduction, is shown. Small individual forest areas in the original map have been merged with those nearby. The maps in Figure 6.21b show the geology of the Netherlands. The original map shows the surfaces of the Holocene, Pleistocene and formations from other periods, like Trias or Jura. In the map on top, several formations have been grouped together. In the example of the forests, cartographic common sense could be used to merge the individual patches. However, with a geological map, just grouping what is close is not enough. An understanding of the geological timetable and classification system is required. For instance, Holocene and Pleistocene

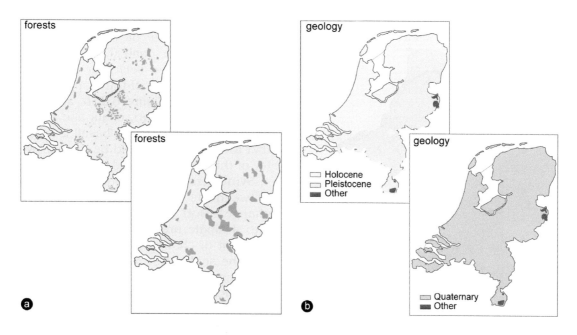

FIGURE 6.21 (a) Graphic generalization versus (b) conceptual generalization

can be grouped together since in a geological classification, both are from the Quaternary. Grouping the small patches classified as 'other' with Pleistocene is not allowed in a geological context. In other words, conceptual generalization requires knowledge of the map contents. To generalize these maps, one depends on principles of the discipline involved. Their classification system changes, and this results in a different structure of the legend as well. It should also be realized that though one can subdivide generalization into several sets of processes, these processes usually interrelate. It is not only simplification or only replacement. Often one process is necessary as a direct result of another process. For instance, if a road has to be enlarged in order to remain visible after scale reduction, then several houses along this road will need to be replaced; otherwise, the road symbol will cover the houses.

Figure 6.22 illustrates the processes involved in graphic generalization: simplification, enlargement, displacement, merging and selection. For each of these actions, three illustrations are given: the original map detail, a generalized map detail before scale reduction and the generalized map detail after scale reduction. Simplification, sometimes called 'smoothing', should reduce the complexity of the map. The illustration in Figure 6.22a shows a river that has a very sinuous nature with many bends. After generalization, the character of the river should be preserved: a meandering river should still be recognized as such. Enlargement (Figure 6.22b) is sometimes needed; otherwise,

symbols would disappear, or would no longer be legible after scale reduction. This action affects roads. To keep a road symbol legible, it has to be enlarged. It should be realized that the road in the generalized map, after scale reduction, would be much too broad. A road at a scale of 1:10 000 could be 10 m wide, while the same road, represented by a similar symbol, on a 1:50 000 map would have a width of 50 m. Displacement is usually the result of other generalization procedures. It is also a critical procedure since one should take care that, for instance, a symbol representing a house is not placed along the wrong line symbol. Figure 6.22c shows the need to displace a house because of the enlargement of a road symbol. In Figure 6.22d, some individual houses are merged to form a built-up area. Selection, as demonstrated in Figure 6.22e, is the process of selecting representative map objects from a set of similar objects, the rest of which are omitted in the resultant map. It is important to note that omitting symbols should not disturb the overall impression of the phenomenon's distribution. Selection is necessary; otherwise, the map image would become too cluttered. The illustration shows how some islands along the coast are left out.

The procedures involved in conceptual generalization are explained in Figure 6.23. They are merging, selection, symbolization and enhancement/exaggeration. Figure 6.23a shows that merging of symbols cannot be done without expertise since it has consequences for the legend as well. Some

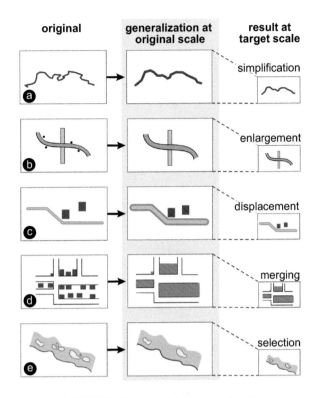

FIGURE 6.22 Graphic generalization:
(a) simplification; (b) enlargement; (c)
displacement; (d) merging; (e) selection

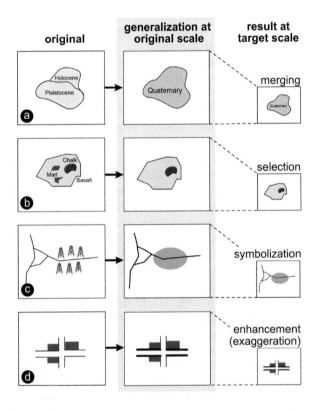

FIGURE 6.23 Conceptual generalization: (a) merging;
(b) selection; (c) symbolization; (d) enhancement

symbols will disappear from the legend, while a smaller number of new units might appear. Selection in the context of conceptual generalization stands for selecting legend classes and requires knowledge of the mapped phenomena. In the example of Figure 6.23b, a lithographic map has symbols for marl, chalk and basalt. Although small in areal extent, the basalt is so characteristic that to leave it out would be to destroy the character of this volcanic island.

Symbolization denotes that the relation between the symbol and the space it represents changes. Dots (e.g. a group of oil rigs) will change into a single area symbol (e.g. an oil field). The moment of change depends on the original scale and the scale after reduction. Generalization can also result in a map where some symbols attract too much or not enough attention. These symbols have to be enlarged or reduced in size. A main road through a village could become insignificant after scale reduction. So it has to be enhanced by using thicker lines to attract attention proportional to its importance in the map image. Figure 6.24 illustrates some of the generalization procedures executed when generalizing topographic maps from a scale of 1:25 000 to 1:50 000. Compare also with Figure 6.25.

6.4.3 Generalization Processes and Tools

The previous section explained the principles of generalization. How does it work in practice? Several cartographers have struggled with this question and suggested conceptual models in an attempt to solve the generalization problem. Among them are Brassel and Weibel (1988) and McMaster and Shea (1992). Their model, in a slightly adapted version, is used to explain the practice of generalization and the tools available. McMaster and Shea developed their method explicitly for digital cartography. Figure 6.26 summarizes their approach. They deconstructed the generalization process into three tasks, translated into the questions of why, when and how to generalize? 'Why generalize' has been discussed at the beginning of Section 6.4.1.

'When?' is related to a cartometric evaluation of the original map data in relation to the generalized map. A large reduction in scale of the DCM will cause legibility problems related to graphic congestion and the coalescence of graphic symbols. These problems can be avoided by the application of, for instance, threshold values or minimum distances between graphical objects. Whenever objects come too close to each other or whenever a specified graphical density (number of objects within a 10

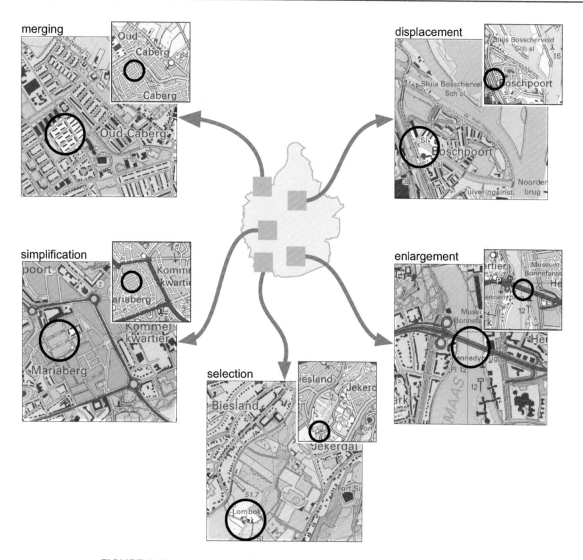

FIGURE 6.24 Examples of graphic generalization from topographic maps 1:25 000 to 1:50 000 (Kadaster Geo Informatie)

by 10 cm² on the map, on paper or on the screen) is surmounted, generalization algorithms can be started up. McMaster and Shea describe this as the transformation control. It consists of the selection of suitable algorithms and of the suitable parameter values to apply to them.

'How?' is related to the tools that can transform the geometric and attribute component of the geospatial data, to generalize effectively the original map data. Here the tools, or rather the algorithms, that execute the actions of the graphic and conceptual generalization are discussed. The generalization of attribute data (classification and symbolization) will be discussed in Section 7.3. Many generalization algorithms do exist. Most of them are related to graphic generalization actions, in a vector or raster data environment. This is not strange since these actions can only be translated into an algorithm

when one can observe regularities in them. For most aspects of graphic generalization, this can be effectuated. An extensive overview of these algorithms is given by McMaster and Shea (1992). However, until recently, most of the algorithms available can only solve relatively simple isolated problems that are linked to a single basic graphical element like lines. Some of these algorithms can still be relatively complex. Let us look at an example related to the selection action. The starting point is a database of the European road network, which contains all motorways, major and minor roads. When part of it has to be displayed at a reduced scale, many roads would have to disappear. It would be simple to suggest the selection of just the motorways. However, this would result in a European network full of gaps. Often, near frontiers, motorways stop and change into major roads, which cross the border. In a

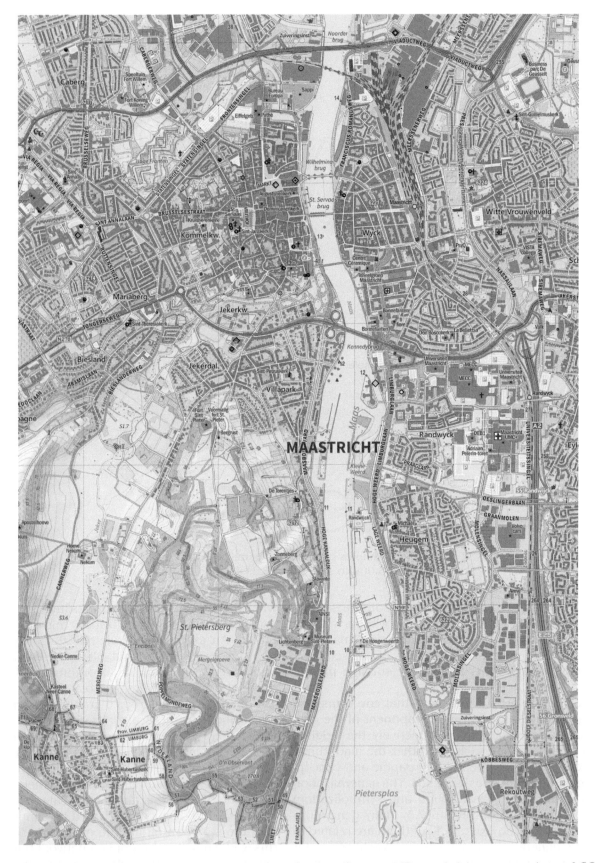

FIGURE 6.25 Maastricht 1:50 000, 2019, open topography map (Source: J.W. van Aalst, www.opentopo.nl CC-BY)

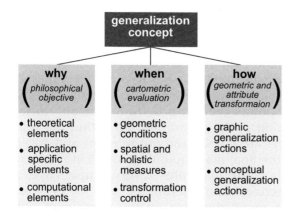

FIGURE 6.26   A conceptual model of generalization (after McMaster and Shea, 1992).

simple example is related to the display of towns based on the number of inhabitants. After scale reduction, it would be easy to preserve only those towns with over 100 000 inhabitants. However, in very densely populated areas like the German Ruhr, the map would still be cluttered with symbols and text. Cartographers then would apply other criteria as well to omit some more towns in order to keep the map legible. On the other hand, lots of areas with only a few inhabitants would remain empty. Normally, a cartographer would keep some smaller towns in these areas. Again, special conditions would have to be incorporated in the algorithm to make it do just this. Kadaster, the National Mapping Agency in the Netherlands, managed to include a fully operational automated generalization process 1:10.000 to 1:50.000 in their map production process (Stoter et al., 2014).

manual situation, the cartographer would incorporate these roads because of their importance to the overall road network. In the algorithms, extra conditions have to be incorporated to do this. Another

The generalization algorithms available in most GIS packages are related to line simplification. Figure 6.27 shows the result of two algorithms.

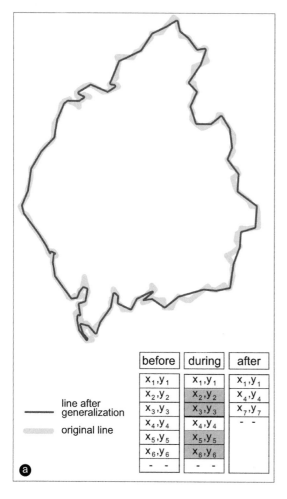

before	during	after
x_1,y_1	x_1,y_1	x_1,y_1
x_2,y_2	x_2,y_2	x_4,y_4
x_3,y_3	x_3,y_3	x_7,y_7
x_4,y_4	x_4,y_4	- -
x_5,y_5	x_5,y_5	
x_6,y_6	x_6,y_6	
- -	- -	

line after generalization

original line

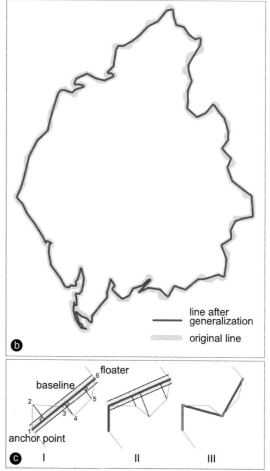

line after generalization

original line

FIGURE 6.27   Line simplification algorithms: (a) nth point; (b) Douglas–Peucker; (c) the principle of the Douglas–Peucker algorithm

In Figure 6.27a, the *n*th-point algorithm, which retains every *n*th point, is explained. It follows a simple approach and does not consider any relations between neighbouring points. The grey line represents the original data, and the black line the generalized data. In the example, *n* is set to three. During execution, only each third point is retained as can be seen in the table below the map. The user can define *n*, and with it the magnitude of generalization. The result will depend on the density and homogeneity of the points along the line. A large value for *n* will result in a strongly generalized line, and very likely in the loss of original line characteristics. An algorithm that considers the line characteristics is the Douglas–Peucker algorithm. It has been accepted as one of the best generalization algorithms available and has been implemented in many GIS packages. The algorithm considers the whole line during the generalization process and eliminates points in an iterative process. It is efficient but relatively slow in processing (Douglas and Peucker, 1973). An example is shown in Figure 6.27b. Again, the grey line represents the original data, and the black line the data after generalization. The algorithm works with a baseline and a tolerance zone. The baseline connects the beginning and end points of a line (here 1 and 6). The first is called the 'anchor point' and the second, the 'floater'. The tolerance zone, defined by the user, should guarantee that the line characteristics are preserved. The smaller the zone, the better the characteristics

are preserved. The distance perpendicular to the baseline is calculated for all points between the floater and the anchor. The point with the largest distance (point 2 in Figure 6.27b II) will be saved and will function as a new floater point. For the remainder of the line, this process is repeated. All points that, at a certain moment during execution of the algorithm, fall within the tolerance zone are considered not relevant and are left out in the final result (points 3 and 5 in Figure 6.27b). This figure illustrates a very simple case, and in practice, many iterations have to be executed. The result depends on the number of points in the original line, their distribution and density, as well as the width of the tolerance zone. Online open tools like https://mapshaper.org have these algorithms available and allow the user to drop their files execute line generalization.

Generalization in the raster domain means generalization of the attribute data since the algorithms available simplify the image. The simplest approach is to define the dominant grid cell or pixel attribute for a region and give this value to all pixels in the region. This approach does not preserve the character of the mapped phenomena, and more clever algorithms have been developed. Most of the raster generalization tools originate from image-processing disciplines. Often the algorithms work with some kind of filter matrix, which is moved over the whole raster data set. Figure 6.28 shows how these filters operate. The filter in Figure 6.28a(II) has a low value for the

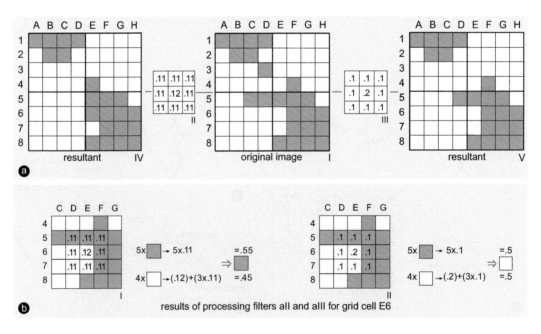

FIGURE 6.28 Area simplification algorithms: (a) the effect of two different filters; (b) the effect of the two filters on a single pixel

central pixel, which gives the surrounding pixels more weight in the process. The filter in Figure 6.28a(III) will have a less generalizing effect on the final data set because the value of the central pixel is high. The result of the generalization process is given in Figure 6.28a(V), while in 6.28b, the effect of both filters on a single pixel is shown. The size of the filter also influences the result.

It has not been possible to translate conceptual generalization processes into algorithms with any great success. The reason is that such processes require knowledge of the mapped theme. Several experiments with knowledge-based generalization systems have been performed (Buttenfield and McMaster, 1991, Müller, 1991). Müller tried to apply a knowledge-based system to the generalization of topographic maps in the scale range between 1:25 000 and 1:250 000. Figure 6.29 shows how the graphic and conceptual generalization actions fit in here. Topographic surveys have clear generalization rules; otherwise, they would not be able to produce uniform map series that cover a whole country. Müller notes that linear elements tend to dominate the map image more, the further the scale decreases. It has also been noted that the number of object classes decreases as well, and roads become disproportionate in size with respect to other line symbols. Of all geometric relations, topological relations are preserved the longest, whereas shapes, distances, etc. are affected. In the experiment conducted, rules and observations like the ones mentioned in Müller's system were applied. This was done under the assumption that the data would be available in raster or vector format and that the algorithms needed would be available.

While going through such a scenario, one still finds many unsolved problems. Most rules point out what should not be done, but do not say what should be done. And if they do, they do not say how to do it.

Many problems still exist, and several potential solutions are only available in experimental environments. Some GIS packages offer the user an interactive generalization environment. The user can select a specific area or a specific data layer, decide which algorithm to use (e.g. for line generalization) and can subsequently set parameters. Results can be directly viewed on the screen, and when results are not as expected, the user can change the algorithm or its parameters. It is also possible to use different parameters or algorithms for different parts of the study area.

6.5 RELIEF

6.5.1 Introduction

In mapping the terrain, the cartographer has to deal with relief data. Again, there is a struggle to transform three-dimensional reality onto the two-dimensional plane of the screen or paper. Relief display should result in a map that provides a geometrically accurate view of the terrain and its shapes (morphology). The method of relief display depends on the purpose of the map. A map intended to give a global impression of terrain requires a different approach compared with a map from which one wants to determine heights within accuracy limits of 10 cm. A map to be used for building a dam will contain different height information from a tourist map for a skiing area.

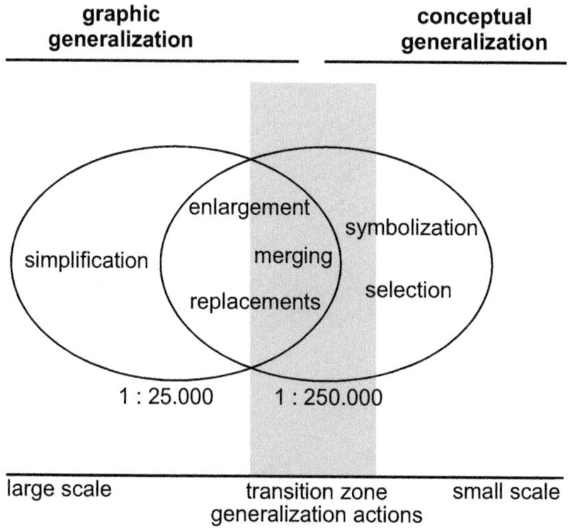

FIGURE 6.29 Knowledge-based generalization (after Müller, 1991)

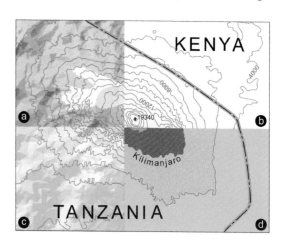

FIGURE 6.30 Absolute and relative terrain representation, Mount Kilimanjaro: (a) contour lines and hill shading; (b) contour lines and height points (absolute); (c) layer tints (relative) and hill shading; (d) contour lines and layer tints

It is possible to represent terrain in absolute or relative terms. In Figure 6.30, both methods are shown. Absolute heights can be displayed by contour lines or height points, as shown by the graphic representation of the north-eastern side of Mount Kilimanjaro in the figure. Their value is determined above or below a reference plane. Relative height indicates whether a certain location is higher, equal or lower than other locations, which can be effected by layer tints, sometimes enhanced with hill shading, as is shown for the western side of Kilimanjaro (Figure 6.30). In mapping relief, cartographers have employed many graphical techniques. The oldest relief maps portray the terrain with simple symbols. Mountains were drawn in aspect or sketched. Relief mapping further developed via hachuring without three-dimensional stimuli, to a systematic hachuring, later followed by hill shading to enhance the relief impression. Figure 6.31a shows hill shading by the Swiss cartographer Imhof. He introduced the *luftperspektivische Geländedarstellung* to represent relief (Imhof, 1982, 2007). It is based on the natural effects of atmospheric colours in the mountains and is seen as one of the best methods of indicating map relief on a geometrically accurate orthogonal map, while still preserving the third dimension. The shading is printed in tints of violet, and a yellowish tint is added to the slopes facing a fictitious light source. The relief impression is further enhanced by adding more contrast in the higher areas. This technique is employed in Swiss topographic maps. Since this method of relief representation is a skilled and (as with most other relief representation methods) laborious activity, cartographers have tried to achieve the same effect with computers.

Figure 6.31b shows a detail from the *Grote Bosatlas* (2001). In this map, layer tint mapping is applied. All areas between certain contour lines get a specific colour, which ranges from dark green for low areas to reddish-brown for high areas. A different approach to relief mapping entails representations such as block diagrams, perspective views and panorama maps. In these maps, the third dimension is not projected on an orthogonal base map. Figure 6.32 gives an example of a panorama map of the Atlantic Ocean produced for a National Geographic Map by H.C. Berann. A review of most relief representation techniques is given by Imhof (1982, 2007).

6.5.2 Digital Terrain Models

To create relief displays by computer, one needs data to start from. The key phrase here is digital terrain model (DTM). DTM is usually defined as a numerical representation of terrain characteristics. When dealing with the altimetric aspects only, they are often called 'digital elevation models' (DEMs). Both DTMs and DEMs can be seen as DLMs (see Section 1.2) that can be manipulated in a GIS for surface operations, or to create DCMs. There are several applications that require a more general approach to digital terrain modelling allowing for the incorporation of three-dimensional geospatial objects that are application-dependent and may be called 'topographical', 'geographical', 'geological', etc. At the same time, the functionality of existing GIS software packages, like Google Earth, is being expanded to combine relief data with planimetric data of topographical and thematic coverages. To cover all applications, the definition should be

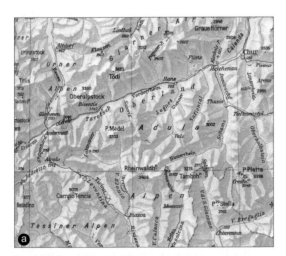

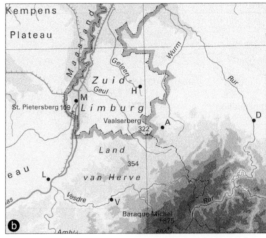

FIGURE 6.31 (a) Hill shading (detail from the Schweizer Weltatlas, courtesy Kanton Zürich; courtesy EDK) versus (b) layer tint mapping (detail from the Bosatlas, courtesy Noordhoff Atlas Productions)

FIGURE 6.32 Terrain shapes and structure: National Geographic Map of the ocean floors (courtesy National Geographic Society)

expanded: a DTM is a digital three-dimensional representation of the terrain surface and selected zero-, one-, two- and three-dimensional geospatial objects that are related to this surface. The use of three-dimensional maps can be very effective in explaining spatial relationships. For instance, when mapping the Earth's surface, DTMs can give an explanatory insight into its forms. Using the possible ways of looking at the terrain in an interactive environment, by changing view angle and azimuth, the sometimes difficult interpretation of the contour line pattern of a topographic map or chart can be avoided. The height information can be combined with, for instance, land use information.

DTMs have found a wide range of applications. They are used in civil engineering for determining the earthwork cut-and-fill volumes, landscaping and creating a visual impression of the environmental impact of civil engineering projects. In topographic mapping, they are used to visualize terrain forms. In geological and geophysical mapping, they visualize surface and underground structures. They are also in use in navigation simulation, for instance,

to train pilots. Last but not least, the military applications have to be mentioned. Here, DTMs provide information on visibility from a specific point, while slope information is used to plan the most suitable route, and some missile guidance systems use DTM information for navigation.

In order to collect data for DTM building, the same data collection techniques that were discussed in Chapter 2 are used. Existing maps that contain contour lines and spot heights can be digitized. One should note that the quality of the DTM can never surpass that of the map the data were derived from. Because contour lines are often a product of interpolation, the quality of DTMs based on digitized data is often less than those collected by photogrammetric, surveying or laser scanning techniques. This last technique provides the most accurate data. The result acquired by photogrammetric techniques depends on the photo scale. Laser scanning results in the so-called point cloud which has to be processed and can result in country-wide very detailed terrain models, as has been effectuated in the Netherlands. The height model has on average eight observations per square metre and a height accuracy of 5 cm (https://www.ahn.nl) (see also Figure 1.6). Next to the technique, the data gathering method applied will influence the usability of the DTM. If one applies selective sampling, all characteristic points in the terrain can be incorporated into the DTM. Another method applied is systematic sampling. Here data are sampled at regular distances. This approach provides no guarantee that characteristic points such as the highest and lowest point can be incorporated since they can be located between the sampled points. Selection of a method will depend on the purpose of the DTM (measurements or presentation), the nature of the terrain (roughness and accessibility) and the available hardware and software.

To structure the DTM data in a DLM, one can choose between a grid approach and a triangulation approach. The first results in a regular network of points covering the study area. Height values are determined at these points. Figure 6.33 shows an example. In the figure, the height values are based on original height measurements as displayed in Figure 6.33a. The distance between the points, in relation to the distribution of the original data, as well as the local or overall interpolation method applied, defines the model's accuracy. An example of local interpolation is illustrated in Figure 6.33b. For each point value to be determined, the heights of the six closest original data points, including a distance weight factor, are used. An example of an overall approach is the calculation of a polynomial function that includes all original points. The grid point height values can be derived from this function. Both approaches have advantages and disadvantages with respect to the result and processing time. In Figure 6.33c, a triangular network is drawn. An important characteristic of this approach is the fact that each of the original data points is incorporated into the model. This offers the opportunity to consider local relief characteristics. The network in Figure 6.33e is the result of the Delaunay triangulation algorithm. It results in a triangular irregular network (TIN). This algorithm is incorporated in most GIS packages (often in a slightly adapted version). Characteristic of a Delaunay triangle is that it has edges with the shortest possible length, and the angle between two edges is as large as possible.

In a GIS environment, the DTMs are used to execute surface analysis. The basic geospatial units of a DTM, grid squares or triangles, have an important function. Two of their attributes, slope and aspect, play a prominent role in a calculation related to the surface analysis (Figure 6.33f and g).

Slope is defined as the amount of change in height over a fixed distance, expressed as a percentage or in degrees. Aspect is the orientation of the geospatial unit with respect to the north and is expressed in degrees.

6.5.3 Terrain Visualization

Questions one is likely to ask when executing a terrain surface analysis range from simple queries such as 'What is the height at this location?' and 'What areas can be found between 200 m and 400 m?' to more complex queries such as 'Which terrain is visible from this point?' or 'How much Earth has to be moved if …?' Most answers can be given in map form.

The shaded relief map is DTM's most prominent derived product. Shaded relief maps used to be products that could only be created by a few very skilled cartographers. Today, this hill-shading technique is available to every GIS user. However, it should be mentioned that, as with generalization algorithms, the users can still make many mistakes since they have to set lots of parameters. In traditional cartography, the sun is put at a certain angle in the north-west to create a shaded relief map that is perceived correctly. Having the light source in the south, for instance (which after all is quite natural in the northern hemisphere), would result in a map on which people would see mountains in valleys and the other way round. This relief inversion is caused by stimuli received by the human brain. In a GIS environment, the user can freely set the location of the light source and so can create maps that give a wrong impression of the terrain.

Figure 6.34a explains how a shaded relief map can be created from a DTM. In the example, a TIN-based DTM is used. For each triangle, a normal vector (the vector perpendicular to the triangle plane) can be calculated. Next, the angle between the normal vector and a vector representing the light source with respect to the observer is determined. The light source is placed in the north-west at an angle of 45° to the plane. Working with normalized data, this will result in a value of between one and zero for each triangle. The next step is to create a grey scale and to define the lightest tint as zero and the darkest tint as one. Now it is possible to match triangle values with a tint from the grey scale. The approach described here is the simplest solution. More advanced algorithms can consider special cases as well. The result of the algorithm is shown in Figure 6.34b. Here, some of the original triangles can still be recognized. This is due to the density and distribution of the original height data

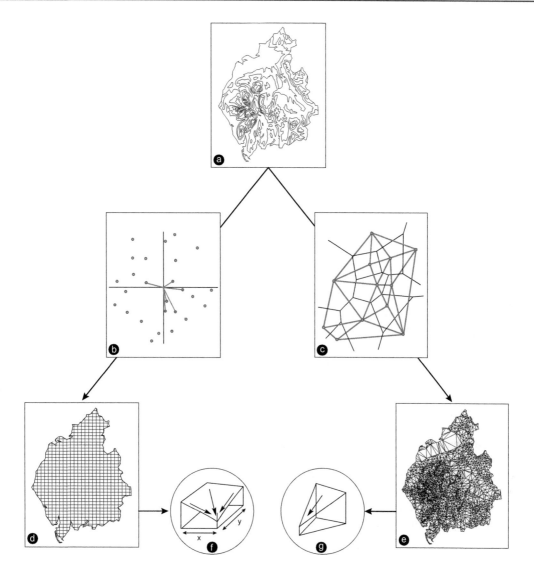

FIGURE 6.33 Data representation from contour lines (a) to a regular grid (d) and TIN (e), basic geographical unit in regular grid (f) and in TIN (g). Between (a) and (d), an interpolation algorithm (b) is applied, and between (a) and (e), a triangulation algorithm (c) is applied

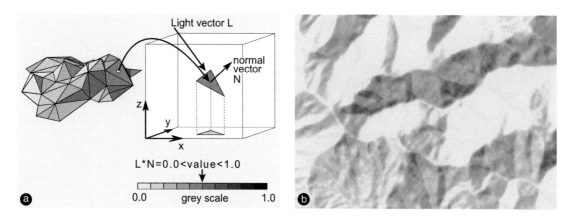

FIGURE 6.34 Principles of hill shading: (a) the (grey) value of a terrain patch is determined by the values of the patch's normal and the position of the light source relative to the patch; (b) an example of a shaded relief map

127

as displayed in Figure 6.33. The option to change the algorithm's parameters, like the position of the light source, is sometimes needed because of terrain characteristics. An example is a mountain ridge with 45° slopes, which ranges from NE to SW. Without interference, the resultant map would be black and white only – something cartographers would try to avoid if possible.

Other derived products are shown in Figure 6.35. On the left of this figure, a profile based on the DTM data is presented, while on the right is a visibility map. Profiles can be used to see what effect the terrain will have on a new high-speed train link, and give insight as to where earthworks have to be excavated. The dark area in Figure 6.35b shows the terrain visible from the selected point, looking northwards. The dashed lines indicate the view angle. One can also set the maximum viewing distance. In the examples, it was assumed that weather conditions would allow for 30 km visibility. More complex, but also possible, is the calculation of the terrain visible from two different points, or the calculation of the terrain from where one could see the location of the observer. As well as having several military applications, this type of calculation for visibility maps is used in environmental impact assessment, for instance, in order to determine where to put high-rise building without visually disturbing the countryside. Most DCMs, from which the figures in this section are derived, neglect the third dimension, although it is available in the DLM. As the panorama maps in Figure 6.36

demonstrate, it is possible to preserve the third dimension in the image. The main map in the figure shows the terrain with other data, such as hydrography, vegetation and roads draped over it. The inset in the upper left of the figure shows a detail of data in two dimensions, the insets at the bottom show the same terrain from different viewpoints, and the upper right inset shows a fully shaded perspective map with terrain features draped over. Figure 6.37 shows a topographic map draped over a detailed terrain model. These days GIS software with digital terrain modelling capacity has draping functions available. Next to maps, satellite imagery can be used as well to create realistic three-dimensional views. Google Earth has made detailed terrain data widely available as backdrop for in imagery and offers user a 3D perspective on our planet. Users can drape their own maps on this terrain as seen in Figure 1.19, but they can also add other 3D feature like buildings.

In order to be able to create maps like those in Figure 6.36, as well as any other three-dimensional maps, the nature of the program functionality available is very important. Figure 6.38 shows a minimum functionality. The functions or utilities can be grouped into four main categories: three-dimensional visualization utilities, cartographic design, cartographic modelling and final display utilities. These should be accessible via a graphical user interface.

Three-dimensional visualization utilities are concerned with the user's view of the map. They include geometric map transformations

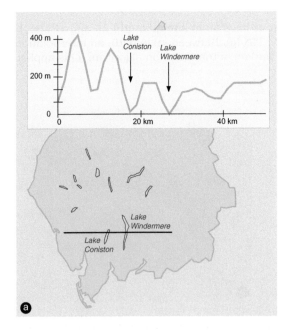

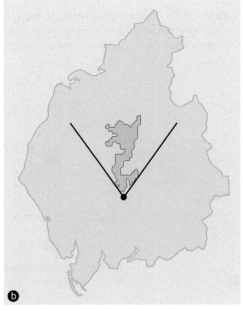

FIGURE 6.35 DTMs and derived products: (a) profiles; (b) visibility maps

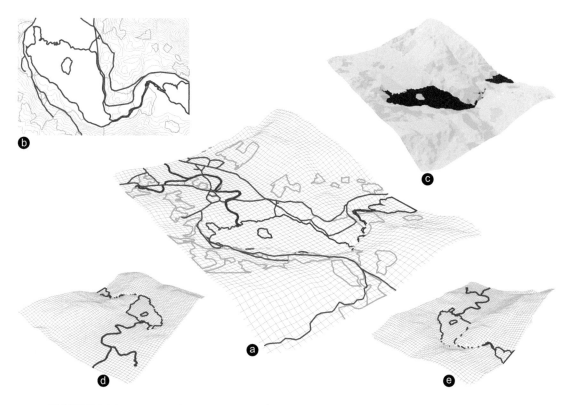

FIGURE 6.36 Perspective maps. Looking at the terrain from different viewpoints and with different data layers draped over. The images in (a), (b), (c), (d), and (e) offer a different angle on the landscape, each with a slightly different content

FIGURE 6.37 Three-dimensional terrain views: (a) topographic map 1:25 000 draped over laser scan-based terrain model (Lutterzand, the Netherlands); (b) new large-scale 3D map series automatically generated from building data set (Level of Detail 1.3), basic topographic data, dense matched aerial photography and laser scan height information from Dutch height database AHN3 (Naarden, the Netherlands) (Kadaster Geo Informatie)

such as rotation, scaling, translation and zooming to position the map in three-dimensional space with respect to the map's purpose and the phenomena to be mapped. Geometric manipulations are necessary since in a three-dimensional image presented on a flat screen, there is the possibility of elements disappearing behind other elements, as is shown in Figure 6.38. This can negatively influence the map's task of information transfer. To avoid this, a proper viewing position should be found by rotating the map around each of the x-, y- and z-axes separately. An important feature is the option to scale the map along the z-axis to find the proper vertical exaggeration. Most software packages that can handle the third dimension has this particular option. During the

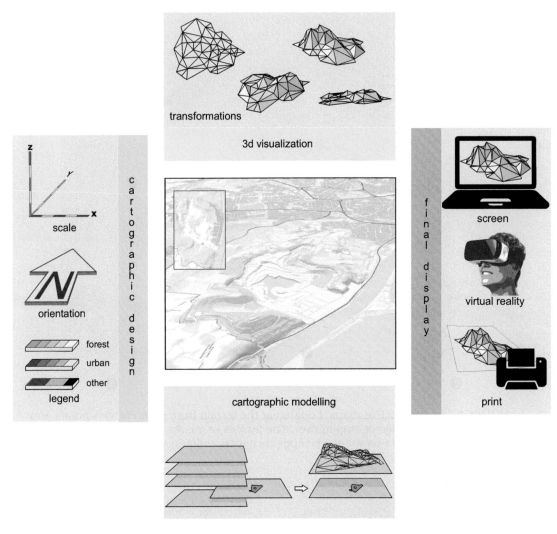

FIGURE 6.38 Basic functionality of a three-dimensional cartographic production unit

mapping process, it should also be possible to use equipment like stereoscope to view the map in 'real' 3D. The visualization of the terrain in a virtual reality (VR) environment will embed the viewer in the terrain and result in a different sometimes very realistic experience. When in the field with a mobile device, augmented reality (AR) can be used to superimpose elements from the topographic database, like object names, on the user's view.

Cartographic design utilities refer to the main design functions, which should include options to choose proper symbology. Among them are the definition of colours, line sizes, fonts, etc. and the positioning of legend, north arrow and scale bar. Information on the orientation of the non-orthogonal map with respect to the more familiar 2D view is also of importance. Referring to this orientation, it should be noted that a direct link

between symbols in the map and those in the legend is necessary. The design process is influenced by 3D perception rules. This means that as well as the use of graphical variables, pictorial depth cues such as shading, texture, perspective and colour should be applied. The relative importance of each of the depth cues depends on the degree of realism of the final image. Other basic operations that might influence the design, such as coordinate transformation, selection, classification and generalization of the data, are assumed to have been executed beforehand.

Cartographic modelling can be seen as manipulating maps or map layers. Tomlin (1990) describes it as a geographical data processing method, but within the framework of this book, its only purpose is visualization. It also provides a link with GIS databases and allows the cartographer or user to retrieve other map data and combine them partly or as a whole with the basic three-dimensional map data already displayed, as seen in the main map of Figure 6.38.

Final display utilities should help the user to produce the final map using known cartographic techniques as well as appropriate computer graphics techniques, both depending on the output medium (e.g. the monitor screen or PDF). From the computer graphics world, complete rendering programs can be implemented. These not only take care of hidden surface removal, texturing and shading, but also include complete atmospheric models for realistic images. These last methods should only be applied if, from a cartographic point of view, they enhance the main task of the map, i.e. information transfer. VR and AR are options too.

6.6 TOPOGRAPHIC DATA: MAPPING AND CHARTING ORGANIZATIONS

6.6.1 Introduction

Users working in a mature GIS environment will find most of the data needed in their own GIS database. If specific data needed for an analysis are not available, the user has to consider other sources. When geometric data are needed, the local map library used to be the first place to go. After finding the map, it would be digitized. This might still be the only way of acquiring the data needed in many cases. However, it is much more convenient if the data can be obtained digitally, and one might find the right provider through the portals of the GDI. Government organizations like those responsible for topographic mapping and planning have made their data available via those portals. For topographic mapping organizations, the sale of digital data nowadays overtakes the sales of paper products. For these organizations, the first task to be performed with computers was to implement the new technology in the conventional mapping process. Their goal was to produce the same maps using the new technology. The first data sets created were unstructured and could only be used to draw maps. Such an approach is not that strange if one remembers their goal: to speed up the production of the paper maps. However, changing information needs in the civil and military market forced the mapping organizations to build structured topographic information systems and make their data available in a format that suits user demands. Organizations such as NATO and the European Union also increased the need for data sets that could cross international boundaries.

Environmental problems, requirements of car navigation systems and DTMs are all examples of topics that do not stop at national boundaries. In Europe, the umbrella association of the topographic mapping organization EuroGeographics is stimulating cooperation and standardization, by making the data of the European National Mapping Agencies interoperable. However, the approach towards the graphical representation of the geodata still varies from country to country, as can be demonstrated by the samples in Figures 6.39–6.44.

This section will study the different topographic mapping activities of several organizations and countries, which each emphasize different aspects of the topographic mapping process. This includes the OSM, EuroBoundaryMap seamless administrative boundary data, the American National Map concept and the approach of the British OS (see their respective websites for recent developments). Regarding data sources, apart from authoritative governmental organizations, there are commercial companies specializing in providing digital geometric data. Some of these have their own data programs, often linked to a GIS consultancy, with data sets of administrative boundaries, road networks, etc. On demand, they can provide any type of digital geospatial data. Attribute data are often (digitally) available from census bureaux. Several GIS vendors have set up data programs as well, to increase GIS awareness and sales. An example is ESRI's living atlas (https://livingatlas.arcgis.com/en/). It is a global network of users and providers of geographical data, and enables the sharing of these data. The data sets include open official governmental data as well as user data and can be directly used in ESRI (online) software or your software (see Figure 6.45). Examples include not only the topographic maps from the United States and the Netherlands, but also adapted generic global data sets for base maps with and without hill shading, boundaries, etc. in different colour palettes. Satellite images from Sentinel 2 and others, as well as aerial photographs from all kinds of regions, are available. All data offered have some sort of metadata description that informs the user of its source, age, coverage, update frequency, etc. This is rather important if one wants to know if the data fit the purpose of use. Many data found on the Internet lack such description, but it should be mentioned, however, that such data come on the Web as they are: no guarantees and no maintenance. But for some, these data would just fit their requirements. That is why products like Google Maps, which lack an authoritative quality stamp, cannot be used for many applications, but are of course very suitable for navigation and

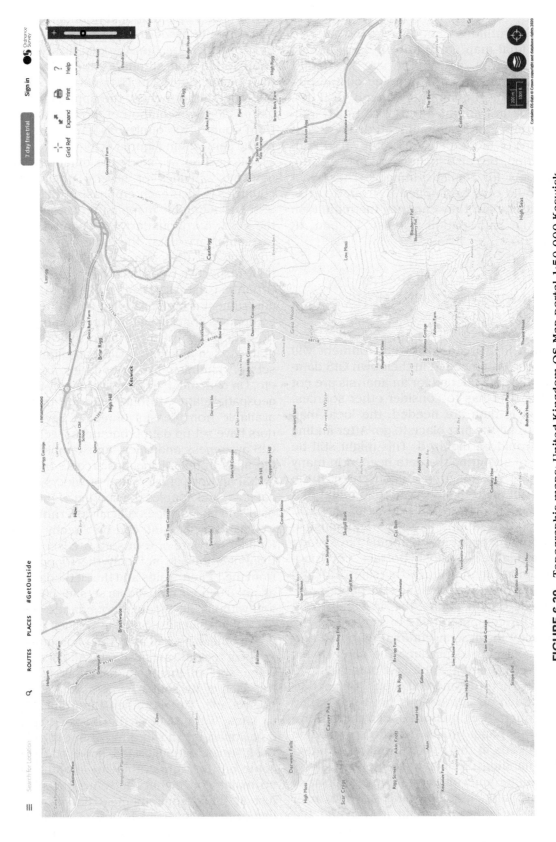

FIGURE 6.39 Topographic maps: United Kingdom OS Map portal 1:50 000 Keswick
https://osmaps.ordnancesurvey.co.uk/) Quotation Reference: CS-122303-Y9X1S8

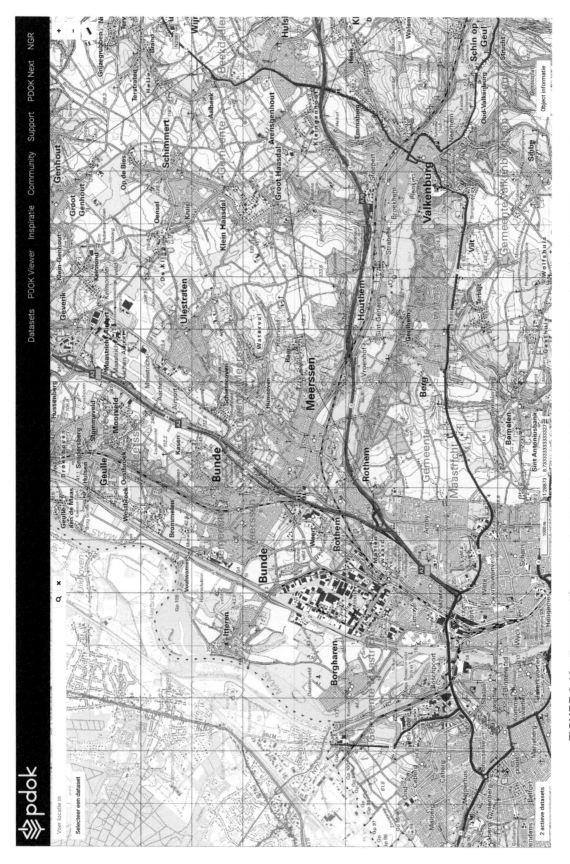

FIGURE 6.40 Topographic maps: the Netherlands TOP50 raster, Meerssen, visualized in PDOK portal, the leading platform for high-quality geodata https://www.pdok.nl (Kadaster Geo Informatie)

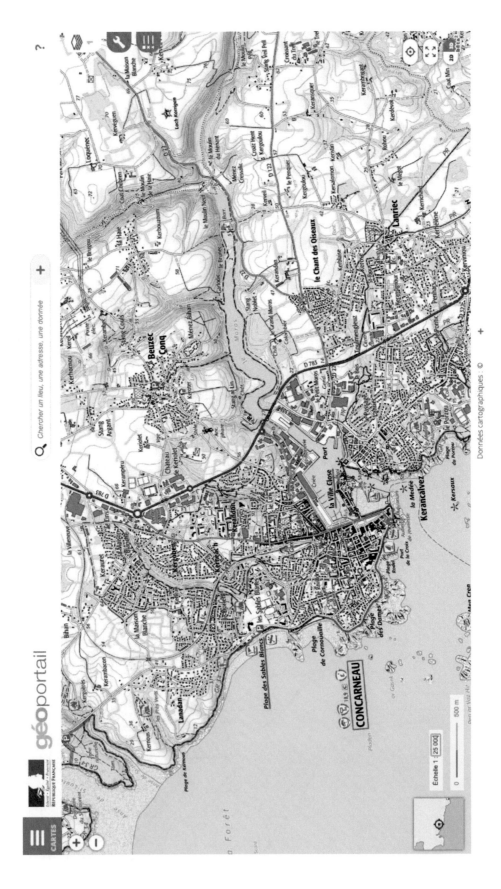

FIGURE 6.41 Topographic maps: France Géoportail 1:25 000 Concarneau (geoportail.gouv.fr / IGNF)

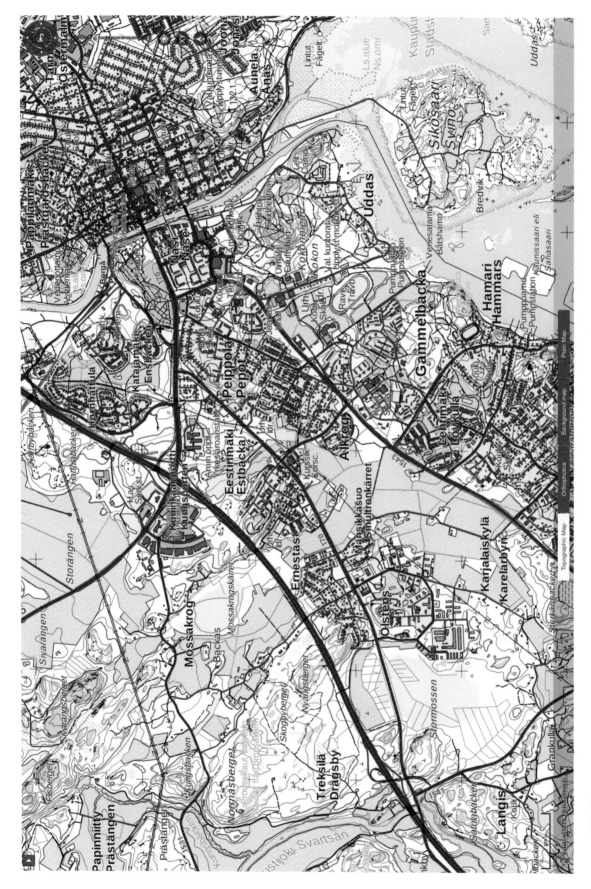

FIGURE 6.42 Topographic maps: Finland L4321 1:20 000 (National Land Survey of Finland Topographic Database 12/2019 – CC-BY)

FIGURE 6.43 Topographic maps: United States IA Dubuque North 20181116 TM 1:24 000. The typical marginal information of a topographic map sheet is found at the bottom (Credit: U.S. Geological Survey Department of the Interior/USGS)

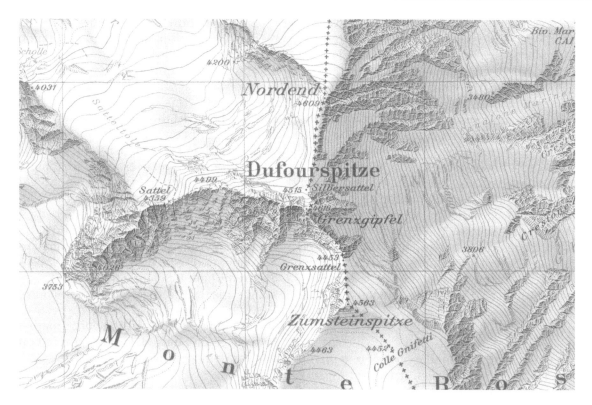

FIGURE 6.44   Topographic maps: Switzerland sheet 2515, 1:25 000

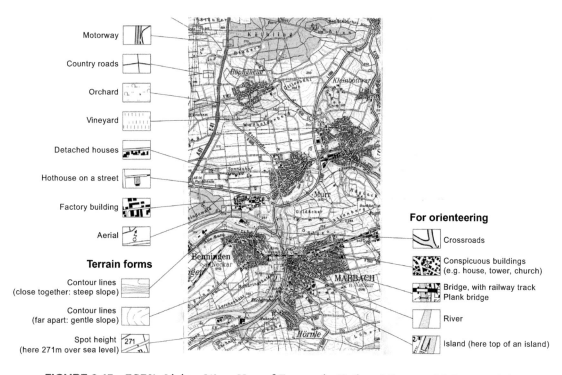

FIGURE 6.45   ESRI's Living Atlas, Map of Europe in National Geographic's map style

quick overviews. Recently, many data have been offered in formats formally decided upon by the ISO (International Organization for Standardization), the Open Geospatial Consortium, like GML Geography Markup Language), and the JavaScript Object Notation-based GeoJSON. Another popular common proprietary format is ESRI's shapefile. All these formats can be used in open software like QGIS.

National Mapping Agencies have started to offer their data in these formats, and it is expected that most data will be available in these standards soon.

As can be seen when comparing Figures 6.39–6.44 and 6.46, there are distinct styles in topographic mapping, and differences in detail. While on German topographic maps (6.46 almost all individual buildings are rendered separately, on the Dutch map (Figure 6.40) to simplify the map image, there is a tendency to render houses as built-up areas. On most topographic maps, colour patches are only used for forests and water, and white is for all other types of land cover, while on Dutch maps, white is only used for arable land. The Finnish map has forests in white and arable land orange. All these topographic maps have contour lines (indicated in blue on the glaciers in the Swiss map), while the Swiss and German ones also have hill shading.

6.6.2 OpenStreetMap (OSM)

OSM is a collaborative mapping project to generate a free and editable world map. It started in 2004 in the United Kingdom where only proprietary data was available. Currently, it has a worldwide detailed coverage and is expanded and updated by volunteers while you read. They crowdsource the data via digitization of satellite imagery or a real photographs, by adding GPS track collected by cycling and walking the terrain, and updating attribute information.

OSM data is used in many web services and exists in different variations. Figure 6.47 shows some of these visualizations. Special mentioning is needed for the Humanitarian OSM team. Via special actions and events, they organize the so-called mapaton, where people in (remote) groups work to map the current situation during, for instance, a disaster.

6.6.3 EuroBoundaryMap

The EuroBoundaryMap is a product of EuroGeographics, the organization of the European National Mapping Agencies. As a database of all European administrative boundaries, it was originally created in 1991 for census purposes. It is a seamless database at 1:100 000. It covers 55 European administrative boundary data sets and includes geometry, names and codes of administrative and statistical units. It also links the statistical Local Administrative Units (LAU) and Territorial Units for Statistics (NUTS) – codes member states of the European Union.

The project might not appear to be so special, but if one realizes that almost all countries have their own national grids and adhere to different hierarchies in their administrative units, it is not that straightforward to match them into a single data set. Some countries have only two levels of hierarchy, while others have six levels. Additionally, the geometry was derived from maps with scales ranging from 1:5000 to 1:600 000. Line generalization has been applied to create boundaries with a similar point density for all countries. On international boundaries, edge matching has been applied. Coastline data have been added because administrative boundaries do not always coincide with the coastline and most attribute data only refer to the land of the particular unit.

Figure 6.48 shows the resulting example of the sample data set provided by EuroGeographics. It shows the area where Belgium, Luxembourg, Germany and France meet. The diagrams in the figure show the hierarchies of the units in the respective countries and the map scales of the source maps used. The data set could be used, for instance, to compare and analyse demographic and economic data.

6.6.4 USGS National Nap Concept

Geospatial data supply is influenced by the policies of the different National Mapping Agencies. These policies may be widely diverging. In the United States, the United States Geological Survey (USGS) considers its geospatial data as public property, which should be made available at a nominal charge to every citizen. This approach has become mainstream, and more and more countries now offer their data as open data.

In the United States, civil topographic mapping is taken care of by the USGS. The major map series (over 55 000 sheets) is published at scale 1:24 000. According to the USGS website, 'The National Map is a collaborative effort among the USGS and other Federal, State, and local partners to improve and deliver topographic information for the Nation'. It has many uses ranging from recreation to scientific analysis to emergency response. A map viewer where the user can select and add layers is available (https://viewer.nationalmap.gov/).

6.6.5 Ordnance Survey

The OS (https://www.ordnancesurvey.co.uk) is the topographic mapping organization in Great Britain (Figure 6.49). It is responsible for the small-, medium- and large-scale mapping of the country. Small- and medium-scale map series are produced at scales 1:625 000, 1:250 000, 1:50 000 and 1:25 000. More than 220 000 large-scale

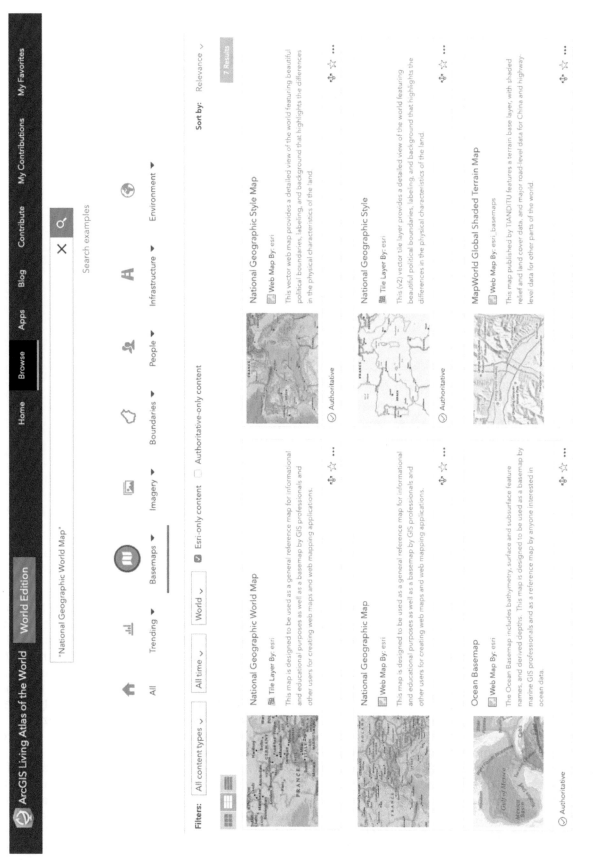

FIGURE 6.46 Legend of a German topographic map. The green patches indicate forests. From 'Tipps zum Kartenlesen' – © Deutsche Gesellschaft für Kartographie, Landesamt für Geoinformation und Landentwicklung Baden-Württemberg (www.lgl-bw.de), 01.2020, Az.: 2851.3-A/1113

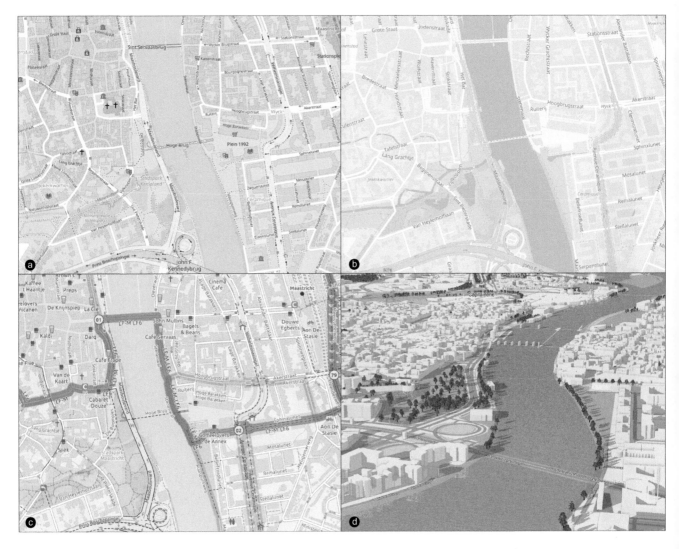

FIGURE 6.47 OSM: (a) basic visualization (https://www.openstreetmap.org); (b) bright visualization (https://maps.omniscale.com); (c) cycle map (www.opencyclemap.org); (d) 3D version (https://demo.f4map.com) – © OpenStreetMap-auteurs / license: CC BY-SA.

maps, called 'National Grid Plans', are published at scales 1:1250, 1:2500 and 1:10 000. To speed up and enhance the production of these large-scale maps, computer cartography was introduced at the beginning of the 1970s. After digitizing more than 40 000 maps, times changed. During the mid-1980s, two keywords had a great impact on the approach: GIS and privatization. The first affected the cartographic approach: cartographers at the OS had to change from unstructured drawing files to topologically structured data. The second had economic consequences: income and savings. The combined impact completely changed the mapping procedures.

In 2001, the OS launched the OS MasterMap, a large-scale map covering the whole country. It consists of the recompiled national grid plans. Each of the 400 million elements in the OS MasterMap database has a unique 16-digit reference number, a 'topographic identifier' (TOID). An example of such code is 0001000006032892, which identifies the Tower of London. The 'map' is seamless and allows the users to select any coverage they need. Via the TOID, other data can be precisely linked to the OS data. The data are also organized in thematic layers. Currently, this product is offered as OS MasterMap Topography Layer.

OS OpenData is today's keyword. Data is offered online via OS Open Zoomstack. According to their website, it 'makes OS open data more accessible, customizable and easier to use. It provides a single, customizable map of Great Britain to be used at national and local levels'. Downloads for GIS applications are available for the whole data set, or parts

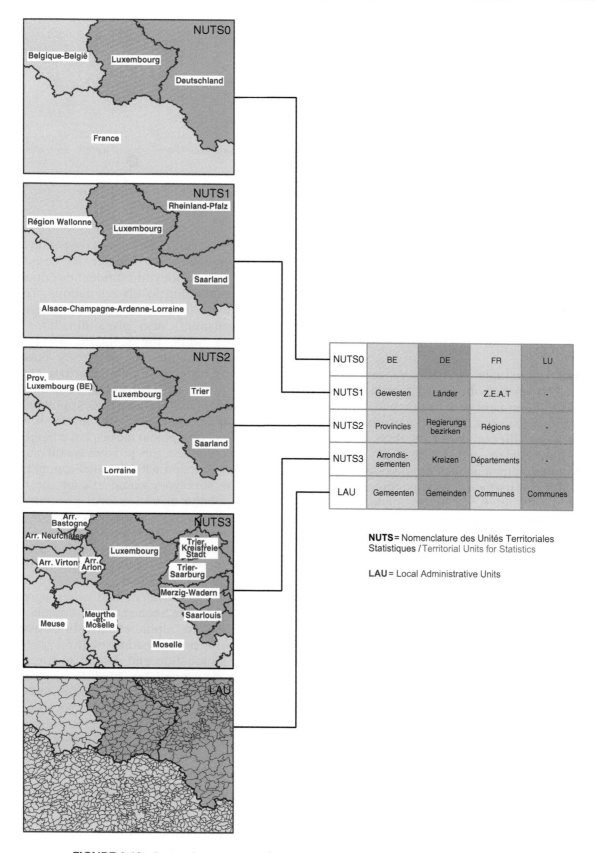

FIGURE 6.48 Luxembourg and neighbouring countries from EuroBoundaryMap data, and the administrative structure of those countries

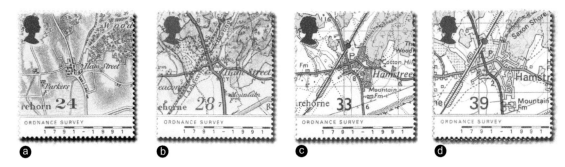

FIGURE 6.49 Development of OS products illustrated by stamps: (a) 1816; (b) 1906; (c) 1959; (d) 1991 (courtesy British Royal Mail)

like rivers, roads, terrain or geographical names separately. The Explorer 1:25 000 and Landranger 1:50 000, the classic paper maps for leisure purposes, are still popular and are available as app on mobile devices with interactive functionality.

6.7 GEOGRAPHICAL NAMES

Geographical names (or geonyms or toponyms) are used to relate both to topographical features in one's environment and to mapped information about those features. As they can also form the link between different data files, their spelling has to be standardized, and this generally is the task of a national names agency, frequently either situated within a national mapping organization or related to it (Figure 6.50). In the United States, this is the US Board of Geographical Names (BGN), in the United Kingdom, the Permanent Committee on Geographical Names for British official use (PCGN) and in the German-speaking area, the Ständiger Ausschuß für geographische Namen (StAgN). For the international exchange of geographical names information, rules have been elaborated by the United Nations Group of Experts on Geographical Names (UNGEGN), which organizes sessions where the standards they propose are voted on, and later sanctioned by the UN Economic and Social Council. On a regional level in Europe, names are processed according to the INSPIRE directives, which aim at a harmonization of georeferenced data, thus allowing to identify, access and use these data throughout the European Union; specifications have been developed for geographical names as well.

The names standardization on a national level consists of defining the official spelling of a name, including all its diacritics (such as ë, å, ç, ñ, š), special letters (such as ß, æ, ð, ij) and the accents and hyphens needed, as well as information about which letters have to be written with capitals instead of in lower case. Information

about the spelling procedures regarding geographical names belonging to a specific country can be found in the Toponymic Guidelines countries must publish according to UN resolutions. The website of UNGEGN (https://unstats.un.org/unsd/ungegn/nna/toponymic/) also gives directions to these guidelines, while the standardization itself process is documented in the freely downloadable Toponymy training manual (Kerfoot et al., 2017) available at https://unstats.un.org/unsd/ungegn/pubs/documents/Training%20Manual.pdf.

Of course, spellings can only be standardized when the geographical names have been collected. In many countries, this process is still ongoing, usually in conjunction with detailed topographical surveys. The surveyors would ask for the names the local population as well as local government is using to refer to geographical objects in its environment. The meaning of the names and their pronunciation would then be recorded as well. These names will then have to be processed, checked regarding their spelling by the national geographical names agency or its equivalent and, when agreement on their spelling has been reached, compiled into gazetteers that are alphabetical lists of geographical names with an indication of the type of feature named, its location and extent. In the data specifications developed by INSPIRE, also information about the language, gender, status (official or not official), the dates the spatial object was incorporated in the data set, the date the name was made valid (standardized since..) as well as the script used for the official name (in Europe, the Roman, Greek and Cyrillic alphabets are in use) is incorporated.

The official name for a geographical object in use in a country where that object is situated is called 'endonym'. Countries or language communities can also have names for geographical objects lying outside that country, and that are different from the official local name versions for these objects; these names are called 'exonyms'. Leghorn is an English

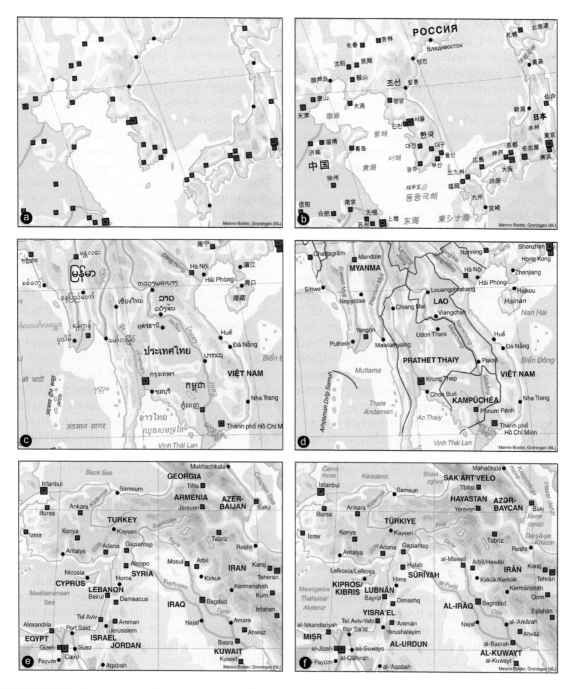

FIGURE 6.50 International Names standardization: maps without names (a) are not really helpful; we need names on them for orientation (b, c). But we should also be able to read those names, and for that purpose, there are standard conversion rules, for instance, for converting names in foreign writing systems into the Roman alphabet (d, e). But even when using these standard conversion rules, we have to make sure that the names have been updated (f) in order to reflect current local use. Maps by Menno Bolder

exonym for the Italian port of Livorno. Belgrado is an Italian exonym for the Serbian capital Beograd. Wolga is the Dutch exonym for the Russian River Волга. Exonyms are usually very much entrenched in the national culture and history of the language communities using them, but they may also be a barrier in international communication, and that

is the reason why UNGEGN tries to restrict their use, and promotes the use of endonyms instead. Exonyms and endonyms are both examples of allonyms, the phenomenon that different names are used to refer to a single topographic feature. North Sea, Noordzee, Nordsee and Vesterhavet are all allonyms that refer to the body of water in between the east coast of Britain and the continent. Contrary to allonyms are homonyms, the phenomenon that different topographical features bear the same geographical name. The toponym Birmingham may refer to a city in Britain, and to at least five settlements in the United States (the most important of which is situated in Alabama). The UNGEGN has published a glossary in which the terminology used in names standardization is explained (Kadmon, 2002, free download at https://unstats.un.org/unsd/geoinfo/UNGEGN/docs/pubs/Glossary_of_terms_rev.pdf).

National standardization of geographical names forms the basis of international standardization. The rules and recommendations developed here refer very much to the conversion of one writing system to another. When both the source and receiver country of geographical names use the same alphabet, the rule is to copy the official spelling of the endonym, including all the diacritics it might have. Use, for instance, the endonym form Bucureşti instead of the English language exonym Bucharest. If source and receiver country have different writing systems, an official UNGEGN-approved conversion system is called for to change the name from one writing system to another (one does not speak of alphabets here, because some writing systems do not use signs for representing individual sounds or phonemes, like in an alphabet, but for syllables or even for complete concepts). Conversion systems can be transcriptions, where the sounds of one language are represented in the writing system of another language (a transcription of the name of the Russian capital Москва gives Moskva, and of the Iraqi capital gives Baghdad). بغداد Opposed to transcription is transliteration, a conversion system in which the letters of one alphabet are converted to letters in another: بغداد gives Baġdād here. In principle, when transliteration is used, the transliterated name can be transliterated back to the former alphabet and result in the original name form, so Baġdād transliterates back to بغداد This is not possible in transcription: the name of the French port Bordeaux will give Бордо in the Cyrillic alphabet as used in Russia, but when transcribed back to the Roman alphabet, this will result in Bordo. The UN-accepted system for converting Chinese names into the Roman alphabet is called the 'Pinyin system'. According to this system, the name of the Chinese capital 北京 gives Beijing in the Roman alphabet; it used to be known under the English exonym Peking.

Standardization of name spellings is in itself not enough; this information as to how to write the names should be distributed as well, in the form of gazetteers or name servers. The United Nations is developing a geographical names database, as are most countries. Currently, many of them have started to provide names information on the Internet. The United States provides the official spelling of its names on https://www.usgs.gov/core-science-systems/ngp/board-on-geographic-names/domestic-names, from the USGS Geographical Names Information System. It also provides names from outside the United States through its National Geospatial Intelligence Agency, the NGA GEONet Names Server (GNS), at http://geonames.nga.mil/gns/html/index.html. A private website based on the NGA material is GeoNames (http://www.geonames.org). There you can search for toponyms worldwide or per country, and for each toponym found, the following information is given: name, type of feature, coordinates, height above sea level, if applicable number of inhabitants, history of the name and sources and dates when an attribute information for the named feature was added to the file, you can zoom in on the feature in Google Earth or in the GeoNames server itself.

The names of features are stored in layers that can be turned on or off: names of administrative entities, hydrographic features, functional areas (cemetery, pasture, forest, etc), settlements, transportation infrastructure items, building functions, physical features (canyon, volcanoes, capes, deserts). Their administrative hierarchy is given as well: Switzerland[CH] » Basel-Landschaft[BL] » Bezirk Sissach[1304] » Ormalingen[2856].

In Europe, under the EuroGeoNames project, a website linking all the national geographical names databases was developed in 2009. It contained information on the geographical names, their feature type, coordinates, the institution in charge of maintaining the database, the language, the data source, a national ID number and the status of the name (whether officially standardized or not). A special service showed the location of the named object on the map. Although this service has been discontinued, the data for this regional gazetteer can be accessed since April 2019 through: https://www.euro-geo-opendata.eu/service/open-regional-gazetteer-service. It is no name server though for the time being. In Australia, toponyms are listed in http://www.ga.gov.au/map/names/.

As soon as names are available in a digitized form, it is possible to query them and also select name parts, for instance, like all names containing the part – dam. Figure 6.51 shows a screenshot from the name server developed by Kadaster, the national mapping and cadastral agency in the Netherlands, on the basis of their 1:10 000 data files, queried for that name part – dam.

FIGURE 6.51   Distribution of all names in the database for the 1:10 000 topographic map coverage of the Netherlands containing the name part – dam.

Increasingly, spatial data exchange will be governed by standards as those developed by INSPIRE or similar bodies, and that is why they are referred to here.

FURTHER READING

Canters, F., and H. Decleir. 1989. *The World in Perspective: A Directory of World Map Projections.* Chichester: John & Wiley Sons.

Gruenreich, D. 1992. "ATKIS - a topographic information system as basis for GIS and digital cartography in Germany. From digital map series to geo-information systems." *Geologisches Jahrbuch A Heft* 122:207–216.

Imhof, E. 1963/2007. *Kartographische Geländedarstellung (Cartographic Relief Representation).* New York/Redlands: Walter de Gruyter / ESRI Press.

Kadmon, N. 2000. *Toponymy. The Lore, Laws and Languages of Geographical Names.* New York: Vantage Press.

Kessler, F., and S. Battersby. 2019. *Working with Map Projections A Guide to Their Selection.* Boca Raton, FL: CRC Press.

Lapaine, M., and L. Usery 2017. *Choosing a Map Projection.* Springer / ICA.

McMaster, R. B., and K. S. Shea. 1992. *Generalization in Digital Cartography.* Washington, DC: Association of American Geographers.

Rhind, D. 1998. *Framework for the World.* Cambridge: Geoinformation International.

Snyder, J. P. 1987. *Map Projection: A Working Manual.* Vol. 1395, *U.S.G.S. Professional Paper.* Washington, DC: U.S. Government Printing Office.

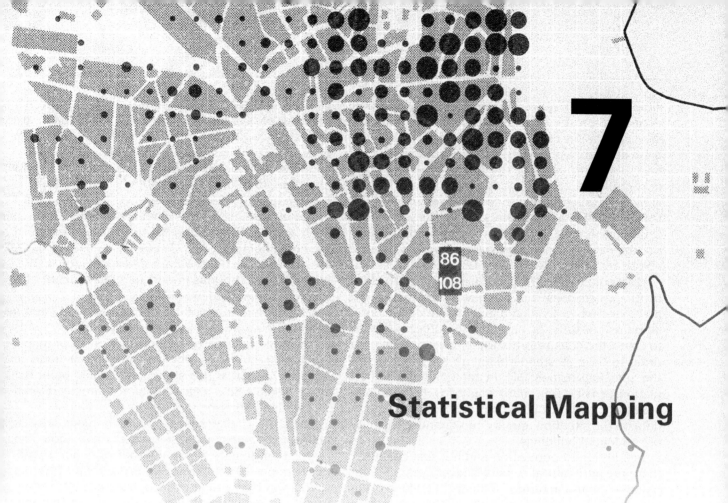

Statistical Mapping

7

In Chapter 6, the locational component of geographical objects or features was discussed. This chapter will deal with the attribute component. Attributes can answer questions like: 'What does one find at a specific location?' and 'What are its characteristics?' The map type that visualizes these attribute data is the thematic map, and its design is discussed in this chapter. Attribute data can be qualitative (language, soil characteristics, geological formations) or quantitative (air pressure, height above sea level, income or election results) (see Figure 7.1). The attributes can change and the objects that have these attributes can change; visualization of these changes is dealt with in Chapter 8. For rendering attribute data, one needs a topographic background: it is no use to present a geological map if one cannot locate the geological formation one is interested in. Attribute data are usually collected from imagery, during fieldwork or statistical surveys.

7.1 STATISTICAL SURVEYS

Statistical surveyors collect socio-economic data, either continuously, in samples or in censuses. In some countries with advanced administrative structures where the population registers of all municipalities have been converted into online facilities, there is a continuous updating of population data.

Depending on the size of the municipality or the subdivisions discerned, other data might be collected by drawing samples from the total population and assessing the present characteristics of these samples: by asking those sampled where and how they spent their holidays last year, or the number of loans they took out or their current employment

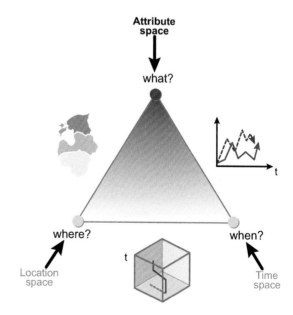

FIGURE 7.1 Working with geospatial data from the perspective of the data's attribute component

situation. This kind of information might be more detailed than the general data currently collected in a population register (name, gender, place and date of birth, nationality, marital status, children and their birth dates or the fact whether they own or rent their accommodation). However, although it is more detailed from a thematic point of view, it is less detailed in a geospatial context; because of its sample character, it might only be correct to show the data on a higher aggregation level as it would not be representative otherwise.

Most countries still have censuses every 10 years. In these censuses, the population characteristics are collected for as high a percentage of the population as possible, by enumeration officers that try to check the census forms when collecting them, in order to have as standardized an interpretation of the questions as possible. In addition to these population censuses, in many countries, housing censuses are carried out, which are meant to gather data on the situation (quality and quantity/capacity) of residential buildings.

In addition to population and housing censuses, there are agricultural censuses (every year or every 5 years), manufacturing industry censuses and company censuses. The scale of the production, the size of the workforce, the turnover or added value or the value of the production assets will be assessed for these censuses.

These statistical data, in whichever way they have been collected, are subject to privacy regulations. If people could not be sure of privacy, the information would not be provided by them. This means that when the data are to be rendered or published in alphanumerical form, they are combined first, so that information on individual, identifiable households, farms, plants or companies cannot be worked out from the data. Another method of safeguarding privacy when data are collected for 1 × 1 km cells is to randomly add numbers (+1, +0, −1) to figures for individual cells, and to suppress data cells with eight households or less or less than 25 people (see *People in Britain*, Census Research Unit, 1980) in a data grid cell, so that it is still impossible for the other households living in such a cell to work out what the characteristics of the household would be. This secrecy is required by the data suppliers, who have given the information in confidence, but has important consequences for the data users: data are combined, and may or may not be averaged, resulting in either overall figures or average figures. Visualized, this generally leads to choropleth maps or proportional symbol maps. Their characteristics will be commented upon in Section 7.5, but at this stage, one should point out some visualization aspects that are directly related to the data gathering techniques.

There is a false suggestion of homogeneity in choropleth maps: values for a phenomenon like population density or the number of general practitioners per 10 000 households are represented in such a way that a suggestion of homogeneity is conveyed: values seem the same throughout the enumeration unit. Look, e.g., at Figure 7.2a: each dot represents the income of a farmer located at the site of a dot. Incomes tend to vary considerably in this example, because of the variation in soils and relief, i.e. gently sloping valley farms on good soils versus small-patched hillside farms on steep slopes and poor soils. When the statistical enumeration officers visit, they will collect the income data (or they might obtain them from the Inland Revenue), which then will be combined (because of privacy regulations) and represented as a choropleth: for

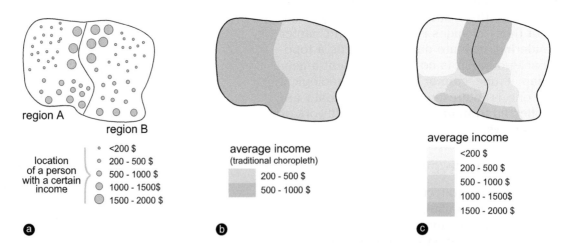

FIGURE 7.2 Farm income figures: (a) reality; (b) represented in the usual choropleth manner; (c) with boundaries adjusted to the phenomenon

every enumeration area, the combined income data will be divided by the number of farms, and these average data categorized (see the choropleth method procedure in Figure 7.21). The resulting image is shown in Figure 7.2b, represented with tints that get darker when the values they represent get higher. So, the extreme local differences in characteristics that geographers would be interested in have been obliterated. This is because the enumeration area boundaries are seldom drawn with a specific phenomenon in mind – they are not relevant for this phenomenon. It would be preferable to base the boundaries on the original data (as found in Figure 7.2a), delineate the areas with similar values and determine averages within these new boundaries, so that the choropleth result would be as in Figure 7.2c. But cartographers seldom are in a position to do this, as the original data are not available to them. However, one should still be aware of this problematical aspect of rendering statistical data.

Another aspect is that when non-area-related ratios are being represented, the impression of the choropleth will be determined more by the size of the enumeration unit than by the actual values expressed by the (grey) value tints: it is the size of the areas that makes the ratios represented in them stand out, while it should be the size of the original numbers they refer to. Consider the example shown in Figure 7.3. When mapped, the large marginal areas in the west and north stand out, suggesting that the problem of large practices is a minor one in Britain. However, when one analyses the data, it is found that the largest number of patients lives in the darkest, urban areas with high patient/GP ratios (see also the example Figure 3.11).

What people see when confronted with proportional symbols related to areas (as in Figure 5.1b or 5.13b) is a ratio between the size of the symbol and the size of the enumeration area it refers to. When areas are compared, one actually compares these ratios and the ratios are considered to stand for the whole enumeration area, again suggesting a homogeneity that does not exist.

In fact, this problem is only offset when the original data coordinates are known (like the addresses of the households being surveyed for the census) so that the data can be represented within regular cells, as in the atlas *People in Britain* (1980), for example (Figure 4.11).

7.2 DATA ANALYSIS

When statistical data are presented for representation, a number of questions need be asked – in

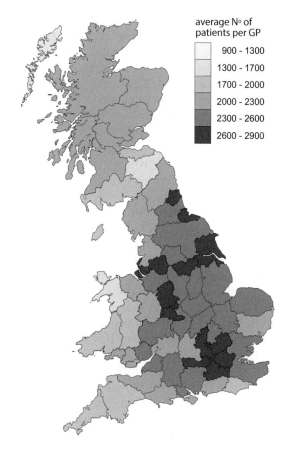

average N° of patients per GP

- 900 - 1300
- 1300 - 1700
- 1700 - 2000
- 2000 - 2300
- 2300 - 2600
- 2600 - 2900

FIGURE 7.3 The influence of differences in enumeration area sizes on data perception in choropleths

order to assess how useful they will be. The answers might be directly available in the metafiles that accompany the relevant file, but it is necessary nevertheless that one realizes the implications of these answers. The questions would revolve around when the data were collected, in which way, for what purpose, for which period of time or to what area they refer to, etc. These answers will indicate the utility, reliability and accuracy of the data. They will indicate whether the data have been collected in a way comparable to similar gathering exercises in the past, so that time series can be realized. The conclusion might be that – as the data were collected with a specific objective in mind – they are biased.

After their assessment of the validity of the data, it is the data characteristics that need be determined prior to further analysis/processing. In the first place, there is the nature of the objects the data refer to: Are these point-like objects, linear objects, areal objects (two-dimensional) or volumetric objects (three-dimensional)? Examples are signposts, roads, fields and mountains.

The next characteristic is the type of change in the data: Is this change gradual or not, and do changes occur abruptly from one place to the next? The smoothness of the change is related to distributions being continuous or discrete/discontinuous. A continuous distribution describes data that can be measured anywhere, such as air pressure or temperature. Discontinuously distributed phenomena can only be ascertained at particular locations and not elsewhere; examples are land cover or vegetation types.

The measurement scale on which the object attributes can be measured comes next. The representation of the data will later depend on this measurement scale, which is why it is important to find out about it in advance (if specific map-related measurement scales are not appropriate, it might be necessary to change the measurement scale of the data). If only a very simple population map is required, and population density figures are available, some relevant threshold value might be selected and the area with lower values represented as uninhabited, while the area with values higher than this threshold can be shown as the ecumene – the human-inhabited world.

The measurement scales are traditionally divided into *nominal, ordinal, interval* and *ratio* scales (see also Section 5.3). Phenomena attributes are measured on a *nominal* scale when differences in data are only of a qualitative nature. Examples are differences in gender, language, religion, land use or geology. The position on an *ordinal* scale of object attributes is determined, when it is known that their values are more or less than another, without knowing the exact distance on the scale between them, or when these concepts have a vague/inaccurate meaning, such as 'cool–tepid–hot' or 'small–medium–large'. Tepid is warmer than cool and hot is warmer than tepid; how much warmer would be unknown, but it is at least possible to establish a hierarchy. Both a hierarchy and the exact distance on the scale are available for measurements on an *interval* scale – with the restriction that it would not be possible to work out relationships/ratios between the measurements other than their distances: 8°C is not twice as hot as 4°C; though, it is 4° warmer than 4°C! Likewise, 2°C is not twice −2°C; though, it again is 4°C warmer. This is because a random temperature value has been selected as the 0°C point, in this case the temperature at which water will freeze when the temperature goes down.

If two locations, A and B, along the Dutch–Belgian border have heights of 4 and 8 m, respectively, above the Netherlands datum (Normaal Amsterdams Peil, NAP), this does not mean that the latter is twice as high as the former. It is only the difference in height between the two points that can be assessed, i.e. 4 m. The datum for heights used in Belgium is 2 m lower, so the heights of these two points on the border would be 6 and 10 m, respectively, in Belgium! (Figure 7.4).

When *ratios* can be expressed, one refers to data that can be measured on a ratio measurement scale. In 1992, the gross national product per capita in Spain was twice that of Portugal so these are ratio data.

There is a difference between absolute and relative ratio data. Absolute ratio data are the result of direct measurements or additions of units. Examples are one's income in money (a specific number of euros or dollars) and the number of children of a family. In the first example, income can have all values in between, with decimals, while in the second case, there can only be discrete values; in other cases, there might be positive and negative values, as and when immigration and emigration data are compared.

Relative ratio data are absolute ratio data related to other data sets, and it is these relative data that often tell us more than absolute data, because they have been put into context. Whether a gross income figure of an equivalent of $40 000 a year is high or not will depend on one's perspective and on how much it will buy. In India (in 2009), it would be the income of a top manager, providing him with a large estate and many servants; a similar income in Alaska might only attract a starting primary school teacher. That is why one should provide some yardsticks and compare the income to the average world income or the average income in developing countries or the average income of managers. Even if it were considered low now, when compared with the remuneration for similar jobs in India in the past, it might seem enormous. So, the relation to a point in the past should be established, and one would be using index values to express these relationships. For example, if in 1910 a certain job paid $800 a year but now pays $30 000, there would be a 37.5 times increase. If the value in 1910 equated to 100, the present index value would be 3750. Expressing data in index values makes them easier to digest.

In addition to the index values that can be used in time series, other relative or derived values can be used in order to give more meaning to absolute data sets: examples are averages, ratios, densities and potentials.

Densities originally referred to the ratio between the population of an area and the resources available to that population. Population density values (number of inhabitants per areal unit, like square

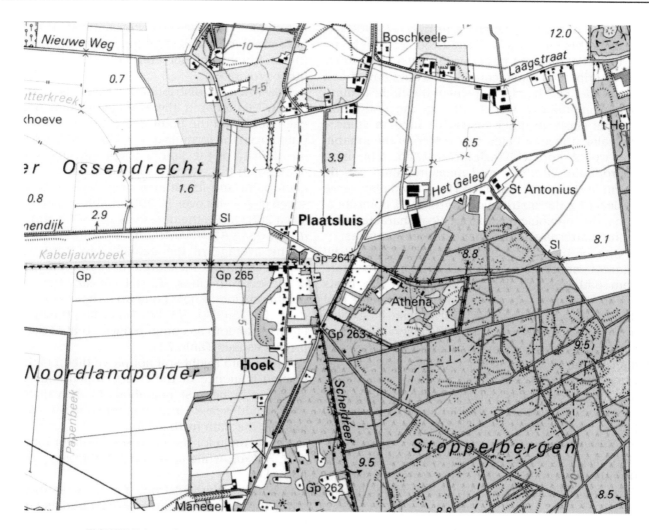

FIGURE 7.4 Height representation on topographic maps of the Dutch–Belgian border
(topographic map 1:25 000, sheet 49G (Kadaster Geo Informatie)

kilometres) originate in the Malthusian era, higher densities referring to smaller means for the population to provide for itself. This concept has lost its economic relevance but has retained relevance as an indicator for well-being or welfare, as high population density figures refer to crowded situations. To retain some relevance as a population/resources ratio, the density concept has been refined and can refer to the number of people in relation to their residential area (residential density) or the number of people in relation to the agricultural area they are cultivating (agricultural density).

Non-area-related ratios can express the relation between any two data sets, as, for instance, between the total number of inhabitants and the total number of general practitioners in a region, or the relation between two subsets of the total population, e.g. between the number of members of the armed forces and the number of those in the teaching profession.

An absolute number of influenza patients of 30 000 in the Netherlands as compared with 28 000 in Luxembourg would not as such be reason for much comment. But if one were to convert these absolute figures into ratios, relating them to the total population numbers of the respective countries, this would result in much more informative data: in the Netherlands (17 million inhabitants), only two in every 1000 people would be ill, whereas in Luxembourg (600 000 inhabitants), 47 in every 1000 people would be smitten, the influenza thus being a real epidemic. Expressed in percentages (realized by multiplying these figures by 100), these values would be 0.2% and 4.7%, respectively.

Averages attempt to characterize the data sets they refer to by one number – and can only succeed in doing so if the variation in measurement is not too high. An average income of $3000 per year is a meaningless figure in a country where 90% of the population has an income around $500 per year and 10% of the population earns over $25 000 per year. In this case, the average cited, $3000, is the (arithmetic) mean. A better or more useful number to describe economic conditions would have been the *mode*, i.e. the class in which the highest number of inhabitants would fall, e.g. the $400–600 income class. In other cases, the median value – the value above or below which 50% of the inhabitants would score with their income – could be more appropriate. Other derived measurements can be found in statistical manuals. They would also serve to describe the distribution patterns of point locations, the topological or hierarchical characteristics of line patterns and the shape of areal patterns.

One of the descriptors of dot patterns is the *nearest neighbour index*. This compares random patterns and actual patterns on the basis of the distance between each point location and the point location nearest to it. These index values can range from 0, when all observations are concentrated in one single point, via 1 (when the distribution is a completely random one), to 2.15, when the distribution of the points is completely regular, all distances between them being equal. Take the province of Drenthe, the Netherlands. From a map of population centres with over 10 000 inhabitants (Figure 7.5), one might get the impression that settlements are more regularly spaced in Drenthe as compared with other Dutch provinces. This might be the result of the distortion in the

potential regular service city network that could have emerged, caused by the lower courses of the Rivers Rhine, Scheldt and Meuse in most of the other provinces.

Now when the distribution of points is totally random within an area, the average distance between each point and the point nearest to it can be computed from the formula $d_e = \frac{1}{2}\sqrt{p}$, where p denotes the density of the points taken into account and d_e denotes the expected mean distance between each point and its nearest neighbour. If there are seven cities with over 10 000 inhabitants in Drenthe (area 2685 km²), the density is 2685/7 = 383.57, so $d_e = \frac{1}{2}\sqrt{383.57} = \frac{1}{2} \times 19.58 = 9.79$ km.

The average of the actually observed nearest distances when measured between the centres of these seven cities, d_o, is 15.66 km. The nearest neighbour index now compares these two values: d_o/d_e, i.e. 15.66/9.79 = 1.6. This value is more regular than random as compared with that of the other provinces (see Table 7.1).

These data might still not prove that it is the disturbing influence of the rivers, which causes this variation in nearest neighbour index values, but they would serve to strengthen one's resolve to analyse the data further.

The formula for the nearest neighbour index is: $R_n = d_o/d_e$, while d_e is computed from $\frac{1}{2}\sqrt{p}$ (see above).

A concept derived from physics describing the attraction between two masses ('bodies') is the *potential*. This attraction force is equivalent to the product of their two masses, divided by their squared distance. In a watered-down version, this concept describes the virtual interaction between the inhabitants of different cities (population

FIGURE 7.5 Population centres with over 10 000 inhabitants in the Netherlands (2015)

TABLE 7.1

Nearest neighbour index values of places over 10 000 inhabitants per province in the Netherlands

Province in the Netherlands	Nearest neighbour index value
Flevoland	2.10
Drenthe	1.60
Overijssel	1.50
Limburg	1.20
Fryslân	1.18
Noord-Holland	1.16
Noord-Brabant	1.15
Gelderland	1.08
Zeeland	1.04
Zuid-Holland	1.01
Utrecht	1.00
Groningen	0.93

potential) or the expected purchases in a market (market potential), etc. The population potential at a point describes the chance there is that people would meet each other, or have contact. This is a chance expressed for each city by the addition of population numbers of other cities divided by their distance to the city where the potential is measured (Figure 7.6). To find the potential value between which one has to interpolate to produce the final map (Figure 7.6b(III)) for each city, the influence of other cities (expressed by their number of inhabitants divided by their distance to this city) has to be added up. The influence of city A upon itself can be expressed by dividing its number of inhabitants by the radius of its geographical extent.

An example of a socio-economic data set analysis is that of statistics on nature, area and number of agricultural holdings in Indonesia (Sensus Pertanian, 2013). As can be seen in Table 7.2, the data made available are on the number and size of individual farms and of agricultural estates/plantations. Not all of these data categories are equally important, however, and if only one map were available to give as good an impression of Indonesian agriculture as possible, choices would have to be made. Data sets related to each other, like average farm size, would require less space

than their separate representation. Individual farms in Indonesia would mostly grow the staple crop, rice, and their average size would show the agricultural density, i.e. the pressure on the land the individual farms would be subject to. Agricultural estates, on the other hand, are sort of extra – they do not produce food for subsistence or for the local market but raise export crops: tea, coffee, rubber, oil palms, sugar. So when asked which would be most important for Indonesia's economy, the individual farms or the agricultural estates, the former would score higher. This importance can be visualized by assigning this aspect the first place in the graphical hierarchy, giving it most emphasis. That is realized by expressing it through area symbols. In addition, using proportional circles, the total agricultural acreages could be shown (Figure 7.7), allowing readers to compare the ratios between these circles and the areas they pertain to on the map. In these circles, the proportion of the area in agricultural estates would increase the interpretation potential, but might make the map more complex and thereby endanger the possibilities for communication.

It is important to note that the representation in this example has actually been a data selection process (as not all the available data have been

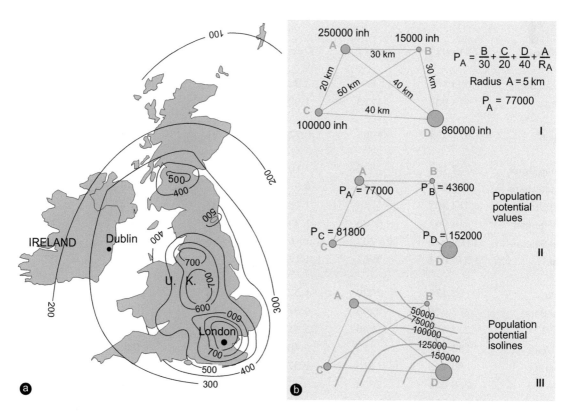

FIGURE 7.6 Potential mapping: (a) an example of a population potential map (after Berry and Marble, 1968); (b) the procedure for assessing the population potential at a point (A)

TABLE 7.2
Agricultural surface and number of holdings in Indonesia, 2013

Province	Individual agricultural holdings			Plantation agriculture			Total surface area (in km²)
	Number of farms	Surface area (km²)	Average area (ha)	Number of estates	Surface area (km²)	Average area (km²)	
Aceh	1 273 825	7043	0.50	90	4142	45	11 185
Sumatra Utara	2 208 491	12 488	0.56	359	11 081	45	23 569
Sumatra Barat	1 313 464	9900	0.75	44	3765	85	13 665
Riau	308 888	2192	0.70	194	14 467	74	16 659
Jambi	384 268	3321	0.86	91	9954	109	13 275
Bengkulu	1 201 322	13 094	1.08	151	14 671	97	27 765
Sumatra Selatan	296 108	2214	0.74	66	4280	65	6494
Lampung	2 205 104	17 954	0.81	55	8043	146	25 997
Bangka-Belitung	75 773	3401	0.44	44	1758	40	5162
Kep.Riau	18 334	113	0.62	5	491	98	604
Banten	1 497 674	9379	0.62	18	414	23	9793
Jakarta	3583	378	0.10	0	1	0	478
Jawa Barat	7778379	48 620	0.62	154	996	6	49 616
Jawa Tengah	10 598 504	55 198	0.52	73	1682	23	56 880
Yogyakarta	1 307 155	4579	0.35	1	30	30	4609
Jawa Timur	12 065 251	68 984	0.57	141	4129	29	73 113
Bali	615 701	3883	0.63	8	813	102	4696
Nusa Tenggara Barat	1 563 287	11 726	0.75	3	852	284	12 578
Nusa Tenggara Timur	2 048 824	11 272	0.55	10	2063	206	13 335
Kalimantan Barat	1 383 861	11 464	0.82	180	10 953	61	22 417
Kalimantan Tengah	449 838	4733	1.05	146	5602	38	10 335
Kalimantan Selatan	971 254	7005	0.72	91	2507	27	9512
Kalimantan Timur	252 164	2896	1.10	116	2189	18	5085
Kalimantan Utara	76 848	750	0.97	19	348	18	1098
Sulawesi Utara	312 967	3714	1.18	37	3392	91	7106
Sulawesi Tengah	514 667	6425	1.24	27	4767	176	11 192
Sulawesi Selatan	2 198 073	25 702	1.16	13	3606	277	29 308
Sulawesi Tenggara	340 044	3987	1.17	11	3712	377	7699
Gorontalo	192 127	3020	1.57	22	409	18	3429
Sulawesi Barat	277 608	2265	0.81	15	1623	108	3888
Maluku	237 445	995	0.41	10	1276	127	2271
Maluku Utara	153 161	735	0.48	0	928	?	1663
Papua Barat	86 139	319	0.37	9	293	32	621
Papua Barat	768 834	2098	0.27	10	344	34	2442
Indonesia	54 942 936	357 110	0.61	2213	126 688	57	483 799

Source: Sensus Pertanian (2013).

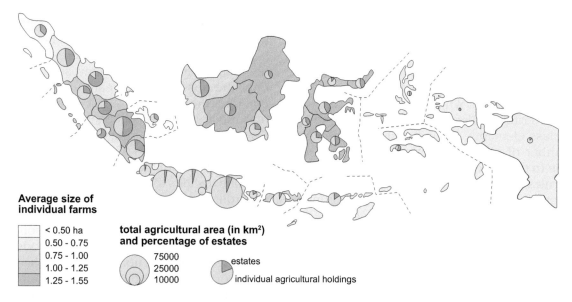

Average size of individual farms

	< 0.50 ha
	0.50 - 0.75
	0.75 - 1.00
	1.00 - 1.25
	1.25 - 1.55

total agricultural area (in km²) and percentage of estates

75000
25000
10000

estates

individual agricultural holdings

FIGURE 7.7   Agricultural holdings in Indonesia, 2013

used) based on the knowledge of the phenomenon (Indonesian agriculture). That is why it is so important that those that use databases for map production have a geographical background that makes them aware of things geographically relevant.

7.2.1 Data Adjustment

In order to make the data selected relevant for comparison purposes, derived figures might have to be adjusted. Fecundity figures, birth rates and death rates are examples here. Birth and death rates refer to the numbers of births or deaths per 1000 inhabitants. Originally, this concept was used to give an indication of the fecundity of the population or salubrity of an area – but this presupposes a 'normal' population structure, i.e. a bell-shaped population pyramid. If there is an over-representation of inhabitants over 65 years, the death rate is bound to be higher than elsewhere; however healthy the climate might be. While in most states of the United States, the death rate has decreased with the heightening of the average life expectancy, it has increased in Florida, because of the enormous influx of retired people with their higher death rate. In order to give a proper image of the actual health-influencing aspects, this figure has to be corrected, for instance, by assessing whether for a specific age group the death rate is higher or lower than in other areas. The fertility rate (the number of births per 1000 women of child-bearing age) is such a refinement over the birth rate, as it takes only the relevant age and gender group into account. The same goes for figures in physical geography, such

as temperature or vegetation. When one would like to assess the effects of geographical latitude on vegetation or climate, one should try to minimize the distorting effects of relief (height, slope and aspect). For temperature, there is an easy rule of thumb, which says that every 100 m shows a decrease in average temperature of 1°C, so that it is possible to reduce all temperatures measured to sea level, thus making all values comparable. Although often other climate effects, like precipitation, are also related to elevation differences, the relationships are not always so simple and linear as is the case for temperature.

The above examples illustrate the importance of data normalization before visualizing them.

7.3 DATA CLASSIFICATION

The risk of the mapping of unprocessed data resulting in unclear visualization is quite high. It is a good cartographic practice to conveniently arrange the data before displaying them. This process is called 'classification'. It can be described as systematically grouping data based on one or more characteristics. Classification will result in a clearer map image, even if it is a generalized image. To be able to classify data, one needs to know what types of data are available, which requires an operational definition of the data. If one is interested, e.g., in the number of ships that entered the port of Rotterdam in 2020, it is first necessary to define what a ship is. Setting these definitions is not easy. Practice has proved that, for instance, with the classification of topographic objects, many different descriptions are in use for the same object (such as lamp post and lamp standard). Even if one agrees on the description, it is still possible that the description might not be completely clear in practice. Clearly, the Golden Gate Bridge fits in the object class 'bridge', but is this also true for a plank over a small stream? When all elements adhere to the same criteria (discrete phenomena), one can count class elements. For instance, one can count the number of oil tankers entering the port. It is possible to count according to quantitative or qualitative rules, such as 10 tankers and five container ships, or 15 ships. When one deals with continuous phenomena like precipitation, the data can be measured: the characteristics of the objects, rather than the objects themselves, can be quantified according to a specific measurement scale, and mapped. Four measurement scales exist, as was mentioned in the previous section.

Classification is helpful to enhance insight into the data. However, to make sense, the number of classes should be limited. Research has revealed that humans can handle up to a maximum of seven classes to get an overview and understanding of the theme mapped at a single glance (see also Figure 7.14). The exact number of classes chosen is influenced by the type of symbolization used, the theme's geographical distribution and the data range (the ratio between the maximum and the minimum data value).

Not everyone is convinced that classification is needed at all times. The American cartographer Tobler (1973) is of the opinion that it is unnecessary to classify statistical data. The major advantage of not classifying the data is that the resulting image is not generalized. For the legends of such maps, he suggests using a continuous grey scale. In Figure 7.22, the lower four maps show the population density in Maastricht municipality with the choropleth method, be it with a different number of classes. The upper map shows the same theme, now rendered by an unclassified choropleth map. The (grey) value of each neighbourhood in the lower maps is derived from the class to which the neighbourhood belongs according to a particular classification method. The general appearance of the maps is different. Extreme values are much better isolated in the unclassified map. However, some cartographers oppose this method since it is virtually impossible to perceive the differences between neighbourhoods that are further apart geographically. The same cartographers also claim that a limited number of grey shades, as in the lower maps in Figure 7.22, improve the legibility of the map. If one has the choice between the two

approaches, one should always ask first: 'What is the purpose of the map?' Is it necessary to be able to determine values for each enumeration area, or is it just an overview one is interested in? It should be mentioned that not many software packages offer the possibility to create unclassified maps. If one remembers Figure 4.6, it should also be noted that most software packages allow one to access the database even during display, and that the user can access the values of each individual geographical unit by pointing at the unit on the map.

If one decides to classify, this process is executed according to the nature of the data. Nominal data are categorized according to taxonomic principles of the discipline involved, such as soil types, climatological zones or geological periods. The map type used to display nominal data is the chorochromatic or mosaic map (see Section 7.5.1). Ordinal data are also based on classifications defined by individual disciplines. Examples are meteorology (temperature: cold, mild, warm) and environmental sciences (forest conditions due to air pollution: healthy, normal, sick). Interval and ratio scales are both linked to quantitative data. Most census statistics belong to this category. Examples are the number of people living in a country's districts, or the percentage of children under the age of 15. Interval and ratio data are displayed on choropleth maps or isoline maps (see Sections 7.5.2 and 7.5.3). The first is primarily associated with socioeconomic data collected for administrative units, while isoline maps show interpolated data derived from physical measurements such as temperature or hours of sunshine.

To reach the best possible classification, several conditions have to be met since not all classification methods are suitable for all situations. Therefore, one should strive for the following.

The final map should approach the *statistical surface* as closely as possible. A statistical surface is a three-dimensional representation of the data in which the height (or z-coordinate) is made proportional to the numerical attribute value of the data (see also Section 7.5.9). In Figure 7.8, the numbers displayed in (a) are used in (b) to give each enumeration area its height. Statistical surfaces offer the user a dramatic view of the data. When functionality (as described in Figure 6.3) is available, the user can even view the data from all possible directions (see also Section 7.5.9). Two types of statistical surface are distinguished: the stepped statistical surface, as displayed in Figure 7.18(c), and smooth statistical surface (Figure 7.18d). Stepped statistical surfaces are derived from choropleth maps, and each of the enumeration districts is clearly visible.

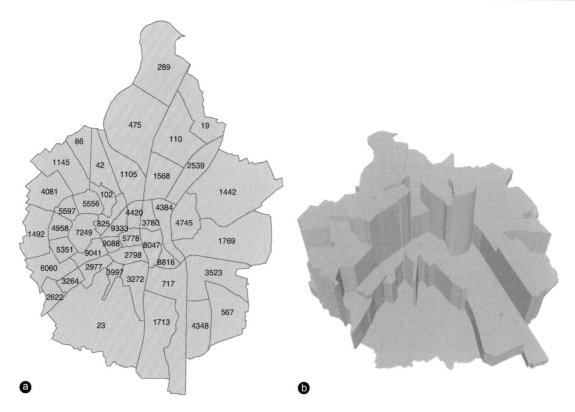

FIGURE 7.8 Population densities of the neighbourhoods of Maastricht for 2018: (a) values for each neighbourhood; (b) the corresponding stepped statistical surface

Smooth statistical surfaces are derived from isoline maps. In this map, the height (z-coordinate) at each location is defined by the intensity of the phenomenon mapped at this particular location.

The final map should display those patterns or structures that are characteristic for the mapped phenomenon. Extreme high or low values should not disappear because of the classification method.

Each class should contain its share of the observed values.

If one adheres to these conditions, the resultant map will give a clear overview of the mapped phenomenon, and it will be possible to determine values at each location in the map. A method used to adhere to the above conditions was proposed by Jenks and Coulson (1963). It can be split into three major steps.

Choose a map type. Since one is dealing with quantitative data, it will be a choice between an isoline map and a choropleth map. In the example elaborated here, it is a choropleth map because the topic deals with population density per neighbourhood of the city of Maastricht (Figure 7.8).

Limit the number of classes. If one looks at the (variable) value of a particular geographical unit, and wants to determine the corresponding value in the legend, it is not possible to use more than seven classes. This seems to be contradictory to the statement made earlier, which said that the map should be as close to the statistical surface as possible. This objective could induce cartographers to discern as many classes as possible, or even to renounce any classification at all. However, it could also be argued that if one intends to grasp all variations in the mapped phenomena in a single moment, the number of classes should be reduced.

Define the class limits. This is the most difficult step in the classification process. Many different methods exist although most software packages offer only three different options. These are either to split the number of the observed values equally over all classes, to have equal class size with respect to the range of the observed values, or to define one's own classification method. This last option allows the user to apply every conceivable method. When class limits have to be defined external to the current software packages, the user should be aware of potential classification methods. The most important classification methods will be explained below. In general, one can distinguish between graphic and mathematical methods. The first type of method will not be found in the classification options of the mapping software.

7.3.1 Graphic Approach

7.3.1.1 Break Points

When all observed values are sorted in ascending order and subsequently plotted in a graph, like the one in Figure 7.9, it is sometimes possible to observe discontinuities. In Figure 7.9, four of these discontinuities are clearly visible: between the observed values number 18 and 19, 24 and 25, 31 and 32 and 38 and 39. This type of graph is, in this context, also known as an 'observation series'. The discontinuities, called 'break points', can function as class boundaries since they are natural breaks in the observation series. It is obvious that not all data sets will show break points, and if they do, the chance that there are enough for the number of classes planned is quite uncertain. In the example, only four class boundaries can be distinguished.

7.3.1.2 Frequency Diagram

A frequency diagram can also be used in a search for discontinuities. If found (Figure 7.10a), they can function as class boundaries. A frequency diagram is especially useful when a large volume of observations is involved. The chances of finding discontinuities increase when the values along the horizontal axis are grouped together. Figure 7.10b shows an example. The existence of break points is strongly influenced by the size of the intervals applied to group the data.

7.3.1.3 Cumulative Frequency Diagram

In a cumulative frequency diagram, as displayed in Figure 7.10c(I), the frequency of the occurrence of the observed values is added up. Changes in orientation of the curve (see the small arrows in

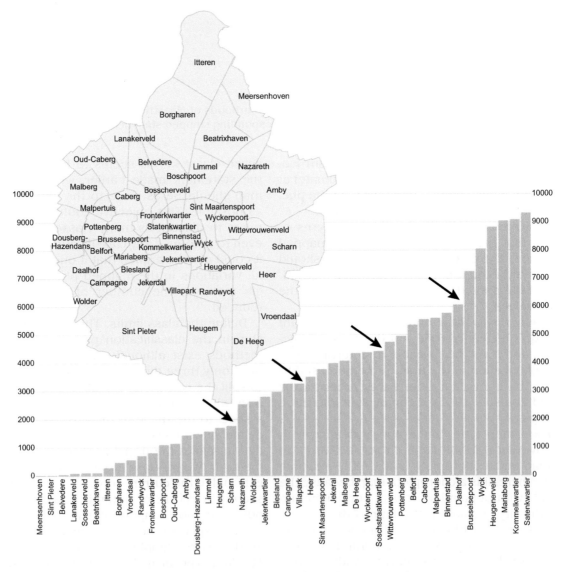

FIGURE 7.9 Observation series, with break points

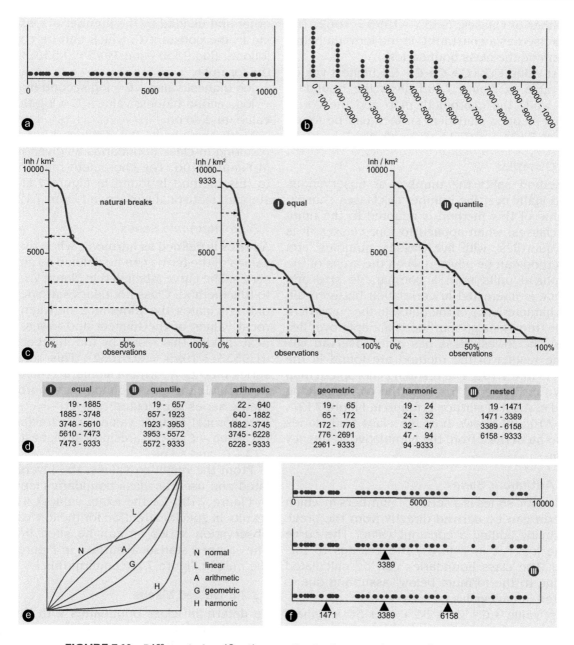

FIGURE 7.10 Different classification methods. See running text for explanation

the diagram) indicate the break points. The values along the vertical axis of those break points are the class boundaries. Again, as with the observation series, there is a limited chance that the number of class boundaries needed is indeed found using a graphic approach.

7.3.2 Mathematical Approach

When looking at Figure 7.9, it is possible to draw a curve along the top of all bars. The shape of this curve can be described by a function. The mathematical approach uses this type of curve to determine the nature and location of the class

boundaries. In Figure 7.10e, curves with several common functions are drawn. The first two methods discussed below can also be executed using a cumulative frequency diagram (Figures 7.10c(II) and (III)).

7.3.2.1 Equal Steps

In this method, the class width is equal for all classes. This method should be applied when the curve created from the observation series is linear. The classes can also be determined graphically, as can be seen in Figure 7.10c(II). If one applies the mathematic approach, the lowest value is subtracted from the highest value, and the result is divided by the number of classes: (9333 – 19)/5 = 1862. The result is used as a constant C in the formula below to determine the class boundaries:

lowest value + C + C + C + C + C = highest value.

The table in Figure 7.10d gives the boundary values, while the choropleth map and the corresponding stepped statistical surface can be found in Figures 7.12a and 7.11a, respectively.

7.3.2.2 Quantiles

This method splits the number of observations proportionally over the number of classes chosen. The name of this method is adapted to the number of classes: when applied to four classes, it is called 'quartiles', with five classes, quintiles, etc. This method can be advised when the areas of the geographical units have a comparable size and when one is interested in correlation between different characteristics of the units. In the Maastricht example, the 41 neighbourhoods are split over five classes. The lowest class has nine observed values. The results of the method are found in the table in Figure 7.10d. The resulting choropleth map is shown in Figure 7.11b, while the corresponding stepped statistical surface is found in Figure 7.12b. Figure 7.10c(III) shows how the class boundaries can also be derived from the cumulative frequency diagram.

7.3.2.3 Arithmetic Series

An arithmetic series is a series of numbers in which each term can be derived directly from the previous term by adding a constant value. The curve with the letter A in Figure 7.10e belongs to this method. The class boundaries can be calculated according to the formula below, assuming one is interested in having five classes:

lowest value + C + $2C$ + $3C$ + $4C$ + $5C$ = highest value.

C has a value of 624 and is calculated as: highest value minus lowest value divided by the number of constants C in the formula (i.e. (9333-19)/15). The table in Figure 7.10d shows the class boundaries, and Figures 7.12c and 7.11c show the choropleth and the stepped statistical surface, respectively.

7.3.2.4 Geometric Series

For geometric classification, the curve labelled G in Figure 7.10e is derived from a geometric series. Each following term is derived from the previous term by multiplying it by a constant C, the ratio of the series. To determine the class boundaries according to this method, the logarithm of the highest value and the lowest value has to be determined. These values are then subtracted from each other and divided by the number of classes, resulting in the constant C, which can be computed as follows: (log 9333 – log 19)/5 = 0.538. C is used in the formula:

log highest value – C = log second highest value
log second highest value – C = log third highest value, and so on.

From the result, the antilogarithm is defined, resulting in class boundaries as given in the table in Figure 7.10d. The choropleth map that belongs to this method is found in Figure 7.11d, and the stepped statistical surface in Figure 7.12d.

7.3.2.5 Harmonic Series

A series is defined as harmonic when the reciprocal values of the terms can be defined as an arithmetic series. The curve labelled H in Figure 7.10e is linked to this method. Class boundaries are defined when one calculates the difference between the reciprocal values of the highest and lowest value and next divides this result by the number of classes ((1/9333–1/19)/5 = 0.01052). This results in the series ratio C. A formula similar to the one applied to calculate the class boundaries according to geometric series is executed:

reciprocal highest value – C = (reciprocal highest value – C) – C = ((reciprocal highest value – C) – C) – C, and so on.

From the resulting values, the inverse is calculated and used as class boundaries (see the table in Figure 7.10d for the exact values). The method results in special attention for the low values in the observation series as can be seen in the table. The stepped statistical surface in Figure 7.12e and the map in Figure 7.11e confirm this.

7.3.2.6 Nested Means

To determine class boundaries with this method, first calculate the average of all observed values. In the Maastricht example, this is 3389. Next, the average is calculated for all values below the overall average and for all values above the overall average (in the example, these will be 1471 and 6158, respectively); see also Figure 7.10f. These three values can be used as class boundaries. This method does not allow one to choose the number of classes freely. There always has to be a multiple of two. In the example, the method is used to define four classes, as is shown in the table in Figure 7.10d. A derived method works with standard deviations added and subtracted from the averages.

Every method results in a different map image, as can be seen in the table in Figure 7.10d and the maps in Figures 7.11 and 7.12. Which method is best? It has already been said that the curve of

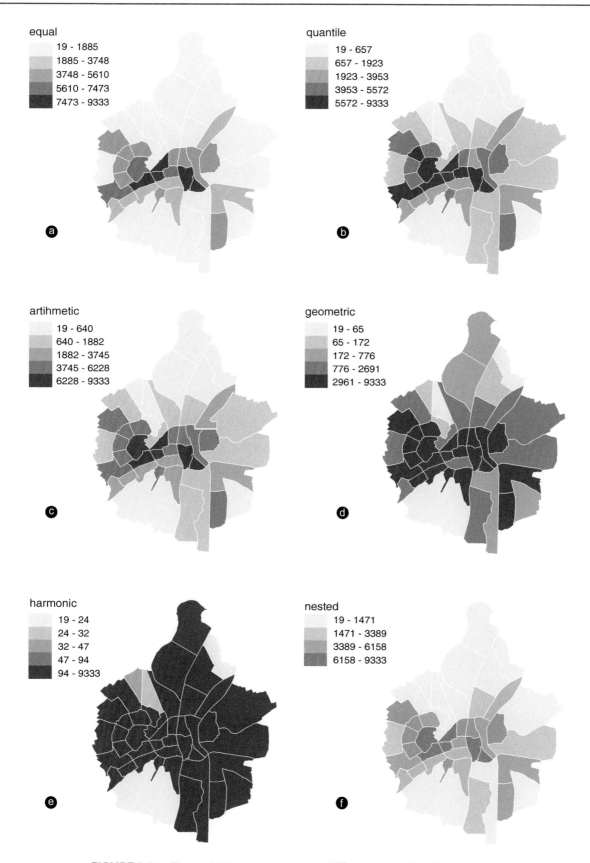

FIGURE 7.11 Choropleth maps based on different classification methods

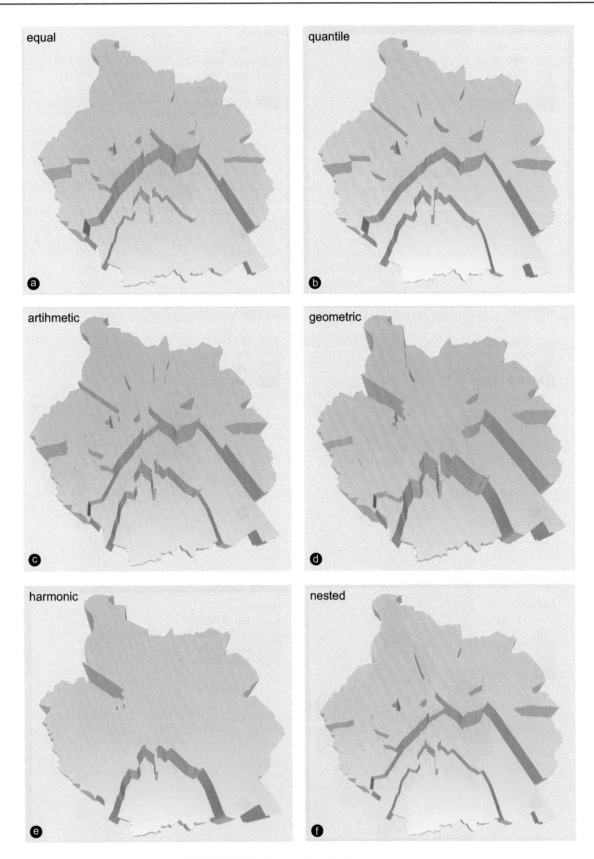

FIGURE 7.12   Stepped statistical surface

one of the functions closest to the observation series will result in the best, i.e. the most accurate, image. However, it is also possible to use the statistical surfaces to get an indication of the accuracy of the method chosen. One has to compare the original statistical surface (Figure 7.8b) with the statistical surface resulting from one of the classification methods. A graphic comparison will immediately show that some maps in Figure 7.11 are not at all similar to the original statistical surface. For instance, the nested means and the geometric method are obviously 'wrong'. For some others, it is much more difficult to judge which one is closest to the original surface. Some calculations have to be made to solve this problem. The attribute value of statistical surfaces in Figure 7.12 is based on the class average. For each of the statistical surfaces, the average values, now given to each geographical unit, are summed and compared to the corresponding values of the original surface. The classification method closest to the original values is most suitable for this data set.

The class boundaries are not the only characteristic that affects the image perceived by the map reader. The visual presentation of the maps in Figure 7.11 is quite important as well. The choice of the (grey) values given to each of these classes will influence the map reader during the map-reading process.

7.4 CARTOGRAPHICAL DATA ANALYSIS

From Section 7.2, it is clear what data aspects are important for communication, how they can be best described statistically or otherwise, or what should be done to make them comparable. This section will describe what aspects of the data (or the descriptors that have been selected for representation) have to be made explicit in order to allow for a proper presentation.

Firstly, the core of the cartographic data analysis will be presented, and this core will be refined and added to later on, in order to show all the aspects that have to be taken into account for a proper presentation.

The core, then, consists of assessing the characteristics of the components of the information and deciding which graphic variables (see Chapter 5) to use for them. Using the graphic variables selected, one should be able to convey to the reader the nature of the components and the differences that have to be shown. The first step in the analysis of the information is finding a common denominator for all the data elements or categories selected for representation. This common denominator will then be used as the title of the map. For the example of Indonesia in Section 7.2, this would be the size of agricultural holdings in Indonesia in 1970. For the data set discussed in this section (see Table 7.3), the common denominator or descriptor of all data elements (also called 'invariant') would be fruit production in the Federal Republic of Germany (FRG) in 2007. It is the common denominators of the data elements that will be used later on for identifying the data set, through their application as map titles.

Then, as a next step, the data variables or components, i.e. those aspects of the attributes that vary or change from data element to data element, should be assessed and their character described.

TABLE 7.3						
Fruit production in Germany in 2007						
Land	Apples	Pears	Cherries	Plums	Others	Total
Baden-Württemberg	352 533	17 490	22 189	30 658	4393	427 263
Niedersachsen	296 458	8304	3436	5205	5186	318 589
Sachsen	93 972	3233	7021	0	4276	108 502
Nordrhein-Westfalen	62 432	4956	1436	2684	34 341	105 849
Bayern	42 908	8346	4991	6443	22 796	85 484
Rheinland-Pfalz	35 721	4305	12 217	19 400	0	71 643
Thüringen	39 131	530	5049	1543	1173	47 426
Mecklenburg-Vorpommern	37 166	175	162	0	3956	41 459
Hamburg	39 990	711	284	0	0	40 985
Sachsen-Anhalt	27 297	488	2844	0	0	30 629
Brandenburg	15 177	407	2310	926	1724	20 544
Schleswig-Holstein	17 054	396	933	0	683	19 066
Hessen	8644	444	315	729	336	10 468

Source: Statistisches Jahrbuch (2007).

For Indonesia (see Table 7.2 and Figure 7.7), these information components would be area used for agriculture, proportion of the agricultural area in estates, average farm size and a geographical component (the provinces). For the soil map in Figure 3.5, the components would be the geographical location of the sample site and the various soil units discerned. Then, one has to determine the measurement scale at which the attributes in these components are measured. If this has already been settled during the data analysis phase described in Section 7.2, so much the better, but as the data aspects one wants to render might change during this visualization operation, and data aspects might have to be redefined in order to answer reformulated communication objectives, these measurement scales might change as well. Anyway, the measurement scales of the components will be nominal, ordinal, interval or ratio scales, or they might be geographical.

In the Indonesia example (Table 7.2), there are four components:

> Area agriculturally used ratio, absolute;
> Percentage in estates ratio, relative;
> Average size of farms ratio, relative;
> Provinces, geographical.

The fruit production example (Table 7.3) has three components:

> Production size ratio, absolute;
> Fruit type, nominal;
> Länder, geographical.

The soil information example (Figure 3.5) has three components:

> Soil type, nominal;
> Groundwater category, ordinal;
> Soil sample locations, geographical.

It is important to assess not only the nature of the components but also their length and range. The *length* of the components refers to the number of classes or categories that will be discerned. In the Indonesia example, there are six farm size classes; in the fruit production example, one discerns five fruit types; and in the soil map example, there are about 15 soil types.

In addition to the length, the *range* of the data (if this attribute aspect is measured on an interval or ratio scale) should be assessed. In the Indonesia example, the total agricultural area per province ranges between 190 and 22 860 km². The average

farm size range is between 0.5 and 3.6 ha. In the fruit production example, it is between 10 468 and 427 263 tons (a ratio of 1:40) per 'land' and between 162 and 352 533 tons (a ratio of 1:2200) for individual fruit types.

Assessing the length and range of the components provides essential information as there is a maximum range of values that can be visualized given a specific representation method.

As indicated in Chapter 5, the maximum number of colours to be discerned on a map, while still providing an overview and the possibility of visually isolating each category, is eight. This number might be increased slightly by adding other graphic variables, like shape (regular patterns). As can be seen from Figure 7.13, comparing proportional circles allows one to have a range between 1 and 2500 (so the largest circle to be discerned can be 2500 times larger than the smallest circle to be discerned, i.e. being different from a dot). If shown as a pie graph, the relation between the smallest sector in the smallest circle and the largest sector in the largest circle would be 1:275. Larger differences could not really be visualized through the pie graph method. If other methods were used, e.g. composite bar graphs, the ratio in overall sizes could not exceed 1:100, while that of subdivisions could not exceed 1:10.

The final aspect of the cartographic analysis of the data to be portrayed is the information hierarchy: one has to determine what aspects are most important, what are least important and what data categories come in between in which order. This has to be determined as this information hierarchy has to be translated into a graphical hierarchy (see Chapter 5).

Now all the building blocks would seem to be available to match the data components to the graphic variables that convey an idea of the kind of differences expressed by the measurement scales: qualitative differences by rendering nominal components, ordered differences by rendering ordinal components, distance differences by rendering interval components and proportional differences by rendering ratio components (Figure 7.14; this is in fact an extension of Figure 5.7).

On the basis of the measurement scale of the data (nominal, ordinal, interval, ratio), one may find in the scheme in Figure 7.14 what graphical variables (size/value/grain/colour/orientation/shape) to select in order to convey a correct idea of these measurement scales. But at the same time, it should be realized that the selection of specific graphic variables affects the image that map users will see consecutively. Depending on this selection,

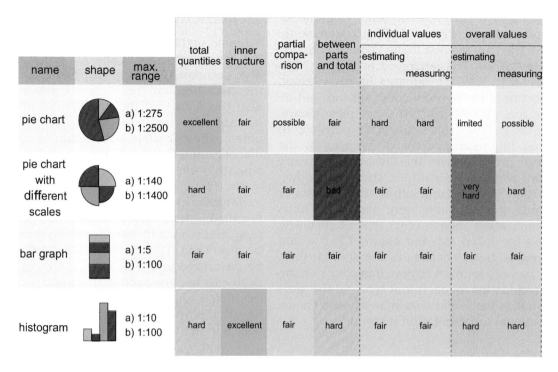

name	shape	max. range	total quantities	inner structure	partial comparison	between parts and total	individual values estimating	measuring	overall values estimating	measuring
pie chart		a) 1:275 b) 1:2500	excellent	fair	possible	fair	hard	hard	limited	possible
pie chart with different scales		a) 1:140 b) 1:1400	hard	fair	fair	bad	fair	fair	very hard	hard
bar graph		a) 1:5 b) 1:100	fair	fair	fair	fair	fair	fair	fair	fair
histogram		a) 1:10 b) 1:100	hard	excellent	fair	hard	fair	fair	hard	hard

FIGURE 7.13   Effectiveness of different diagrams (after Gächter, 1969). (a) within each figure; (b) between figures

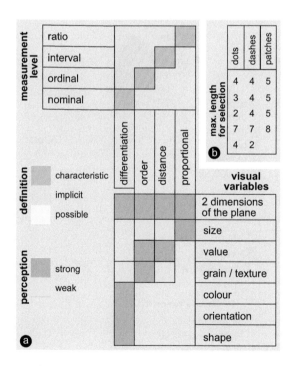

FIGURE 7.14   Cartographic information analysis

users will be able either to differentiate between map objects, to see simultaneously all objects belonging to a specific class or category, or to get an overview of the distribution of the data over the mapped area at a glance. The latter effect is called a 'so high', while the previous refer to either an 'intermediate' or a 'low image level'. In a low-level image, it is impossible to see the distribution over the map of all map objects belonging to a specific class or category.

But there may be some additional considerations before the graphic variables can be applied to render data characteristics according to one of the mapping methods discussed under Section 7.5. After the relation of the data to the Earth's surface and their measurement level has been assessed, and its quality evaluated and locational aspects (shapes and topology) have been taken account of, a preliminary visualization of the data will be effectuated. This preliminary visualization has an analytical function: it will show trends, patterns, etc. to hang on to during further transformations, which might be necessary. Such transformations might be required depending on communication objectives, as well as the audience: more generalized, simpler and less abstract maps need to be produced for a less schooled audience. The results of the data evaluation might be taken into account, requiring a scale reduction, e.g., or an aggregation of the data. Only after all these steps have been gone through will the final map answer the requirements.

Of course, during the data analysis process, one should ask oneself whether it is essential to map the data, and whether it might not be more profitable to graphically portray them otherwise. If the geospatial aspects of the data are well known, they may be visualized otherwise, in order to allow for better comparisons. Figure 7.15 provides some examples. The fruit harvest in the Federal Republic of Germany in 2007 is portrayed in Figure 7.15a (see also Table 7.3): the squares in each region are proportional to its production and are subdivided for the various fruit types (apples, pears, cherries, etc.). From this map, it is possible to get a good impression of the differences in the overall fruit production in the FRG, but it is very difficult to compare, for instance, the pear production in Baden-Württemberg with that in Bavaria or Nordrhein-Westfalen. If that were one of the aims of the visualization, then another type of visualization would be called for, and Figure 7.15b would be the answer. Here, because of the standardized width of the bars for each region, it is easy to compare the production for each fruit type. So the aims of the visualization would also influence the choice of the graphical representation method.

7.5 MAPPING METHODS

Mapping methods are standardized ways of applying the graphic variables for rendering information components. In these methods not only is the measurement scale taken into account, but also the nature of the distribution of the objects (whether they refer to points, lines, areas or volumes; whether their distribution is continuous or discontinuous; and whether their boundaries are smooth or not).

The result of a specific combination of graphic variables according to such a standardized method is called a 'map type'. About nine important mapping methods and nine resulting map types can be discerned. These methods and their results will be described in this section as well as the specific problems in constructing them and the procedures for their production, the transformation possibilities to and from other map types and some general issues in their interpretation. Figure 7.16 shows the

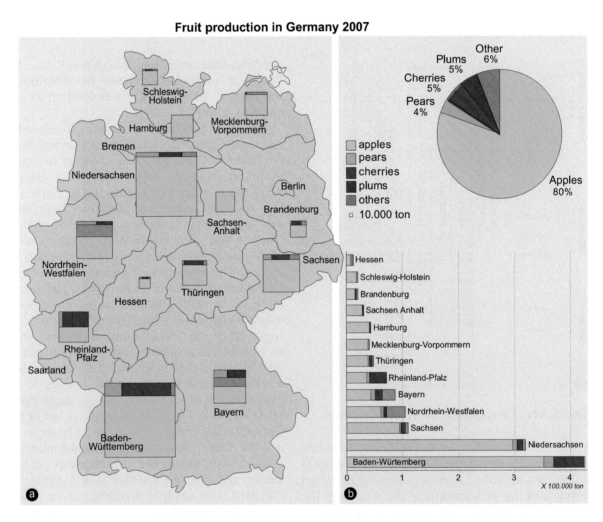

FIGURE 7.15 Fruit production in Germany, 2007

graphic variables	qualitative	quantitative		
	nominal	ordinal/interval/ratio		composite
	variation of hue, orientation, form	repetition	variation of grain, size, grey value	variation of size, segmentation
point data	nominal point symbol maps § 7.5.4 	dot maps § 7.5.7 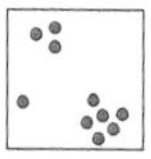	proportional symbol maps § 7.5.5 	point diagram maps § 7.5.6 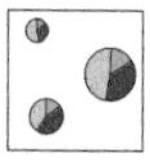
linear data a) lines	nominal line symbol maps 	—	flowline maps § 7.5.8 	line diagram maps § 7.5.6 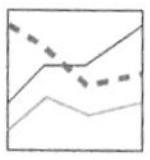
b) vectors	—	standard vector maps 	graduated vector maps 	vector diagram maps 
areal data regular distribution	R.S. landuse maps 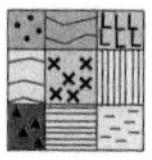	regular grid symbol maps 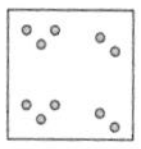	proportional symbol grid maps § 7.5.5 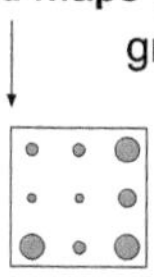 grid choropleth § 7.5.2 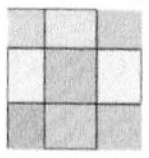	areal diagram grid maps § 7.5.6 
irregular boundaries	chorochromatic mosaic maps § 7.5.1 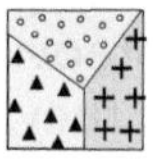	—	choropleth § 7.5.2 	area diagram § 7.5.6 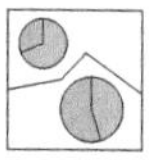
volume data	—	—	stepped statistical surface § 7.5.9 	—
surface data	—	isoline map § 7.5.3 	filled-in isoline map § 7.5.3 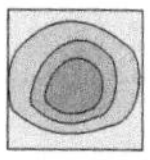	—
volume data	—	—	smooth statistical surface § 7.5.9 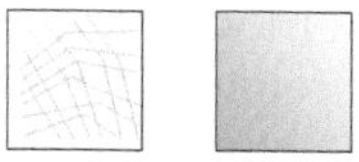	—

(Left margin labels: **discrete** spanning point data through volume data; **continuas** spanning surface data and volume data.)

FIGURE 7.16 *Subdivision* of map types (according to Freitag, 1992), based on the measurement scale, corresponding graphical variables and (dis)continuity of the data. The red section numbers show where the particular map types are dealt with

various map types that will be discerned here in relation to each other. The subdivision is based on characteristics of the objects rendered ((dis)continuity and geospatial reference) as well as their attributes (measurement scale and corresponding graphical variables).

As was mentioned in Section 7.4, it might be necessary, because of reformulation of communication objectives, to select map types other than the one the data have been inventoried in. This would call for transformation of the map type, either by going back to the original data (which would be the best method) or, if these would not be available, by transformation of the source map. The arrows in Figure 7.17 show a number of these transformations: the dot map can be transformed into the choropleth map (see Section 7.5.2) through counting the dots per enumeration area, dividing the number by the area's surface and expressing this ratio through grey tints (see Figure 7.21). The same dot map could be transformed into a chorochromatic map (Section 7.5.1) through setting a threshold value and treating this in a binary way (all enumeration areas above this value have value A, while all others value B). Again, the dot map could be turned into an isoline map (Section 7.5.3) by moving a transparent template with a circle drawn on it over the dot map, and noting the number of dots within the circle at any location. Between the values found thus, isoline boundary values can be interpolated. When these locations are linked, the result will be an isoline map. In turn, this isoline map can be transformed into a statistical surface (i.e. draw the isolines in perspective, then draw any next isoline on a higher plane and drape a grid over it). Another method to produce a statistical surface (Section 7.5.9) from a dot map would be to draw lines at equal distances from dot pairs (Figure 7.18b), compute the area between these lines and assign these areas a height proportional to the ratios between the dot value and the area between the lines. As a next step, these heights can be drawn in perspective (Dahlberg, 1967; see Figure 7.18). This statistical surface would then be used as a data model for checking whether choropleth maps (like that in Figure 7.18e) would be a suitable rendering of the data. This stepped surface data model is the result of a specific view of the data, regarded here as a phenomenon with incremental changes. Another view of the data – as a phenomenon with continuous changes – would lead to a smooth surface data model (Figure 7.18d) and corresponding isoline maps (Figure 7.18f).

In nearly all cases, these map transformations will lead to an information loss, because, when transformed back, the original situation cannot be restored. In many cases, this information loss will be taken for granted if it means that the communication objectives can be reached. More accurate data that do not get through to the map user are less useful than generalized data that do get through. Still, because of this information loss, it would be advisable to get back to the original data.

7.5.1 Chorochromatic Maps or Mosaic Maps

This term has been coined by combining the Greek words for 'area' (*choros*) and 'colour' (*chroma*). So originally, this method rendered nominal values for areas through different colours. But the term is used equally for rendering nominal area values rendered through black-and-white patterns. Both pattern (differences in shape) and differences in colour will give the map reader the impression of nominal, qualitative differences.

The most important condition for the outlook of these patterns or colours is that only different nominal qualities are being rendered and that no suggestion of differences in hierarchy or order is being conveyed. This could be done easily by using different colours or black-and-white patterns, among which none should have more impact (darker, brighter) than others (see Figure 7.19). On the other hand, the colours or patterns selected should be easily discernible one from another.

For printed maps, the use of colour is more expensive; moreover, colours present extra problems because they have associative and psychological values as well; in small areas on the map or screen, it might be impossible to discern a particular colour because perception would be influenced by the colours around it. Saturated colours should only be used for small areas; otherwise, they would dominate the image too much.

When using different patterns instead of different colours, one should preferably select patterns that are comparable in dimension. When both gross and fine patterns are applied, even if they have the same overall percentage black, the coarser patterns stand out.

When chorochromatic maps are being used for the representation of non-area-related phenomena, such as religion or language, the image presented to the map reader might be influenced too much by the actual sizes of the areas taken by specific colours or patterns. The map reader might assume that the sizes of areas might be proportional to the number of peoples with specific qualitative characteristics. Take Figure 7.20, e.g., where dominant languages have been rendered. From this map, a

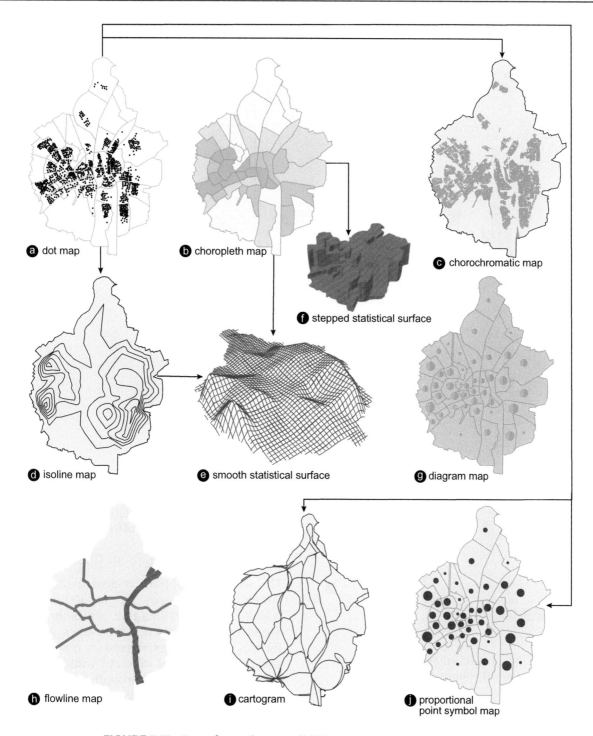

a dot map

b choropleth map

c chorochromatic map

f stepped statistical surface

d isoline map

e smooth statistical surface

g diagram map

h flowline map

i cartogram

j proportional point symbol map

FIGURE 7.17 Transformation possibilities between mapping methods

non-geographically schooled map reader might assume that, as they cover equally sized areas, the numbers of the English and aboriginal language communities might be equal. The high percentage of speakers of aboriginal languages in the interior refers to a very sparsely populated area, whereas population density is much higher along the coasts. A correction, to avoid such inaccurate first impressions, would be to add a diagram showing the actual numbers involved.

A special case of a chorochromatic map would be a grid map in which the dominant phenomenon within each grid cell would determine the assignment of a particular colour or pattern to the cell in order to designate that phenomenon. Remote sensing imagery that has been interpreted is a type of

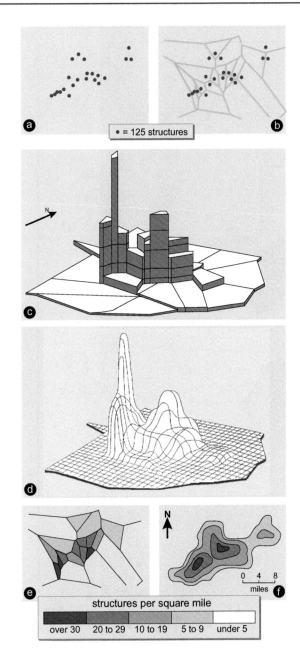

FIGURE 7.18 Transformation of a dot map into a statistical volume (Dahlberg, 1967)

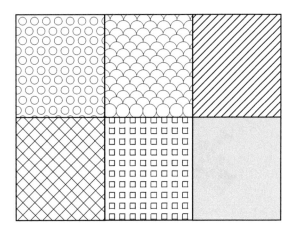

FIGURE 7.19 Patterns with equal (grey) value

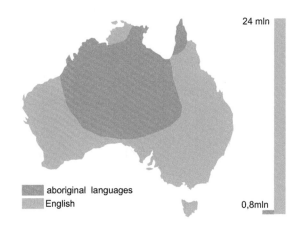

FIGURE 7.20 Dominant language areas in Australia (2016)

7.5.2 Choropleth Maps or Choropleths

The word 'choropleth' is also made up from two Greek words, *choros* for 'area' and *plethos* for 'value'. So it is values that are being rendered for areas in this method. The values are calculated for the areas and expressed as a stepped surface, showing a series of discrete values. As these values are represented through area symbols, they can only be relative values. It is the differences in (grey) value or the intensity of a colour that denotes differences in intensity of a phenomenon, like differences in density. Because differences in value are used, a hierarchy or order between the classes distinguished can be perceived as well. But when correctly applied, percentages or densities that are twice as high are represented by a value that is twice as dark.

Generally, the darker the values, the more intense or the higher the densities of the phenomenon. Another guideline is that the darker the area tints, the less favourable the conditions of the phenomenon.

(grid) chorochromatic map. The basis for such a map would be the images of some bands, which show the intensity of the radiation reflected from the area represented by that grid cell. A combination of these reflection values together determines – through comparison with existing maps or through taking samples from the area – what characteristic this combination of reflection values for the various bands refers to. So the actual interpretation consists of the transformation of a number of grid choropleth maps (see below) into a grid chorochromatic map.

It might sometimes be difficult to combine these two guiding rules. Literacy might be taken as an example: to render increasing literacy percentages on a global map through tints that increase in value would put the map reader aware of these rules on the wrong foot, as the less favourable condition would be represented by lighter tints. In cases like these, one would just change the definition of the phenomenon, and map illiteracy instead of literacy percentages, and all would be well.

There are two main types of choropleth: density maps (e.g. those that portray ratios in which the areas covered are accounted for in the denominator) and non-area-related ratio maps (e.g. the percentage of people over 65 in the total population). From a map-use point of view, it is important to distinguish between these two types, as the visual impression of choropleth maps is governed by both the tint of the areas and their size. When the area is not accounted for in this ratio, this might lead to distorted images (as in the example in Figure 7.3).

The production procedure for both types of choropleth map is shown in Figure 7.21. The starting point will be absolute values for enumeration areas for a specific phenomenon, e.g. the number of people or the number of doctors (Figure 7.21a). Now in order to see whether these absolute numbers are in fact more or less than is to be expected, they are put into perspective by relating them to other absolute figures, like the size of the areas these numbers have been collected for (as in 7.21b), or the number of the total population. So, ratios between these two sets of absolute figures will be determined. The next step will consist of categorizing all these ratios into a limited number of classes (see Section 7.3) as in Figure 7.21c. The limiting factor in the number of classes will be the number of different (grey) values that can be distinguished. The number chosen here is five (see Section 5.3). This range of classes can be extended by adding another colour or pattern, but with seven classes, the maximum has been reached. Finally, all areas that fall into a specific category will be assigned the (grey) value for this category (Figure 7.21d).

The aim of categorizing the ratios for a choropleth would be to improve the possibilities for communication of the information. Generally speaking, through categorization, the image will be simplified, and the existence of trends or patterns will be better visualized. A condition is that differences within classes have been minimized and differences between classes maximized, that the differences between the statistical surface of the unclassified data and that of the classified data would be as small as possible, and that boundaries suggested by the classified model coincide with data boundaries in reality.

As these conditions cannot always be met, a case has been made for unclassed choropleth maps (Tobler, 1973). With present-day printers, area patterns can be generated with a non-incremental increase in the percentage inked, and therefore in (grey) value, so that these percentages inked can proportionally represent the ratios that have to be mapped. Of course, this can also be effectuated for classed data, but the advantage of proportional representation for all data is that no boundaries are suggested because of areas belonging to different classes where in fact only small regional differences occur (see Figure 7.22).

Other corrections against the assumption of boundaries in locations where they do not occur in reality would be either a re-demarcation of enumeration area boundaries or a representation in a grid-cell mode. A choropleth map with boundaries which have been adjusted to the occurrence of the phenomenon has been termed a 'dasymetric map'. If in the dot map in Figure 7.23a areas with a similar density of dots were to be demarcated against another, and density values would be determined anew within these new boundaries, a dasymetric map would be generated (Figure 7.23f). This would be a perfect operation to effect on screen, but no software is able to perform this yet.

As can be seen, by comparing this dasymetric map or the grid choropleth in Figure 7.23b with the original choropleth based on statistical enumeration areas (Figure 7.23d), the density image of the original data has already been improved upon considerably. The area with a high density protruding eastwards in Figure 7.23d, a result of the random demarcation of the enumeration area, which was

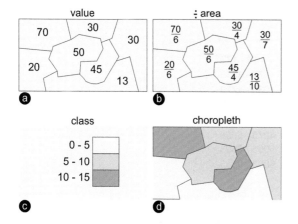

FIGURE 7.21 Production procedure for choropleth maps See running text for explanation

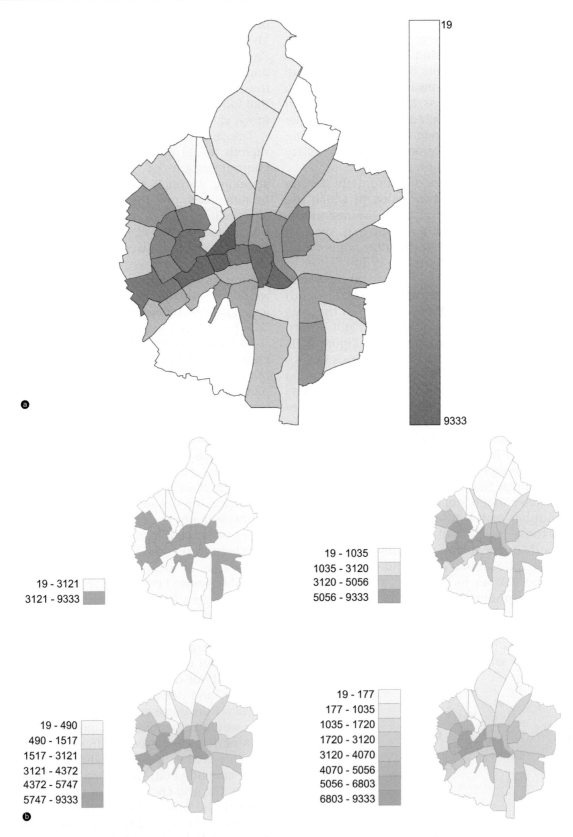

19

9333

19 - 3121
3121 - 9333

19 - 1035
1035 - 3120
3120 - 5056
5056 - 9333

19 - 490
490 - 1517
1517 - 3121
3121 - 4372
4372 - 5747
5747 - 9333

19 - 177
177 - 1035
1035 - 1720
1720 - 3120
3120 - 4070
4070 - 5056
5056 - 6803
6803 - 9333

FIGURE 7.22 Unclassified choropleth map (a) of Maastricht's population density, compared with (b) classified maps that have two, four, six or eight classes. The larger the number of classes, the more the resulting map will approximate the unclassified choropleth

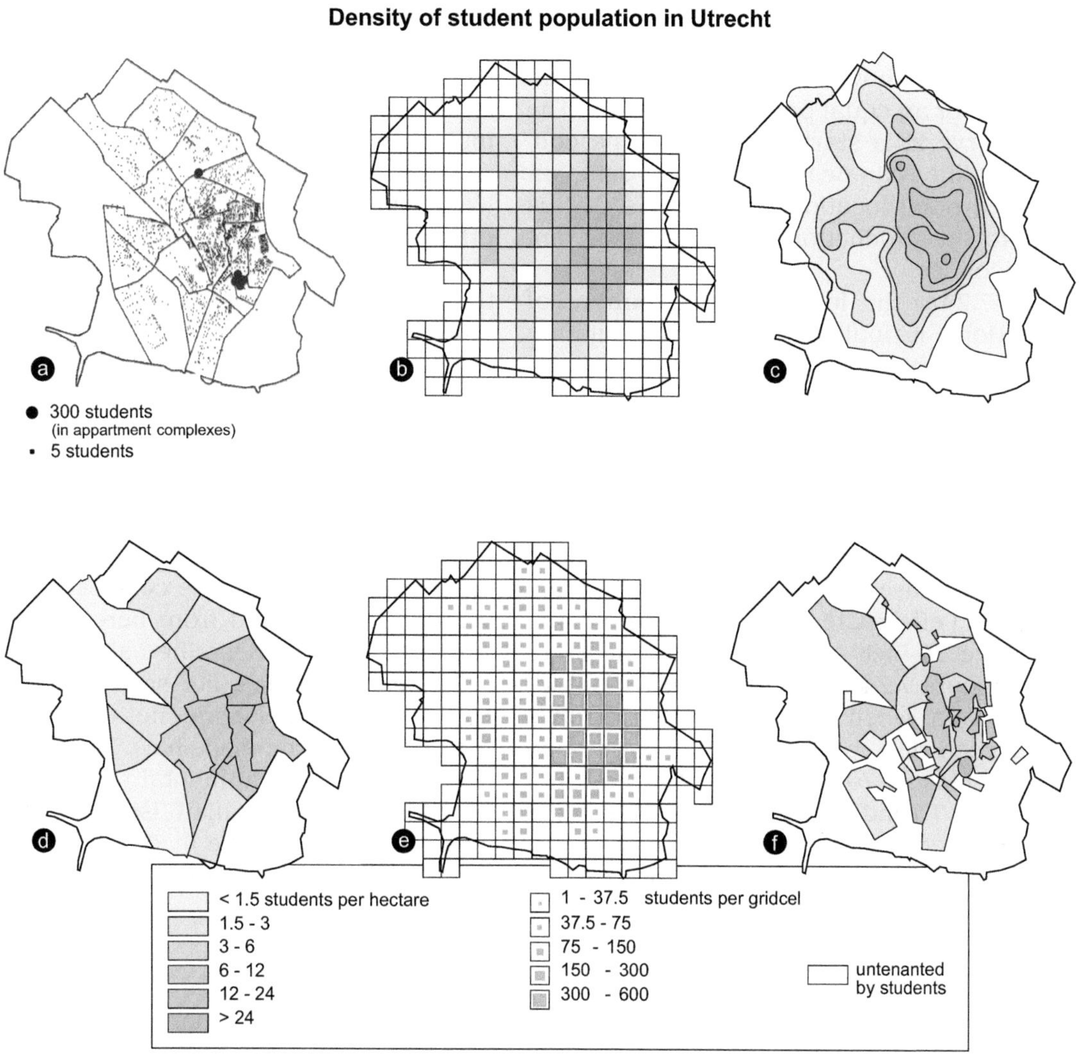

Density of student population in Utrecht

● 300 students
  (in appartment complexes)
· 5 students

☐	< 1.5 students per hectare	
☐	1.5 - 3	
☐	3 - 6	
☐	6 - 12	
☐	12 - 24	
☐	> 24	

☐	1 - 37.5 students per gridcel
☐	37.5 - 75
☐	75 - 150
☐	150 - 300
☐	300 - 600
☐	untenanted by students

FIGURE 7.23 Transformation of a dot map (a) into choropleth maps (b), (d) and (f), a proportional symbol map (e) and an isoline map (c) . The number of dots can be counted per grid cell, per enumeration unit or within newly drawn boundaries adapted to the dot distribution.

not based on the actual phenomenon at all, has been obliterated in Figure 7.23b and f. The false suggestion of homogeneity of the densities within the enumeration areas has been lessened because of the small size of the grid cells.

So a false impression is created by choropleth representation of non-area-related ratios. Two solutions offer themselves: correction of the image by adding weights or changing the base map topography as in cartograms (see Section 7.5.10). The first involves incorporation, as a sort of weight factor, of proportional symbols denoting the actual absolute numbers of the primary data set. In a map of the distribution of doctors, put into context by expressing it as a ratio of the number of doctors and the total population numbers, the number of doctors would be the primary data set, visualized by proportional

symbols, against a background of a choropleth map showing the ratio of doctors and patients.

Despite all these shortcomings, which require corrections or warnings to the map readers, choropleth maps are the type of map most used for the representation of socio-economic data. It is because their construction is relatively straightforward, and they can be computer-generated easily.

7.5.3 Isoline Maps

Unlike choropleth maps, which view the data set they have to represent as discrete values only valid for specific areas, isoline maps are based on the assumption that the phenomenon to be represented has a continuous distribution and smoothly changes in value in all directions of the plane as well. The Greek word 'iso' means 'equal', and an isoline is a line that connects points with an equal value: equal height above sea level, equal amount of precipitation or an equal population density. The values that serve as a starting point for the construction of isolines can be measurement data that apply either to point locations or to areas. Let us first look into the production procedure for point data-based isolines.

In Figure 7.24a, the location of a number of weather stations is indicated, with precipitation data (it is customary for climatological maps to use 30-year data averages). These data are now categorized into a number of classes. In the example, the data, which range from 28 to 104 inches per year, have been subdivided into nine classes (20–30, 30–40, 40–50, 50–60, 60–70, 70–80, 80–90, 90–100 and 100–110, see Figure 7.24b). Now these class boundaries have to be drawn in on the map, and this is being effectuated through interpolation. Let us take the 70-inch boundary first. There are no points in Figure 7.24c with the value 70, but we know such points, when constructed, would have to lie in between data–point pairs with values on the opposite sides of 70 inches. So, to quote the map, in between 73 and 65, for instance. Assuming the change in precipitation values occurs linearly, one would be able to pinpoint exactly the location:

when the two data points with values 73 and 65 (there is a difference in value of 8 inches between them) are linked, a point with the value 70 (3 inches less than the 73-inch point) would lie on 3/8 of the interdistance, reckoned from the 73-inch point. To get other points with the boundary value 70, one would have to proceed in a similar way between other data pairs on the opposite sides of 70.

Because the values increase continuously, more than seven classes would be acceptable in the final map (Figure 7.24f).

When the original data have not been collected or measured for point locations, as for weather stations or heights, but for areas, the first steps of the production procedure would be different. Data collected for regular grid cells (these can be population numbers), as in a grid map (Figure 7.25a), are totalled per grid cell and then assigned to the grid-cell centre (7.25b). These centres are then used as the data points, and from here, the procedure is similar to the one described above (Figure 7.25).

There has been quite some opposition to this application of the isoline method, to the point that area-based isoline maps have been called 'pseudo-isoline maps' in continental Europe (in the United Kingdom, they are called 'isopleth' (*iso* for 'equal' and *plethos* for 'value')). The point made against them is that discrete data (data valid for specific

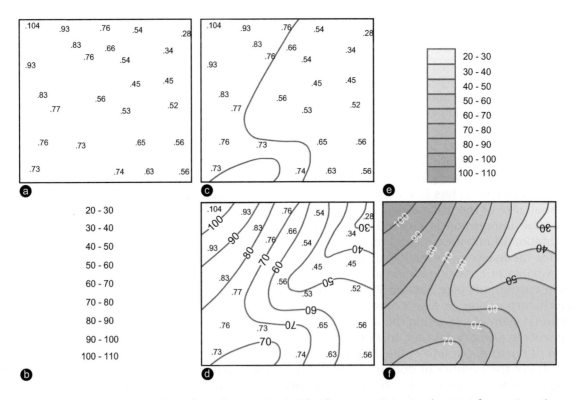

FIGURE 7.24 The production of a point data-based isoline map. See running text for explanation

66	72	80	96	95
63	69	78	86	93
58	64	70	84	89
52	57	60	75	84
45	50	55	70	77
39	44	50	62	74
35	40	50	60	71
38	43	53	58	66

a

66	72	80	96	95
63	69	78	86	93
58	64	70	84	89
52	57	60	75	84
45	50	55	70	77
39	44	50	62	74
35	40	50	60	71
38	43	53	58	66

b

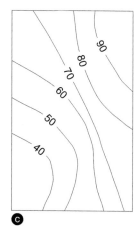

c

d

FIGURE 7.25 The production of isopleth maps through interpolation between grid-cell centre values. The process is described in the running text

enumeration areas) are treated as if they were continuous. But this is all a matter of definition. If population density is not regarded as the ratio between the number of people living in a specific area and the size of that area, but as the number of people within a standard area size, like a circle with a surface of 1 km², which can float over the area, then the concept could refer to a 'continuously changing value'. This floating template (to be perceived as a circle drawn on a transparent material with which a population dot map is scanned) method was used to produce the map in Figure 7.23c.

The prime issue regarding data point-based isolines is the representativeness of these point locations regarding the phenomenon they describe for their surrounding area. The more homogeneous the surrounding area, the more representative the data points.

In contrast to choropleth maps, isoline maps show us trends; they show clearly in which direction values for the phenomenon being represented are increasing or decreasing. Because of this characteristic, they are very well suited for comparing different phenomena and assessing whether there are correlations between these phenomena. In this regard, they perform better than choropleth maps.

7.5.4 Nominal Point Data

Nominal data valid for point locations are represented by symbols that are different in shape, orientation or colour. There is a general division between figurative and geometrical symbols, the figurative ones being used when associations might ease recognition of the symbols. For more abstract phenomena, geometrically shaped symbols are used. Figure 7.26 shows some associative

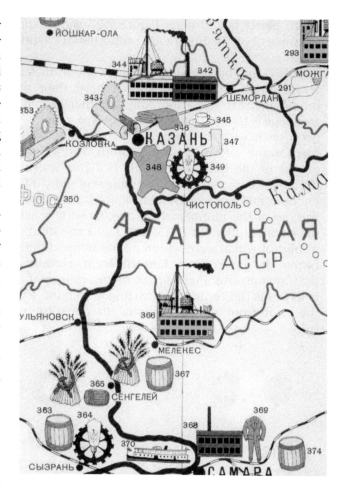

FIGURE 7.26 Detail of a map for the first Five Year plan in the USSR (1929) by V.V. Ermakov and T. Cholodnyj. The main targets to be realized are rendered by associative figurative symbols

figurative symbols. The more elaborate their shapes, the more they will be subject to a tangle on the map, resulting in severe legibility problems.

These figurative symbols probably will be the first map symbols map readers will be confronted with during their education, as figurative symbols (such as grain sheaves for agriculture, wheels for manufacturing or cows' heads for animal husbandry) usually dominate cartographic material for primary schools and tourists. With geometrical symbols as in Figure 7.27, better map legibility is coupled with less recognizability.

7.5.5 Absolute Proportional Method

Discrete absolute values, valid for point locations or for areas, can be represented by proportional symbols. For this purpose, figurative symbols are not well suited as their shapes are so complex that it is very difficult to compare their dimensions. Geometrically shaped proportional symbols are much better suited here. Figure 7.27 shows some examples of geometrical proportional symbol maps.

This is an appropriate place to mention that different graphical languages each have their own grammar. Next to cartography, another graphical language is that of *isotypes* (using associative symbols, each with the same standard value), developed in the 1930s by Otto Neurath. The principal aspect of the grammar for his isotypes was that differences in value would be represented by differences in number of symbols. Figure 7.28 shows an isotype of the number of Europeans with or without a constitution. In 1793, 30% had a constitution, a third of which were within a monarchy (different forms of government had different colours in the original). Another aspect of the grammar for isotypes is the central vertical line, which indicates the shift from unconstitutional to constitutional rule.

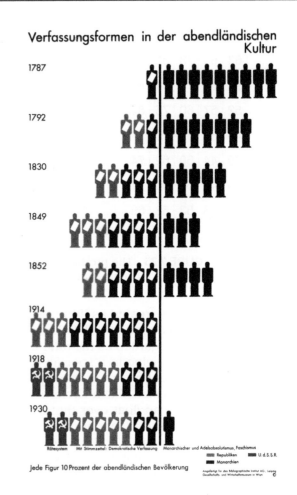

FIGURE 7.28 Isotype of the proportion of Europeans with a constitution (from Neurath, 1930)

The same principle of showing differences in value by different numbers of symbols cannot be used on maps. Figure 7.29a and c show this: the large number of symbols obliterate too much of the map, and thereby threaten to block the geographical background, the very reason for showing the data on the map. Therefore, different values in cartography are represented by symbols differing in size. The areas covered by these symbols will be proportional to the values they have to represent.

The primary considerations for these symbols will be legibility and comparability. Whether symbols are legible or not against the background of the base map depends on the contrast and symbol density. Whether proportional symbols can be compared easily will depend on their shapes. Proportional symbols that vary only in one direction, like columns (Figure 7.29b), score well in people's ability to compare the sizes they represent, much better than proportional circles. On the other hand, these circles would dominate the map image less, they would not monopolize certain directions,

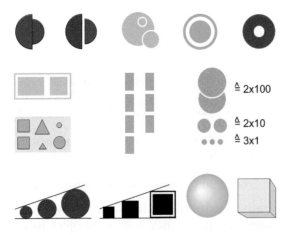

FIGURE 7.27 Geometrical proportional symbols

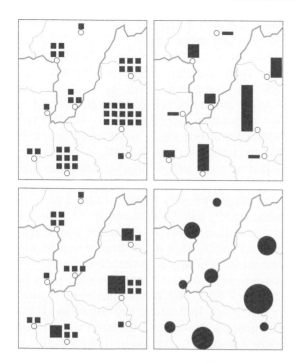

FIGURE 7.29 Comparison of the disturbing influence of different proportional map symbols upon the map (from Imhof, 1972)

and it would be easier to apply them within the areas they relate to. As many map readers underestimate continually proportional circles when comparing them, graded circles can be used instead, which only have a limited number of symbols to denote specific size categories.

The range concept, introduced in Section 7.4, is very important in the context of proportional symbol maps, as it refers to the ratio of the highest and lowest value that can still be represented proportionally, without impairing legibility. As could be seen from Figure 7.13, the range between the smallest proportional circle symbol that can still be perceived and the largest one that can be applied to the map without disturbing the map image too much is 1:2500. As the dimensions are proportional to the square roots of these values, this denotes the difference between a 1 mm diameter circle and a 5 cm diameter circle. When one tries to visualize a similar range through proportional columns, and the smallest value is rendered through a 1 mm high column, the highest value would have to be rendered through a 2.5 m high column, which obviously would be impossible. The only way to render such extremes would be by introducing a threshold value below which all values would be represented by an asterisk or so, thus reducing the range.

It has been proposed to use three-dimensional symbols, in order to increase the range that could

be represented on a map. The idea was that by constructing three-dimensional symbols, the dimensions of these symbols would be proportional to the cubic roots of the values, instead of to their square roots as in two-dimensional symbols. When compared with proportional circles or squares, the space taken by 3D symbols will thus be reduced considerably. Perception research has shown, however, that with 3D symbols simulated on paper maps, it is not the 3D model but the surface the model actually covers on the map that is taken into account, thus leading to serious underestimations of the actual values represented. Thus, the use of 3D symbols is only advocated here in situations in which a better 3D image can be generated, either through anaglyphs or through stereo images (Kraak, 1988).

When used as symbols for areas, proportional symbols tend to be perceived as ratios: it is the ratio between the part of the area covered by the proportional symbol and the area itself which is taken account of. This phenomenon is most clear when proportional symbols are rendered in grid cells. Here, the impression generated by them is similar to that of grid choropleths (Figure7.23b and e).

The production procedure to follow in this special case (of using proportional symbols for designating area characteristics in a regular grid) can be explained on the basis of Figure 7.30. In this map

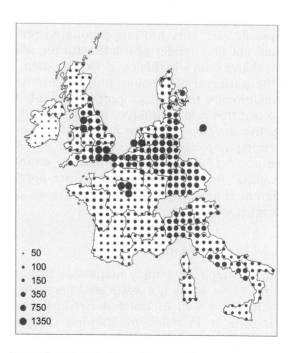

FIGURE 7.30 The use of proportional symbols in regular grids: housing densities in Western Europe. Numbers of houses in thousands per (virtual) grid cell

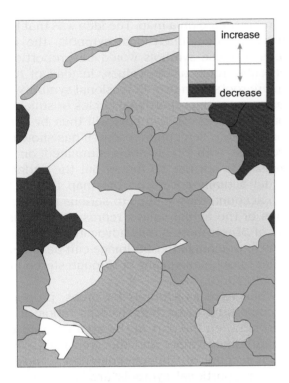

FIGURE 7.31 Diverging map

(housing density in Europe), absolute area figures are related to the number of regular grid nodes that fall into the respective areas: if 1 800 000 houses (eastern France) belong to an enumeration area in which 12 nodes fall, each node will have to represent 150 000 houses and will be proportioned accordingly. As each node also relates to an area of a specific size, this grid proportional symbol map would not only render absolute data but also render relative data – in this case, density data.

The portrayal of absolute positive and negative developments through proportional symbols can also use this grid technique. Relative positive and negative developments are portrayed by binary or diverging maps (see also Section 5.3.2 and Figures 5.16 and 5.17). Figure 7.31 is another example of the latter: here, opposing colours are applied, in different tints, proportional to the more or less diverging values.

7.5.6 Diagram Maps

Diagram maps are simply maps that contain diagrams. Their use is not advocated here. Diagrams function very well in isolated circumstances, on their own or in pairs, in allowing comparisons between figures or in visualizing temporal trends. But when represented against the background of a map, there are usually too many distracting circumstances: the map background, geographical

names, the fact that they are not situated on the same line any more, etc.

Nevertheless, diagrams are much applied to maps. One discerns between *line diagrams*, in which the temporal trend in a phenomenon is indicated (such as yearly temperature or precipitation averages); *bar graphs*, in which the length of the column has been subdivided proportionally between various characteristics; *histograms*, in which a number of contiguous columns are used; *area diagrams* (like pie graphs), in which the area of the diagrams has been subdivided; *flow diagrams*, which will be covered in next subsection; and *areal diagrams*, in which the whole map area has been subdivided according to the percentages of the various characteristics discerned.

An example of the areal diagram map is given in Figure 7.32b. The basis shown in Figure 7.32a is a dot map, indicating the location of two language groups, A and B. In the area shown, there are 29 000 A and 56 000 B, so in this area, the A group forms (29 000/(29 000 + 56 000)×100 =) 34% of the population. This then should be shown in the areal diagram map. If the area is subdivided into 100 grid cells, 34 of them would be designated A and 66 would be designated B. Though this would enormously distort the actual geographical image, by rendering the correct proportions, geography might be helped a bit by locating the designated grid cells in such a way that at least the actual distribution patterns are simulated somewhat.

The phenomenon of diagram maps is not to be confused with maps that have diagrams added to them in order to substantiate the categorization that has been applied. In a choropleth map, the classification can be sustained by showing in a histogram the number of observations that fall within each class. In a scatter diagram or triangular graph, a categorization on the basis of three characteristics that together make up for the whole data set can be sustained.

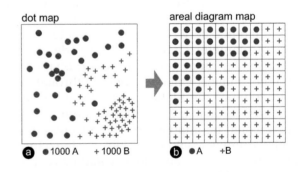

FIGURE 7.32 Production of an areal diagram of an area's language composition. The process is described in the running text

The negative advice at the start of this subsection is based on the consideration that a map is not the proper environment for diagrams because of too many distracting graphical cues. The only advantage of adding diagrams to maps would be to put them into their geographical context. However, with a proper reference to their actual location, series of diagrams would be much better placed next to each other outside the map. Here because of their adjacency, they can be better interpreted. In a multimedia package, they could be consulted outside the map environment.

When cartographers or researchers produce diagram maps, it is mainly for analytical purposes. The data are being analysed in their proper position, which will help the analysis – but as an inventory map, it would not be suited for communication. A good example is the composite migration diagram shown in Figure 7.33. Here, the positive and negative net migration is visualized for five 10-year periods (1900–1950) for cantons in western Belgium. For each individual canton, the diagram shows admirably the trend in the population development, but together the diagrams are too complex to show a clear regional trend, which is why the map has been analysed. As a result of this analysis, a number of migration types have been discerned. There are cantons with a continuous net immigration, with a continuous net emigration; there are cantons that lost many inhabitants in the second decade of the previous century (why?); there are cantons that lost people early in

that century and gained later. Nearly all cantons can be assigned to one of these types and visualized accordingly (as in Figure 7.33b). The resulting image shows much better the actual distribution of the various types and therefore of the migration trends.

7.5.7 Dot Maps

Dot maps are a special case of proportional symbol maps as they represent point data through symbols that each denote the same quantity and that have been located as well as possible in the locations where the phenomenon occurs. In the case of a population dot map, when each dot represents one person, it would be possible to produce a dot map with absolute correct locations of the dots. Whenever people have been aggregated because the representative value of the dot is not one person but 5 or 100, it becomes impossible to show the locations of the persons represented correctly and approximations have to be made. A solution would be to put the dot in the centre of the addresses of the inhabitants it represents, or in their gravity point. The dot locations have to be selected in such a way that they characterize the actual population distribution.

So dot maps show patterns: for instance, concentration and dispersion of the population distribution, in a population dot map, or subgroups of the population, as the students represented in Figure 7.23a.

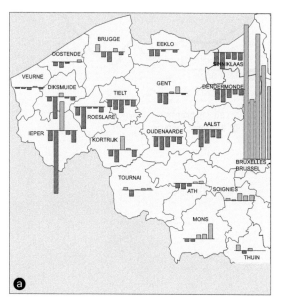

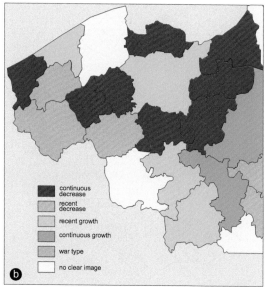

FIGURE 7.33 (a) Composite diagram map of the population development in western Belgium between 1900 and 1950 (Atlas van de Nationale Survey), for analytical purposes and (b) chorochromatic representation of the same information for communication purposes

A population dot map is usually generated according to the following procedure (see Figure 7.34): population data will be available for enumeration areas (Figure 7.34a), and generally, additional information will be available in the form of topographic maps, land use maps or remote sensing imagery. On the basis of this additional information (Figure 7.34b), it will be possible to determine which areas are uninhabited, which are sparsely inhabited and which have a high population density (as the contiguously built-up urban areas). So on the basis of the data and the additional information, the enumeration areas will be broken down into smaller units considered homogeneous as regards their population distribution, with the population numbers that are supposed to apply to them (Figure 7.34c). The final step will be the translation of the values to be represented into graphical form, by choosing a dot with a specific size and a specific representative value (see Figure 7.34d) and by applying the dots regularly over the areas considered homogeneous. Without the additional information, one would only

be justified in applying the number of dots calculated for each area regularly over that area; with the additional information, one can be more specific.

When the size of the dots is too large, a situation will quickly occur in which the dots have a tendency to coalesce or merge, so that individual dots can no longer be distinguished. Though this is not serious for a restricted number of locations (after all, it never is the intention of dot maps to show densities or to have the dots counted), it is better to avoid this if they would coalesce over larger areas.

When the representative value of the dots is too large, few dots will have to be applied to the map, but in sparsely inhabited areas, large tracts would go without any dots at all, so no pattern will show here. When the dots have a small representative value, many dots will have to be located, and there will again be the danger of merging. When the differences in density are just too large for rendering both rural and urban densities, one may be compelled to use an additional dot size for larger urban values (Figure 7.34d).

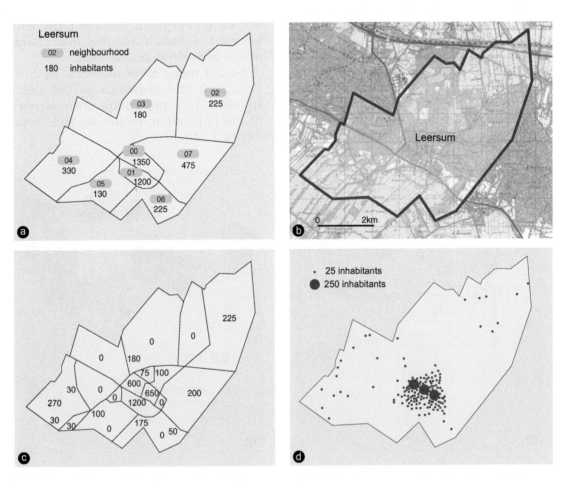

FIGURE 7.34 Production procedure of a dot map: (a) administrative units; (b) topographic map – source Kadaster Geo Informatie; (c) combining administrative units with the topography; (d) the resulting dot map

Because of these considerations, the actual dot size and representative values have to be tested out for a few different areas in order to check whether the required impression will result.

A number of software packages are equipped with the facility to produce dot maps. But, as these work on the basis of random generators of the dot locations within the areas specified, there is not much use in applying them, as the resulting image will bear little resemblance to the actual distribution within the areas.

7.5.8 Flow Line Maps

This is one of the few map types that simulates movement. Movement can be simulated on static paper maps in a number of ways: by using graphical variables that give the reader an ordered impression (through differences in size or in value), by showing a number of situations adjacent in time next to each other (the filmic method, as in a comic strip) or by using symbols that are associated with movement. Flow line maps use the third way, as they use arrow symbols. This is a most useful symbol in cartography, as arrows indicate both the route along which a movement occurs and the direction along the route (by the way in which the arrowhead points); also, the volume transported along that route can be shown by the relative thickness of the arrow's shaft. So far this definition is valid for flow line maps, flow line diagrams and vector maps (see also Figure 7.16).

Vector maps only show the size of the forces that occur at specific points in specific directions. Wind velocity maps are an example. Flow line maps show the specific route of the movement or transport as well, but are not further subdivided. Flow line diagrams are further subdivided, and show, for instance in the amount of goods transported from A to B, the proportion of those goods that have been transported by boat, by dividing up the arrow shaft longitudinally into different sections denoting different transport modes.

The impression given by proportional arrow symbols is one that is governed by both the length of the route and the thickness of the arrow, i.e. the amount transported. This product of thickness and length really is a transportation achievement impression rather than an impression of the amount of goods transported (Figure 7.35a). If it is only the amount of goods transported from all over the world to a given location that has to be visualized, this can be done better by expressing these amounts proportionally at their places of origin (and eventually linking these symbols to the destination by arrows showing the direction, as in Figure 7.35b). The amounts can be compared in this way more easily than from arrow symbols pointing in all directions. Whenever traffic in both directions along a route is in balance, there is not much point in adding arrowheads, and thereby, the symbol changes into a band-like symbol, still proportional to the amount of goods or persons transported along it.

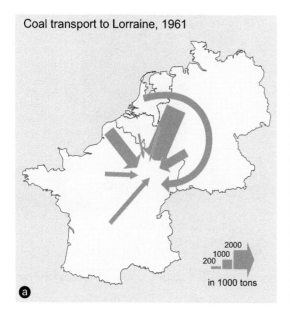

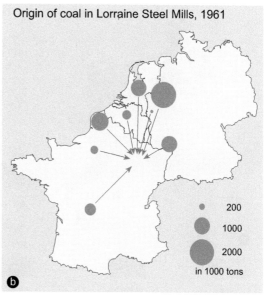

FIGURE 7.35   (a) Flow line map showing transportation achievements; (b) map for comparison of the actual coal quantities transported from various areas to Lorraine

7.5.9 Statistical Surfaces

The three-dimensional representation of quantitative data such as used in choropleth and isoline maps for analytical purposes can help in their two-dimensional representation. Such a three-dimensional representation may be called a 'data model' or a 'statistical surface'. In the case where the theme being mapped is height above sea level, the data model will simulate tangible reality (Figures 7.8b, 7.12, 7.17e). Where it portrays other phenomena, the data model will serve as a yardstick for assessing whether, through classification procedures, a correct view of the data has been generated. Of course, the data model can be used on its own for data communication, but it has some disadvantages here: it is not generalized through classification and therefore might present too complex an image and, because of its relief, some areas might be obliterated by peaks in the three-dimensional data model. Because of its perspective, it would be impossible to read exact values from the map. There would be advantages as well: the image generated by a three-dimensional data model is a very dramatic one, which will be remembered for a long time, and it would present a good overview of the general trend of the data.

7.5.10 Cartograms

Cartograms provide another correction to the false impression choropleths of non-area-related phenomena might provide (see Section 7.5.2), is the adaptation of the size of the enumeration areas in such a way that they have been made proportional to the number of the secondary data set. In our medical example from Section 7.2, with the number of patients per general practitioners, the enumeration areas would have to be made proportional in size to the number of inhabitants or patients. In this way, unfavourable ratios could be weighed against the number of people affected in this unfavourable way. Disproportionately small medical practices in marginal areas could not dominate the resulting image, because the number of people affected in this way, and therefore the area rendered with this light grey tone (compared with Figure 7.3), would be relatively small. An example of such an 'area proportional to' or APT cartogram is shown in Figure 7.36, in which areas are proportional to population data. Figure 7.36 has been constructed starting from a central area (France) outwards; one can also start from the original outline of the area and subdivide the interior surface proportionally (as in Figure 7.17h).

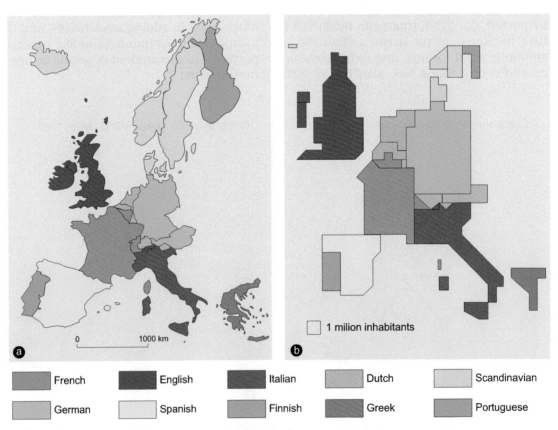

■ French	■ English	■ Italian	■ Dutch	■ Scandinavian
■ German	■ Spanish	■ Finnish	■ Greek	■ Portuguese

FIGURE 7.36 Languages in Western Europe: (a) in geographical space; (b) as an 'area proportional to' cartogram

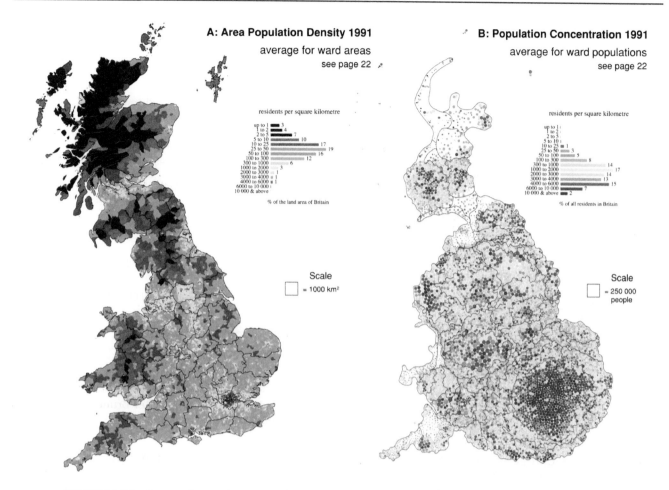

FIGURE 7.37 Comparison of a population density choropleth of Britain and a population density cartogram of Britain (Dorling, 1995). A new social atlas of Britain (John Wiley and Sons, UK)

Figure 7.37 presents another revealing example portraying the population density of Britain. It compares a choropleth and a cartogram. The map at left shows that half of Britain has densities below 50 inh/km². From the map at right, one perceives that over 80% of Britons live in areas with population densities of over 1000 inh/km². Which map gives a better impression of population density?

7.5.11 Chorèmes

Chorèmes, introduced by Roger Brunet in the 1980s, are elementary spatial structures that are each symbolized by a specific graphical shape. Combined, they aim at schematic representations of chosen regions that try to represent the complexity of those regions. Each of those shapes stands for a spatial phenomenon (such as attraction, dispersion, concentration), and by combining these shapes, the territorial dynamics or those regions can be characterized. In analog static cartography, it was difficult to represent the (economic, demographic, etc.) forces that impact specific regions,

and especially for educational purposes, chorèmes tend to make them visible. By combining a number of chorèmes, each standing for a specific spatial phenomenon that influences a region, a specific model of that region is constructed. Such a specific model can be regarded as a hypothesis; whether it is valid or not depends on its power to explain the spatial differences and dynamics that occur in that region. The shapes that are used for chorèmes are combinations of point, line area and network symbols such as plus and minus signs that denote demographic expansion or shrinkage; lines that may indicate barriers, contact zones, fronts or links; and arrows that represent attraction, spatial tendencies or bridgeheads. Hierarchies may be represented by denser shading, or size differences in point symbols.

Brunet was a member of the Reclus research group, a geographical think tank for the French government in the 1980s, and the best-known specific model it developed was the 'blue banana', a model that characterized the spatial forces at play in Western Europe (see Figure 7.38). The central

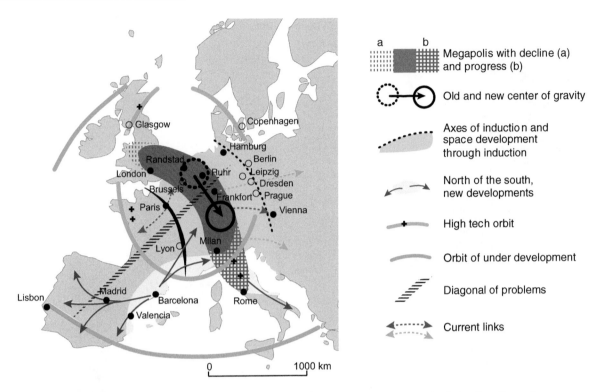

FIGURE 7.38 The economic field of force of Western Europe, Redesign after Brunet, 1989 and 2002 (https://www.mgm.fr/PUB/Mappemonde/M202/Brunet.pdf)

'dorsal' in this European field of force was formed by the banana-shaped link between the British Midlands and Italian Lombardy, with the highest population density and state of economic development. That dorsal was moving in a south-easterly direction from the Randstad/Ruhr area to southern Germany. East and west of that dorsal, its economic influence was felt through induction, and some parallel economic axes (Paris–Lyon–Marseille and Hamburg–Saxony–Vienna) emerged. The plusses stand for regions with a surplus young population that needs jobs, and some of them were accommodated by the green high-tech orbit where new jobs were created in green environments. Farther from the European heartland, shown by a blue arc, was a zone of underdeveloped areas. The checkered diagonal from Estremadura in Spain towards the Baltic linked underprivileged regions in this map.

French geographers were particularly worried by the fact that the central dorsal was bypassing most of their country, and they advised French governments to stimulate new economic activities that linked 'the north of the south' to this central dorsal, for instance, by developing the necessary infrastructure.

The fact that the 'blue banana' (rendered pink in Figure 7.38) had such an impact on economic thinking was proof of the visual attraction and power of these chorèmes and the specific models built from them.

The construction of a specific model from a number of chorèmes is demonstrated in Figure 7.39. First one has to identify the central forces at play that shape an area. For the Netherlands, it is its delta location (Figure 7.39a) where the River Rhine has traversed the ice age-deposited sandy moraines, and where lagoons were formed behind rows of dunes (hostile to marine traffic), filled up by peat and frequently submerged leading to clay deposits (b). In this landscape, in a marine climate (c), a classical centre/periphery structure developed with zones of more labour-intensive horticulture and dairy farming closer to the urban centre and arable farming further away. At a distance from the centre but linked to it by a network (f), a number of regional subcentres (e) developed, some of which industrialized (g), as the poor sandy soils were unable to sustain a growing agricultural population. This structure used to be overlaid by religious differences (mainstream protestant north-west, catholic south-east, separated by a fundamentalist protestant zone (h)) but these differences tend to disappear, on the resulting specific model of the Netherlands (Figure 7.39i) they are already hardly noticeable. The core area, or Randstad, consists of a number of adjacent settlements, almost

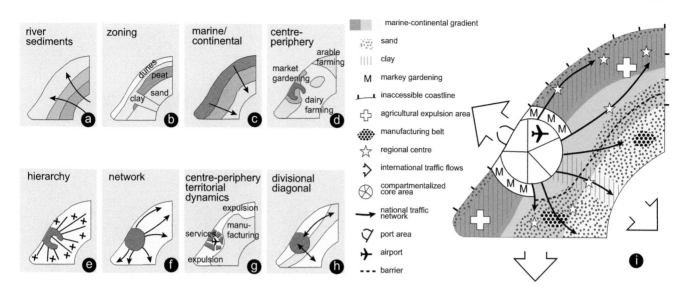

FIGURE 7.39 Construction of a specific model of the Netherlands (i) from a number of chorèmes (a–h)

contiguous, that each specialized (services, administration, transportation (ports and airport)), so the core is compartmentalized. The arable farming areas in the north and south-west have become population expulsion areas, because of mechanization of agriculture. The coastline is hostile to maritime traffic except where the dunes have been traversed with canals. The River Rhine is the main link with the European hinterland.

FURTHER READING

Brewer, C. 2015. *Designing Better Maps: A Guide for GIS Users*. 2nd ed. Redlands: ESRI Press.

Dent, B. D., J. Torguson, and T. W. Hodler. 2008. *Cartography: Thematic Map Design*. 6th ed. Boston, MA: McGraw-Hill.

Field, K. 2018. *Cartography*. Redlands: ESRI Press.

Kimerling, J. A., A. R. Buckley, P. C. Muehrcke, and J. O. Muehrcke. 2016. *Map Use: Reading, Analysis, Interpretation*. 8th ed. Redlands: ESRI Press.

Krygier, J., and D. Wood. 2016. *Making Maps: A Visual Guide to Effective Map Design for GIS*. 3rd ed. New York: Guildford Press.

Robinson, A. H., J. L. Morrison, P. C. Muehrcke, A. J. Kimerling, and S. C. Guptill. 1995. *Elements of Cartography*. New York: John Wiley and Sons.

Slocum, T. A., R. B. McMaster, F. C. Kessler, and H. H. Howard. 2008. *Thematic Cartography and Geovisualization*. 3rd ed. Upper Saddle River, NJ: Pearson.

Tyner, J. A. 2014. *Principles of Map Design*. New York: Guildford Press.

FIGURE 12.2 Construction of a cartographic model: the structural plan of a cartographic-modelling process.

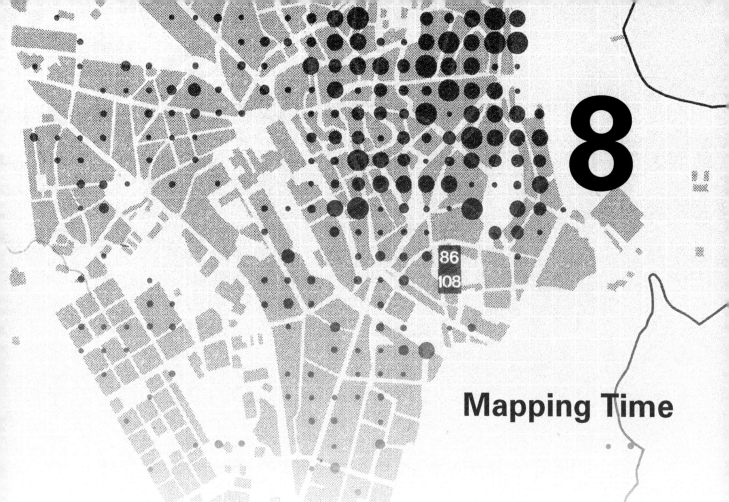

8

Mapping Time

8.1 INTRODUCTION

From the previous chapters, it has become clear that maps are needed when one intends to understand geospatial patterns and relationships. Most maps limit themselves to a single snapshot in time. However, the study of geographical processes or events cannot be successful without considering time as well, since many of the most important challenges facing society today, such as global climate change, economic development and health, require the detection and analysis of changes and trends to support problem-solving. In particular, the temporal component of the data helps in understanding the regionally varied impacts of global climate change, like sea-level rises, or in tracing the diffusion of infectious diseases. This chapter looks at problems like these from the temporal perspective (see Figure 8.1), but without forgetting the lessons learned in Chapters 6 (location space) and 7 (attribute space).

Depicting events requires an inventive cartographic design approach to keep the maps clear and understandable. A classic example is Minard's 1869 map showing Napoleon's campaign in Russia (Robinson, 1967; Rendgen, 2018). This map, presented in Figure 8.2, is considered by many (Tufte, 1983) to be one of the best map designs ever. It simply but effectively visualizes the dramatic

losses during the campaign. It shows the path of Napoleon's 1812 attack on Moscow. The symbol representing the route of Napoleon's army varies in thickness depending on the number of soldiers involved. These numbers decrease from over 500 000 at the start to under 10 000 at the end

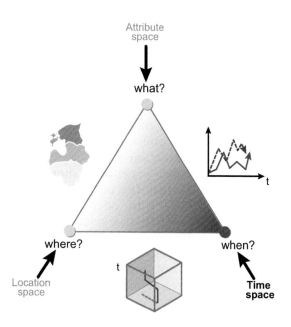

FIGURE 8.1 Approaching geospatial data from a temporal perspective

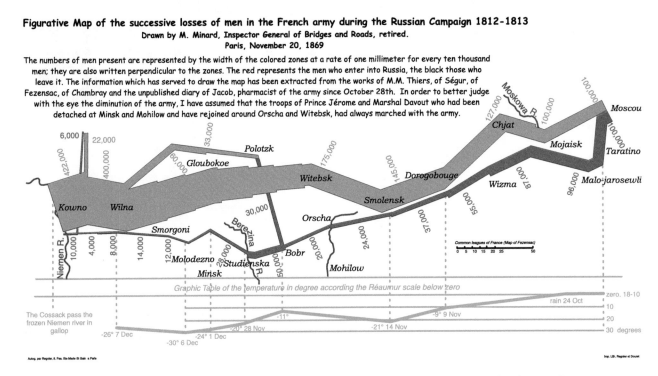

FIGURE 8.2 Minard's map from 1869 showing Napoleon's 1812 campaign in Russia

of the campaign. To explain that it was not just losses due to battles, a graph below the map gives the temperature during the retreat from Moscow. It shows lows of almost −40°C. Many have tried to 'improve' Minard's approach with today's technology such as interactive animations and the space–time cube (Kraak, 2014).

However, not all maps displaying events that stretch over time will be this clear. Generally, such designs tend to become rather complex, due to the amount of data or the length of the time period to be covered. A solution is to split the single map into a set of maps (sometimes called a 'small multiples') physically displayed consecutively, to be read as a story. The individual maps will be less cluttered, but for the reader, this requires greater skills to combine the information found in the individual maps into an event, especially when one needs many maps to display the process (Kousoulakou and Kraak, 1992). With advancing technology, animation seems to be the solution. However, in understanding the process represented by the animation, the reader/viewer should have interaction tools available. If not, the animation is even more limited than the set of maps, where the reader at least has the freedom to move from one image to another in retrieving information. The above solutions to visualize the time dimension are oriented towards presentation, to inform the viewer about

an event that happened or to show a scenario that might take place in future. With the abundance of geospatial data, for instance, acquired by satellite monitoring, or via GPS (Global Positioning System), there is also a need for exploration (Kraak, 1998). Cartographic exploration requires different solutions involving options for interaction and dynamics (see Chapter 10).

The notion of time is not as straightforward as one might think. It seems everyone knows what it is but yet no one can really describe it. The Oxford Dictionary gives as definition: 'the indefinite continued progress of existence and events in the past, present and future regarded as a whole'. In the geosciences, it is all about events and change. An event can be continuous, like the ever-changing temperatures at a meteorological station, or discrete, like municipal boundary changes (see Figure 8.3). Time can be measured objectively or subjectively. A train trip will take 1:30. A lecture is scheduled for 1 hour, but could be experienced shorter or longer depending on how interesting it is. Time is considered to be linear or cyclic. Our system of counting years is an example of linearity of time. Returning days and nights or the seasons are examples of cyclic time. Still it is more complex since it is not always clear to what time one has referred to.

Most prevalent in the literature are world time (world time is the timescale of reality, i.e. the

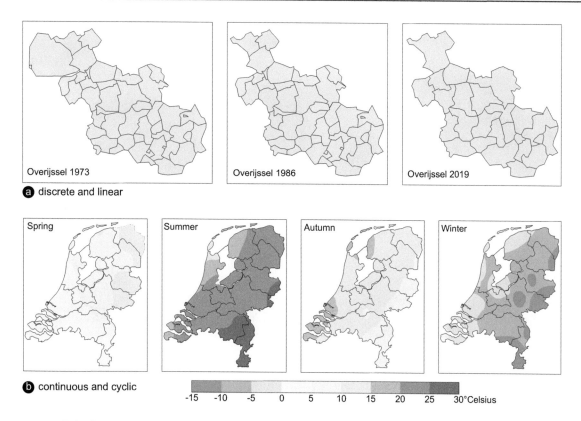

FIGURE 8.3 The nature of time: (a) linear and discrete time – changing municipal boundaries of Overijssel Province, the Netherlands, in 1973, 1986 and 2019; (b) cyclic and continuous time – temperature in spring, summer, autumn and winter

moment an event takes place in the real world), database time (the moment the event is captured in the database) and display time (the moment an event is displayed in a map – see also Langran, 1992; Peuquet, 2002).

The objective of maps is to support decision-making, and the maps discussed in this chapter should be able to answer temporal questions. MacEachren (1995) classified possible questions concerning spatiotemporal data into seven query types, addressing the existence of an entity (if? or whether?), its location in time (when?), its duration (how long?), its temporal texture (how often?), its rate of change (how fast?), the sequence of entities (what order?) and synchronization (do entities occur together?). These questions could be asked both with linear and cyclic time in mind, and at different scales as well. For example, when do traffic jams start? What are the temporal patterns of the traffic jam when looking at daily, weekly or seasonal cycles?

In practice, the most common map reading tasks are to detect changes over time, such as the temperature change between day and night, the extent a car has moved during rush hour or the expansion of a city between two points in time. To inform about the impact of, for instance, a flooding, online newspapers use before/after maps. These allow the reader to move the image/map of current situation on top of the previous situation. The simple temporal questions, such as if, how long and what order, can be answered through typical temporal maps. This should result in detecting trends and patterns, which are higher-order goals that lead to understanding of the data. For the more complex questions, a more advanced working environment is needed as will be discussed in Chapter 10.

Time is inherent in all maps, but it is not always that obvious. Figure 8.4a shows an example. It displays the network of Icelandair. This airline advertises itself having a route network that connects many European cities with as many North American cities through the hub at Keflavik Airport in Iceland. The animated version of this map shows all planes leaving Iceland in all directions at the same time ((https://www.youtube.com/watch?v=Duk39GHvOiQ). It looks fancy but does not explain how it works. One has to analyse the airline's timetable (arrivals and departures) to understand what the map tells. The timeline in Figure 8.4b shows two periods of flight activity during the day. In Figure 8.4c, these flights have

a

Movements Icelandair
Origin and destination - Wednesday November 13, 2019

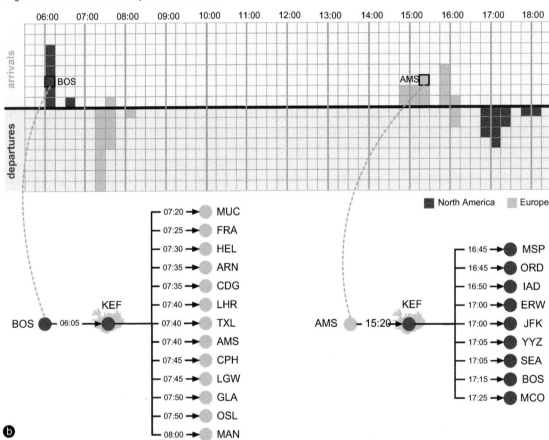

b

FIGURE 8.4 Icelandair's international route network: (a) Keflavik as hub between Europe and North America; (b) timeline with arrivals and departures during the day based on the location of origin and destination explaining the connectivity

been coloured by location (origin and destination). Flights from North America arrive early in the morning, and for instance, a traveller from Boston can jump on any plane with a European destination. A similar, but reversed pattern is visible for the afternoon. A passenger from Amsterdam can select flights to many North American destinations.

Another example of 'hidden time' is shown in Figure 8.5. The top of the figure visualized the

number of domestic flights in Iceland by (a) the origin-destination table and (b) the corresponding flow map. Obviously, the most frequent flights are between Reykjavik (RKV) and Akureyri (AEY). But what if you plan a trip to Grimsey Island (GRY) from Reykjavik (RKV)? For this, we have to refine the temporal unit from a weekly to a daily basis; otherwise, you might not find a connection. In Figure 8.5c and Figure 8.5b, flow map has been

Origin Destination of weekly connections

	AEY	BIU	EGS	GJR	GRY	HFN	HZK	IFJ	KEF	RKV	THO	VEJ	VPN
AEY				3					13	33	0		5
BIU										6			
EGS										22			
GJR										2			
GRY	3												
HFN										9			
HZK										12			
IFJ										13			
KEF	13												
RKV	33	6	22	2		9	12	13				12	
THO	5												0
VEJ										12			
VPN	0												

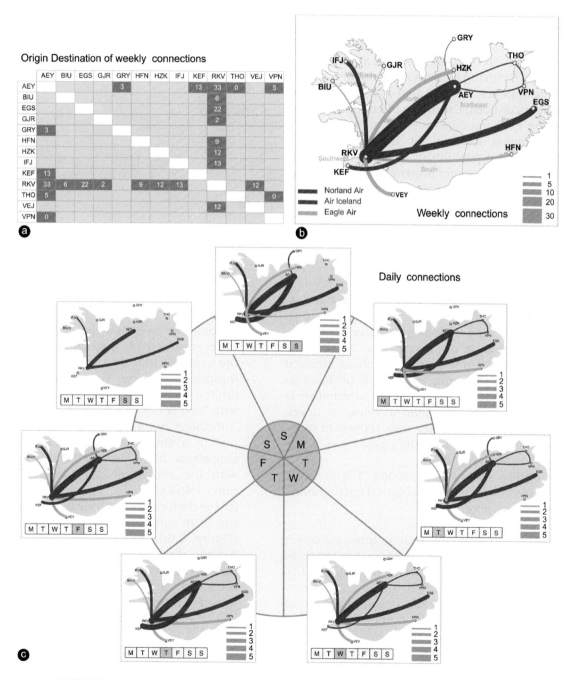

FIGURE 8.5 Visual temporal analysis of the domestic flight networks on Iceland: (a) weekly origin/destination table; (b) corresponding weekly flow map of domestic flights on Iceland; (c) daily frequency of domestic flights in Iceland

split into daily maps. Assuming the Norland flights from Akureyri to Grimsey connect with the Air Iceland flights from Reykjavik to Akureyri, it would be possible to reach Grimsey from the capital on Sundays, Tuesdays and Fridays. Figure 8.5c also reveals flying on Saturdays will not give you many options. The seven maps could be generated into an animation.

8.2 MAPPING CHANGE

Mapping the time dimension means mapping change (Figure 8.6a), and this can refer to change in a feature's existence, such as appearance or disappearance. A calving iceberg appears, and after it melts, it disappears. It could also change in geometry, attributes or both. Examples of changing geometry are the evolving coastline of the Netherlands, or the changing location of Europe's national boundaries or the position of weather fronts. The changes in a land parcel's ownership or in road traffic intensity are examples of changing attributes. Urban growth is a combination of both. The urban boundaries expand, and simultaneously, the land use shifts from rural to urban. If maps have to represent events like these, they should suggest a change. This implies the use of symbols that are perceived to represent change. Examples of such symbols are arrows that have an origin and a destination (see also Section 7.5.8). They are used to show movement, and their size can give an indication of the magnitude of change involved (Figures 7.7–7.35). Specific point symbols such as crossed swords (battle) or lightning (riots) can be used to represent dynamics as well. Another alternative is the use of value (tints). In a map showing the development of a town, the darker tints represent older built-up areas, while lighter tints represent newly built-up areas (Figure 8.6b).

Based on the above observations, it is possible to distinguish between three temporal cartographic depiction modes (Figure 8.6):

- → *Single static map*: Specific graphic variables and symbols are used to show a change in order to represent an event. In Figure 8.6b, the value has been used to represent time. Darker tints indicate older developments and lighter tints newer developments.
- → *Series of static maps*: The single maps represent snapshots in time. Together the maps make up an event. Change is perceived by looking at the succession of individual maps depicting the event in successive snapshots.

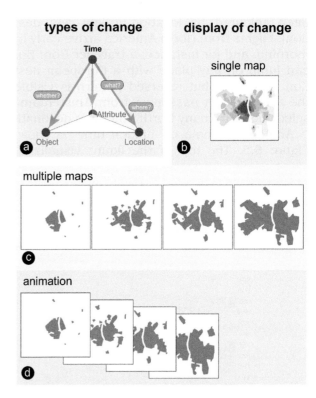

FIGURE 8.6 Mapping an event (urban growth of the city of Maastricht, the Netherlands): (a) types of change; (b) by a single map; (c) by a series of maps; (d) with an animation (a simulation)

It could be said that the temporal sequence is represented by a spatial sequence, which the user has to follow, in order to perceive the temporal variation. The number of images is limited, however, since it is difficult to deal with long series (Figure 8.6c).

- → *Animated map*: Change is perceived to happen in a single frame by displaying several snapshots after each other. The difference with the series of maps is that the variations introduced to represent an event have to be deduced not from a spatial sequence but from a real movement on the map itself (Figure 8.6d). Within this category, one can further distinguish between interactive and non-interactive animations. The most notorious example is the spinning globe ubiquitous in web pages. The interactive animated map will be elaborated in the next section.

Alternative graphic representations of time exist as well. In these cases, not only geographical space but also time space is used to present an event. Examples are given in Figures 8.7 and 8.8.

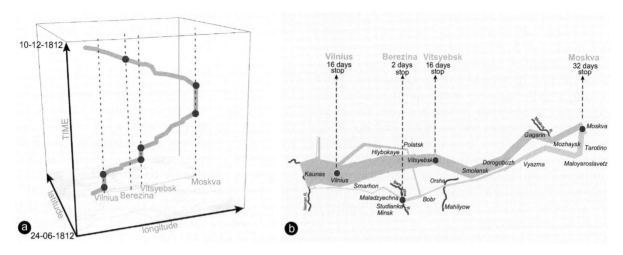

FIGURE 8.7 Mapping Napoleon's Russian campaign (1812): (a) a space–time cube; (b) flow map of Napoleon's march showing the strength of his troops

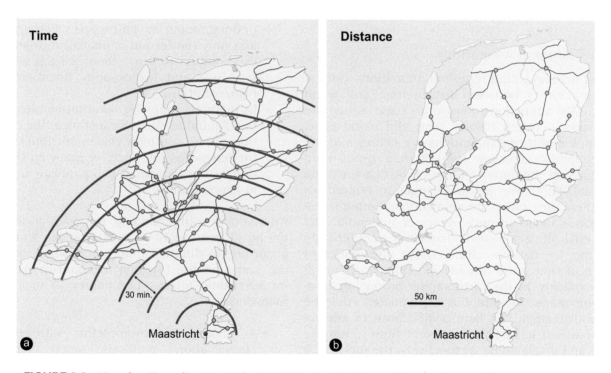

FIGURE 8.8 Mapping the rail network in the Netherlands (a) as a time cartogram (for the year 2010) with travel times from Maastricht (using isochrones) and (b) in its correct geographical layout

The map in Figure 8.7a is a so-called space–time cube. The bottom of the cube represents geographical space, and the event is drawn along the vertical time axis. The cube represents Napoleon's march on Moscow as depicted in Figure 8.7b. Figure 8.8 shows a kind of cartogram. Geographical space is distorted, based on travelling time. The particular example shows travelling time by public transport from the town of Maastricht to other destinations in the Netherlands.

8.3 ANIMATION

Cartographers have paid attention to animation since the 1960s. However, early work only allowed for the manual cartoon-like approach, and experiments were made with either film or television. During the 1980s, technological developments gave rise to a second phase of cartographic animation, with the first computer-produced imagery. Currently, a next wave of cartographic animation is going on, created and enabled by GIS technology. A historical overview is given by Campbell and Egbert (1990). Some early research trends in the application of animation to display changes can be found (Dransch, 1997; Ormeling, 1996; Peterson, 1995). Because of the need in GIScience (geographical information science) environments to deal with processes as a whole, and no longer with single time slices, it is not just visualization methods and techniques that have to be taken into account but also database issues (data storage and maintenance, database design and map user interfaces as well). Current mapping and GIS software offer animation options.

As mentioned before, animations can be very useful in clarifying trends and processes, as well as in explaining or providing insight into spatial relationships. Cartographic animations can be subdivided into temporal and non-temporal animations.

8.3.1 Temporal Animations

When dealing with a temporal animation, a direct relation exists between display time and world time. An animation's temporal scale would be the ratio between display time and world time. Examples of these animations are changes in the Netherlands' coastline from Roman times until the present, boundary changes in Africa since the Second World War, or the changes in yesterday's weather. Time units, the animations' temporal resolution, can be seconds, years or millennia. The GIS environment also distinguishes another type of time, i.e. database time. These three different types of time were already recognized, although not explicitly, by the cartographer producing topographic maps. Topographic map updates would be a good example, as here a difference of several years would exist between world time, database time and display time (respectively, the moment a new road is built, the aerial photograph taken and the final map printed). Temporal animations show changes in the locational or attribute components of spatial data, as shown in Figure 8.9a(I) and (II). For a proper understanding, it is important that the user can influence the flow of the animation. Minimum functionality requires options to play with the timeline: forwards, backwards, slow, fast and pause. In sports, such as running or cycling, athletes can relive their activities via interactive animated playback of the routes followed. This is often linked to all kinds of other statistics collected during the activities. In the Netherlands, the National Mapping Agency has created an interactive animation of 200 years of topographic mapping.

The history of every detailed location in the country can be relive d here (https://www.topotijdreis.nl).

8.3.2 Non-Temporal Animations

Display time in non-temporal animations is not directly linked with world time. The dynamics of the map are used to show spatial relationships or to clarify geometrical or attribute characteristics of spatial phenomena. Here interaction is necessary as well, if only to allow the user to answer the question 'How was it?' Non-temporal animations can be split into those displaying a successive build-up of phenomena and those showing changing representations of the same phenomena (Figure 8.9b and c).

Examples of animations with successive build-up include the following:

* Understanding a landscape. For instance, first only the terrain is displayed, followed by the addition of other themes such as roads, land use and hydrography (location) as in Figure 8.9b(II).
* In thematic mapping, alternating classes are highlighted to show, for instance, the distribution of low and high values (attributes) using the legend as interface, or showing the location of the individual geographical units like in Figure 8.9b(I).

Animations with changing representations (rendering different data or rendering the same data according to different map types, which is in fact the same as the toggling mechanism described for electronic atlases in Chapter 9) include the following:

* A display of choropleths with different classification methods used (attribute) (Figure 8.9c(I));
* Displaying a particular data set by changing the cartographic method of representation, for instance, by showing the same data subsequently in a dot map, a choropleth, a stepped statistical surface and an isoline map (location/attribute) (Figure 8.9c(I));
* Maps with blinking symbols to attract attention to certain map objects, object categories or their attributes;
* A simulated flight through the landscape, as a result of continuous changes in the viewpoint of the user (location);
* The effects of panning and zooming in or out in animation (location and attribute).

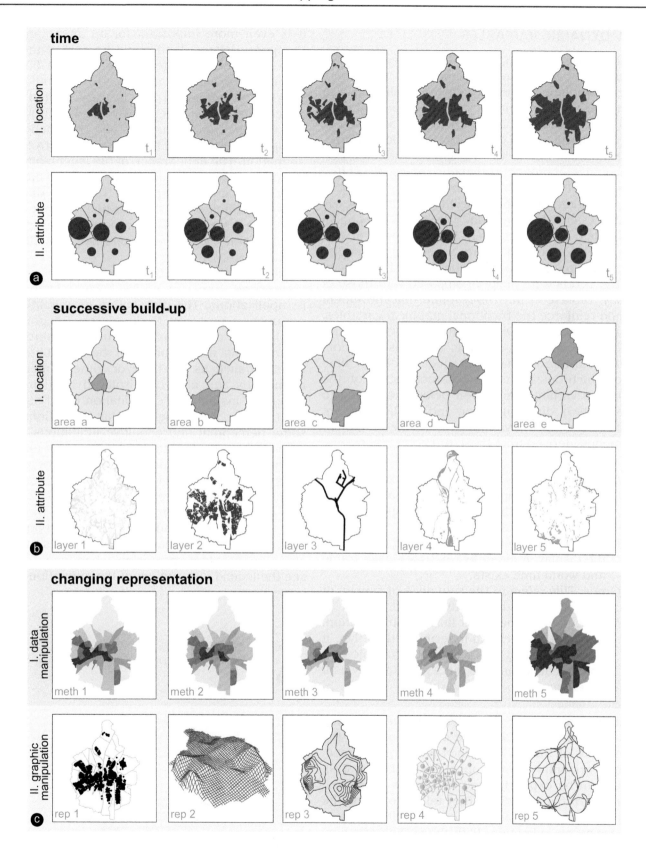

FIGURE 8.9 Classification of animated maps: (a) temporal maps with locational (I) and attribute (II) change; (b) successive build-up according to location (I) or attribute (II); (c) changing representation because of data manipulation (I) or graphic manipulation (II)

8.4 DYNAMIC VARIABLES

A question cartographers have to deal with is, 'How one can design an animation to make sure the viewer indeed understands the development or trend of the phenomena?'. The traditional graphic variables, as explained in Chapter 5, are used to represent the spatial data in each individual frame. Bertin, the first to write on graphical variables, had a negative approach to dynamic maps. He stated (Bertin, 1967, 2011): 'movement only introduces one additional variable, it will be dominant, it will distract all attention from the other (graphical) variables'. Recent research, however, has not sustained this statement. Here, we should remember that technological opportunities offered at the end of the 1960s were limited compared with those of today. DiBiase et al. (1992) found that movement would reinforce the traditional graphical variables. In this framework, DiBiase and MacEachren introduced the so-called dynamic variables: duration, order and rate of change, frequency, display time and synchronization. The characteristics of the dynamic variables are given below. According to Blok (2005), duration, order, and display time are the most important variables, while the others are somehow derived from these three.

> *Display time*: This is the time at which some display change is initiated. The display date can be linked directly to the chronological date to define a temporal location.
> *Duration*: The length of time nothing changes in the display. A direct link between each frame and world time exists.
> *Order*: This refers to the sequence of frames or scenes. Time is inherently ordered.
> *Frequency*: Frequency is linked with duration. Either can be defined in terms of the other. It is worth treating it as a separate dynamic variable because humans react to frequency as if it were an independent variable.

Duration represents the length of time nothing changes at the display, while order deals with the sequence of frames or scenes (see, for instance, each horizontal small map series in Figure 8.9). They can be explicitly used to express an animation's narrative character. They tell a story, and the dynamic variables can be seen as additional tools to design an animation. With those, one can control all visual manipulations.

Both dynamic variables could also be used in the legend of the animation. Although all maps should have a legend to explain their contents, it is even more important for an animation. The (temporal) legend itself could be part of the user interface. Such an interface is required because an animation without an option to manipulate the flow of the animation will be very limited in its effectiveness. The legend as part of the interface will not only help to understand the phenomena mapped but also allow for a dynamic control of the animation. The appearance of the legend interface will depend on the nature of the temporal data and the type of queries expected. Temporal data can be cyclic (think of seasons) or linear (think of history). The first might require a kind of dial to move through time, while the second needs a slide bar (Figure 8.10).

Although the effect of animation is not yet fully understood, one can notice a clear increase in its applications. The distribution of animations is no longer a problem. Media players integrated in the browser can be used to run the animations. Cartographic animations can be distributed via websites such as YouTube or Vimeo. Interactive animations can also be displayed in a browser, using, for instance, a combination of HTML5 and D3. However, it is recognized that animations have some perceptual and cognitive limitations. Users might sometimes fail to notice important changes in the animation. In addition, viewers of animations might be overwhelmed by the amount of images (frames) they receive, and because of this, they would be unable to remember what they saw. This problem is called 'cognitive overload'. The dynamism of the animation also amplifies the split attention problem, occurring when one tries to see the legend and content of the animation at the same time.

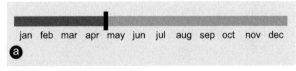

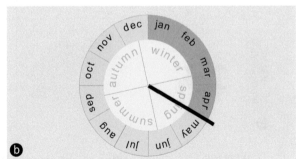

FIGURE 8.10 The animation interface: (a) for linear time; (b) for cyclic time

FURTHER READING

Dransch, D. 1997. *Computer animation in der kartographie: theorie und praxis.* Berlin: Springer Verlag.

Kraak, M. J. 2014. *Mapping Time: Illustrated by Minard's Map of Napoleon's Russian Campaign of 1812.* Redlands: ESRI Press.

Langran, G. 1992. *Time in Geographic Information Systems.* London: Taylor & Francis.

Parkes, D., and N. Thrift. 1980. *Times, Spaces, and Places.* Chichester: John Wiley & Sons.

Peterson, M. P. 1995. *Interactive and Animated Cartography.* Englewood Cliffs, NJ: Prentice Hall.

Peuquet, D. J. 2002. *Representations of Space and Time.* New York: The Guilford Press.

Rendgen, S. 2018. *The Minard System: The Complete Statistical Graphics of Charles-Joseph Minard.* Princeton, NJ: Princeton Architectural Press.

Tufte, E. R. 1983. *The Visual Display of Quantitative Information.* Cheshire, CT: Graphics Press.

Tufte, E. R. 2006. *Beautiful Evidence.* Cheshire, CT: Graphics Press.

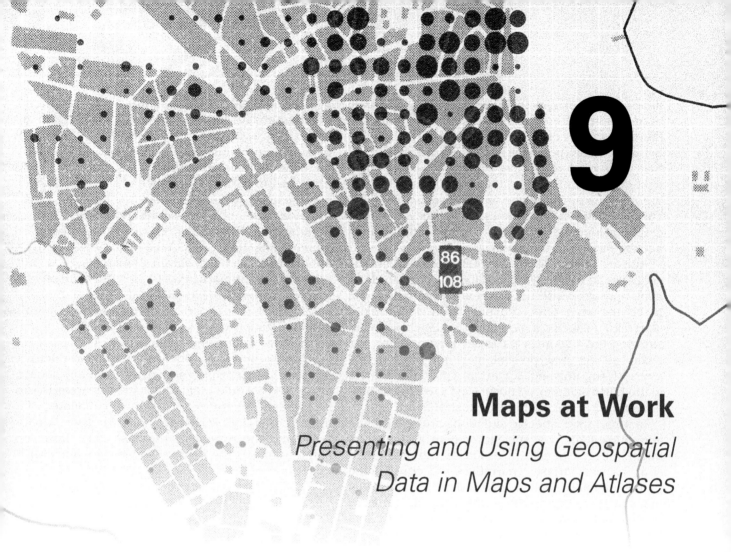

Maps at Work
Presenting and Using Geospatial Data in Maps and Atlases

9.1 INTRODUCTION

Cartography includes not only the design and production of maps but also their use (see Section 1.1), and in order to get the relevant information from maps, users apply specific map-use strategies. They also have to be aware of the intentions and techniques of those that designed the maps and atlases they use. Atlases are intentional combinations of maps or data sets, structured in such a way that specific objectives are reached. In a way, atlases are similar to rhetoric: if a number of arguments are presented in a speech in a given sequence, a specific conclusion is reached. So if one combines a number of maps on specific themes in a specific sequence, this would be done with the objective that the user arrives at a specific conclusion. Atlas editors have certain points to make, and in order to do so, they select and process specific data sets for specific areas. Objectives of atlases may include the introduction of children to their environment (see also Figure 6.15), or to access global information in a reference atlas, but it may be just as well to provide awareness of specific environmental

problems or to evaluate the availability of good factory sites. Objectives and structures together may be called the 'narrative of an atlas', and the way in which this narrative is directed may be called the 'atlas scenario'.

The objective of an electronic school atlas of Sweden, e.g., would be to communicate basic spatial information about the country in such a way that it would keep the pupils interested. In order to effectuate that, the scenario could be the simulation of a flight of geese from south to north over the country, thus providing a gradual overview (as the geese would see it from above) of the country's geography with points indicated where adventures can be had (to keep up the interest) but also allowing for free roaming in order to discover patterns of one's own. The narrative would tell the pupils about the changes in the landscape and the way people earn their living during the series of seemingly criss-cross but actually northbound flights.

Atlases work with a number of 'tools' in order to structure the information. An information hierarchy is arrived at by the use of the sequence tool and the scale tool: more important themes or areas are shown earlier in the atlas or at a larger scale than less important ones. By showing a number of thematic maps consecutively, a causal relationship between these themes is suggested. Specific areas that are regarded as most important can be highlighted by zooming in on them in an inset map. In the margins of the atlas maps, there can be references to cities in other continents at the same geographical longitude or latitude, or to related themes depicted on other maps.

Atlases have a structure that is evident in the sequence in which the information is presented. An atlas may start with the world and gradually zoom in on the area under consideration, or it might start with a specific region and gradually zoom out, showing the region in its larger environment. Or, again, atlases may present their information in a matrix form, starting, for a Eurocentric audience, in the Earth's perceived north-west (North America) and ending in the south-east (Australia).

In a narrative, specific situations are highlighted and presented in a specific sequence; narrators might hide a specific message or moral in their story as well. Atlases' narratives are constructed by the (default) sequence in which the individual maps are presented, within which specific areas or themes get more or less coverage. The data sets for an atlas are selected and processed in order to answer the editor's objective. The processing of the data has as its first aim to make the various data sets comparable, at the same generalization level, letting them be valid for the same date, preferably at the same areal aggregation level, etc.

Access and navigation are the most important aspects when dealing with atlases. Working with atlases is one of school pupils' first contacts with information systems. Learning to access atlases means learning to work with indexes and registers, and understanding the atlas structure. Going from large-scale local maps to small-scale global maps is a preferential structure used in school atlases, for instance. This structure can be provided by attention and sequence: attention, again, is translated into the number of pages and the larger scales used for specific areas, and sequence refers to the order in which the various regions are presented.

The ability to compare maps or data sets is one of the essential characteristics and objectives of atlases: comparison of a map of a known area with that of an unknown one highlights the differences and similarities. By processing the spatiotemporal data, atlas editors see to it that individual maps can fruitfully be compared with the other maps contained. These comparisons can be of a thematical/topical nature (e.g. comparing illiteracy and

average income for the countries of a continent), a geographical nature (e.g. comparing settlement patterns in the United States with those in China) or a temporal nature (as in Chapter 8). To be fruitful, the documents compared must be on the same scale, with similarly detailed base maps available, generalized in a similar manner. The settlement densities of areas should reflect, if possible, population densities. For relevant comparisons, all maps at a specific scale must have been drawn by applying the same generalization rules. It is only relevant to compare maps when the representation of these maps has indeed been standardized (see Figure 9.1). Apart from generalization, this standardization covers the use of symbols, representational values and scale series. As the part of Java shown in Figure 9.1 has a population density ten times higher than the part of Germany shown, one would assume that if maps of both areas would have been generalized to the same degree, far more settlements would be shown for Java than for Germany. The reverse is the case here, however, and that makes the comparison less meaningful.

9.2 PAPER ATLASES

Aside from their possibilities to compare or their accessibility levels, paper atlases can be differentiated on the basis of their objectives. These objectives are embodied in the different existing atlas types. Paper atlases may be differentiated into reference atlases, school atlases, topographic atlases, topical atlases (which represent only one particular theme for many areas, such as the *FAO World Atlas of Agriculture* or the *Atlas of War and Peace*) and national atlases. This last category can be defined as atlases that contain a comprehensive combination of high-resolution geographical data sets that each completely cover the same country. Topographical atlases are either complete sets of topographical maps, bound in book form, or collections of typical landscapes, as exemplified on details of topographic maps. Whatever the paper atlas type, paper atlases isolate in the sense that their maps show only a particular area at a time, at a particular scale, pertaining to a specific date or period in time and to a specific topic.

As was stated in the previous section, by showing a number of thematic maps consecutively, a causal relationship between these themes is suggested. This is the starting point for what are called 'geographical narratives' here: longer or shorter series of maps that together explain geographical phenomena. A short geographical narrative is presented on an atlas spread (two facing pages

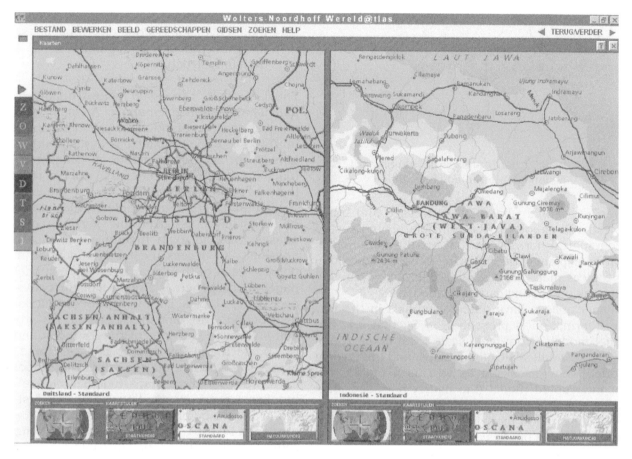

FIGURE 9.1 The Wolters Noordhoff Wereld@tlas contained the desirable functionality to compare two areas with the same scale on screen; zooming in on one would automatically result in zooming in on the other simultaneously (courtesy Noordhoff Atlas Productions)

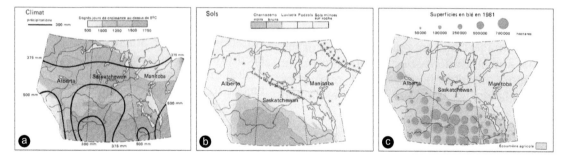

FIGURE 9.2 Geographical narrative: causal relationships as expressed through the juxtaposition of topical maps of the same area: (a) climate (precipitation and temperature); (b) soils, as determinants of (c) the wheat-growing area (from InterAtlas. Les resources du Québec et du Canada, Montréal 1996)

in a paper atlas) in the *InterAtlas*, a school atlas produced for Québec (1986). In Figure 9.2, one can see an example of such a geographical narrative: the upper map shows the climate in central Canada and especially the two foremost aspects for wheat growing: the length of the growing season (the number of consecutive days with temperatures above 5°C) and the amount of precipitation. The next maps show the soils that would allow for good wheat harvests, and the map to the right is the result of the former two: the actual spatial extent of the wheat-growing area. This extent is explained by the threshold values depicted in the two former maps, and comparison of the three maps explains not only the present spatial extent but also the areas where wheat growing might be extended in future. The other maps and graphs on the atlas spread show where the wheat is going

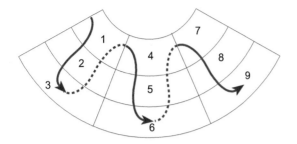

FIGURE 9.3 Matrix atlas structure: ordered sequential interrupted presentation

after harvesting: where it is processed and transported and exported. The principle behind such a geographical narrative is that the previous map(s) (partly) explain(s) the following ones; juxtaposition of topical maps in a sequence is usually understood as presenting causal relationships.

Paper atlases have different models in the way the information is presented; in Section 9.1, two radial models (from large-scale local maps zooming out to small-scale global maps, and the reverse model) were discussed as well as a regular linear interrupted model (the matrix one, see, e.g., Figure 9.3). Another model that can be discerned is the confrontation model: each atlas spread showing confronting physical and socio-economic maps of the same area and, by doing so, showing the different degrees to which the physical world has been affected by humans – a critical issue in this time when humankind is striving for sustainable development. In the same way, historical atlases compare the political situation of the same area for different periods, remote sensing atlases juxtapose satellite imagery and topographic maps of the same areas and natural resources atlases might contrast maps of yet-to-be-developed resources (like virgin oil fields) with maps of already-exploited resources.

9.3 ELECTRONIC ATLASES

9.3.1 Electronic Atlas Types

If the atlas platform is no longer a book but a screen, then it should be called an 'electronic atlas'. If paper atlases are considered intentional combinations of maps, then not all electronic atlases might fit this definition. Unless they would have a default sequence in which the maps would be presented, some could better be defined as intentional combinations of specially processed spatial data sets, together with the software to produce maps from them. There are several types of electronic atlases to be discerned:

- ➡ View-only, stand-alone electronic atlases;
- ➡ Interactive, stand-alone electronic atlases;
- ➡ Electronic atlases integrated into a spatial data infrastructure.

The latter two, discussed below, qualify for the data set definition given above. As these electronic atlases are either stand-alone products, distributed on DVD's or USB sticks (cf. Moellering's virtual map type 2), or accessible online (cf. Moellering's virtual map type 3, see Section 2.1), we can subdivide them as done in Table 9.1.

View-only, stand-alone electronic atlases can be considered as electronic versions of paper atlases, with no extra functionality, but with the possibility to access the maps contained at random, instead of the linear browsing that occurs in paper atlases. There is already a distinct advantage over paper atlases, i.e. the cost of production and distribution. They are much cheaper to produce, and it is much easier to distribute them (and thus to update them) than paper atlases. Some extra aspects that ease their use might be the possibility to view different maps together on the same monitor screen, by dividing it up. The first electronic atlas to be produced, the *Atlas of Arkansas*, produced in 1987 by Richard Smith, was the forerunner of this atlas type, distributed on a number of floppy disks. Other examples are the DVDs accompanying each volume of the (paper) national atlas of Germany (*Bundesrepublik Deutschland Nationalatlas*, 2000–2006).

View-only, web(-based electronic) atlases still are, same as the previous category, electronic versions of paper atlases, but as they are distributed online, they are easier to access, at least in a digital environment with web access, and they are easier to update as the updated maps don't have to be distributed offline to the users. The atlas of public health of the Netherlands (*Atlas vz-info*) is

TABLE 9.1

Types of electronic atlases

	Stand-alone electronic atlases	Web-based electronic atlases = web atlases
Static/view only	x	x
Interactive	x	x
Interactive + integrated	?	x

The x indicates that examples of these categories discerned do exist (these are described in the text underneath), probably with the exception of the category in the lower left field, for which no examples were found.

an example. https://www.volksgezondheidenzorg. info/onderwerp/atlas-vzinfo/alle-kaarten. The option 'alle kaarten' in this atlas gives access to a list of all maps it contains, and as a step towards interactivity, for a number of topics, maps of related topics can be juxtaposed.

Interactive stand-alone electronic atlases are intended for a more computer-literate audience. These are atlases that will allow their users to manipulate the data sets contained. The principle here is that there are no true maps: each map is a specific selection of data, processed in such a way as to come as near as possible to the essence of the theme's distribution, but that will always be biased by subjective elements. In an interactive environment, users can change the colour scheme used for one of their own liking; they can adjust the classification method or extend the number of classes at will. An example that allows this interactivity is the *Atlas of Switzerland 3* (see Figure 9.4), the electronic national atlas of Switzerland produced by the Swiss Federal Office of Topography (2002). A special item in this atlas is the functionality to generate terrain models and panoramas from every location designated within Switzerland, in any required direction, vertical exaggeration or sun angle. Any thematic maps contained in the atlas can be draped over the 3D models; in Figure 9.4, land cover and internal rail traffic have been combined with the relief. Any item on the map can be queried as well.

Apart from the map object query function, in many interactive electronic atlases, either standalone or online, data sets can be combined so that the atlas user is no longer restricted to the themes selected by the cartographer for the atlas. Computations can be effectuated on areas, themes or themes within specifically determined boundaries, and much of the GIS functionality would be available here. Still, the major emphasis will be on assessing the spatial information and the visualization of the result. The definition by Elzakker (1993) of electronic atlases refers mainly to atlas type: 'An electronic atlas is a computerized GIS related to a certain area or theme in connection with a given purpose, with an additional narrative faculty

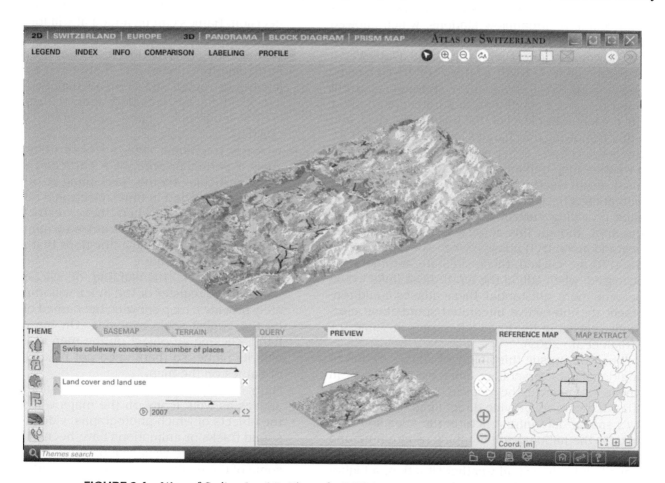

FIGURE 9.4 Atlas of Switzerland 3. View of a DTM-based block diagram with land cover and rail traffic superimposed (Reproduced by permission of swisstopo (BA19106))

in which maps play a dominant role'. As these electronic atlases tend to become more complex, the term 'atlas information systems' can also be used for them.

Interactive web [online electronic] atlases have most of the advantages available in stand-alone ones, but can be updated continuously. On the other hand, as it would take more time to download information as compared to when taken from a DVD, the maps need be less complex in order to minimize their transfer time. In principle, maps would be transferred in a more generalized state, allowing smaller areas to be presented at a lower generalization level, when zooming in.

Finally, *interactive web atlases integrated in a spatial data infrastructure* have a number of interactive online electronic atlases that already have links with the spatial data infrastructure: *The National Atlas of the United States* (http://www.national-atlas-program-has ended-any-data-still-available?qt-news_science_products=4#qt-news_science_products), which was discontinued in 2014, used to have a link to additional data servers: it contained, for instance, a map on rivers with real-time streamflow stations, which showed water gauges. If such a water gauge was clicked on, it would connect with the USGS WaterWatch website (now (2020) https://waterwatch.usgs.gov/?m=flood&r=in&w=map), showing current water resources conditions, and immediately show the present water level at that point in the river, and a graph showing the changes in water level there for the last 24 hours. Other national web atlases have a weather map option; when clicked, they show the most recent weather forecast for the country. Many web atlases have additional links to other relevant sites providing current information measured on the spot, though this service is provided as well in stand-alone DVD atlases: when specific URLs on the DVD are clicked, these sites are automatically accessed, when still in the air. If these links would become more substantial, these atlases could represent the interactive integrated stand-alone electronic atlases in the lower-left corner in Table 9.1 (Kraak et al., 2009).

When we pursue this ability of atlases to link to additional data servers, we foresee a near future in which, for instance, the maps in the online national atlas of a country provide direct links to that country's spatial data infrastructure: the atlas soil map links to the national soil mapping programme and detailed soil maps of the area centred on; climate maps link to the actual weather situation, a higher education participation map links to the relevant statistics for the clicked-on municipality; highway system maps link to the actual traffic density map as monitored by the relevant sensor system. Other sensor systems would be tapped by the atlas air pollution map, and the nation's relief map would be linked to a climate change scenario, forecasting the effect of sea-level changes on the land. In 2020, experiments were under way both in Spain and in Ukraine of linking a national atlas portal to the nation's SDI (spatial data infrastructure).

9.3.2 Electronic Atlas Functionality

Though the access to electronic atlases is of course limited to Mac or PC locations (with Internet for the online atlases), they provide a number of advantages over paper atlases worth noting:

- Customized maps can be produced on them. The atlas contained in the *Microsoft Encarta Premium 2009* encyclopaedia was a good example. In the special legend window, all object categories one wants to include in the map can be clicked on and an inset detail map will show immediately the effect of this decision. Names can be added at will for each object category (Figure 9.5).
- Geometrical information can be provided immediately: distances between points, lengths of routes and areas of regions, however defined, are provided upon request, as are the geographical coordinates of the cursor position.
- Whereas traditional atlases isolate (they only show a particular area, at a particular scale for a particular theme, pertaining to a specific date or period in time), electronic atlases have the power to shed these restrictions and to move beyond map frames by panning, zooming and presenting animations that show developments over time.
- Search engines. Highlighting or clicking a name in the register or the index will immediately provide a map showing the named object on the largest scale available. On the other hand, clicking an object on the map will provide its name: so there need not be a clutter of names on the map.
- Many electronic atlases are multimedia atlases as well. Icons on the maps will show the objects of which photographs, video clips, sound tracks or animations have been stored.
- Atlases may have visualization options in that some of them allow one to aggregate thematic data to larger regions or that allow one to change relative to absolute renderings of the

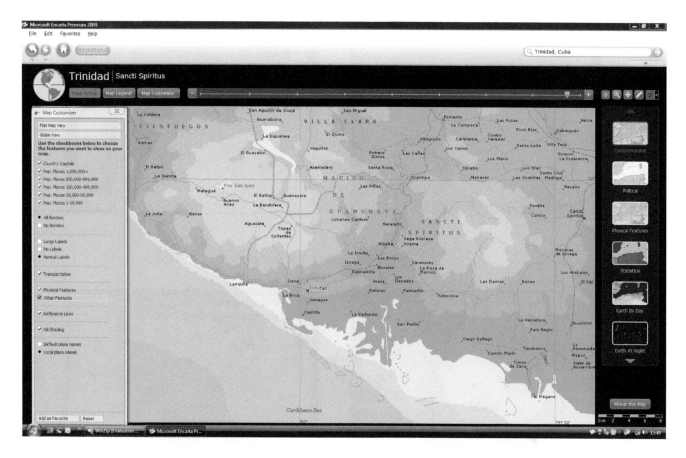

FIGURE 9.5   Legend category selection window, in the atlas in *Microsoft Encarta Premium* 2009 (Used with permission from Microsoft)

data sets, or vice versa, providing new views and possibly new insight into geographical phenomena.

The combination of database and graphical user interface (GUI) and other software functions developed to access the information in an atlas information system is different from a GIS: special care is taken to relate all data sets to each other, to allow them to be experienced as related, and to let them tell, in conjunction, a specific story or narrative. There will usually be a central theme (e.g. what has happened to the environment in the last 20 years, for instance, or the question whether all inhabitants of this country have equal access to its resources).

The final and perhaps most important contribution of web atlases is their ability to geographically order data, and by doing so function as user interfaces to information contained on the websites concerned. People searching for topical data on specific areas might access the national atlas website of a country, select the relevant topic and zoom in on the area required, with the result that the system provides them with all the files that answer the

geographical and topical requirements. These files could either be in the national atlas database or be in other databases to which the user can be linked. This power of managing spatial information is another unique aspect of maps.

Electronic atlases (and indeed all atlases) are only useful if their users have a clear idea of its overall possibilities and structures, how to access the information they want and the way to get back to the starting point. In order to realize this, they must have 'maps' of these electronic atlases at their disposal (Figure 9.6) and the electronic atlas should have a function showing where its users are on this map. Generally, we can say that the more interactive and integrated the electronic atlases are, the more complex it becomes to navigate through them.

Similarly, to the subdivision of paper atlases based on their objectives, a differentiation of electronical atlases can be devised. Paper national atlases were the first to be developed into electronic national atlas information systems and reference atlases followed. Nowadays, the emergence can be seen of earth sciences atlas information systems, physical planning atlas information systems,

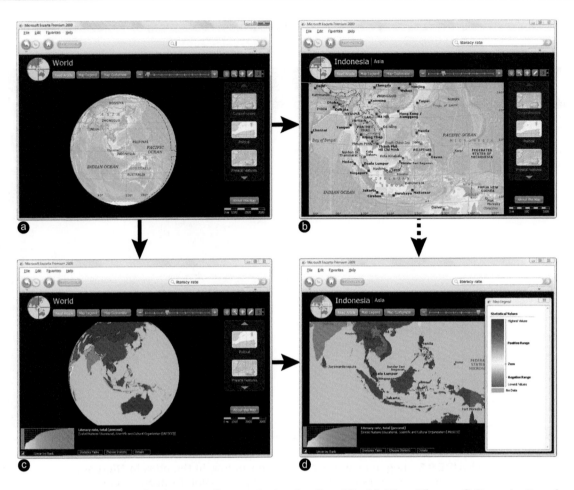

FIGURE 9.6 Proposed navigation map for an electronic atlas (*World Atlas Microsoft Encarta Premium*, 2009): one may zoom in from (a) the topographic globe to the (b) larger-scale topographic maps or move to the (c) thematic globe, and (d) zoom in there. Alternately, toggling between topographic and thematic maps of the same area is possible. (Used with permission from Microsoft)

socio-economic atlas information systems and historical atlas information systems.

A good example of the latter is the *Centennia Historical Atlas* (see Figure 9.7), an electronic atlas showing the changes in the political geography of Europe since AD 1000. The spatial and temporal (with a monthly increment) data on every boundary change have been stored in the file, and each of them can be visualized. The program can be put into a motion mode, which will result in an animated picture of historical developments, like Napoleon's incursions into Russia. Textual descriptions of the events are provided, of course, and it will be only a matter of time before other elements, such as imagery or even excerpts from historical motion pictures, are added as well.

The overall objectives of the various types of atlas information system can be translated in the relation of each data set contained in the relation of each data set it contains to the other data sets contained. These relationships are generally

elucidated through the use of metaphors and materialized in the form of specific structures and scenarios. These scenarios form the starting point for the design of GUIs.

The GUI for the school atlas of Sweden mentioned in Section 9.1 would have to provide functions for taking off and landing and for changing the weather conditions. The structure of the data would have to allow for access at all points, for providing information about all points or linear objects on the map overflown and for links with other media at specified locations.

The various electronic atlas types discerned will each have different scenarios and structures. For the strictly reference type, the scenario might be to simulate a complete impartiality, by not pre-programming any Eurocentric viewpoints or, in order to boost the user's interest, it might be directed at comparisons between different regions, and to quantify these comparisons. Socio-economic atlas information systems should have an inbuilt

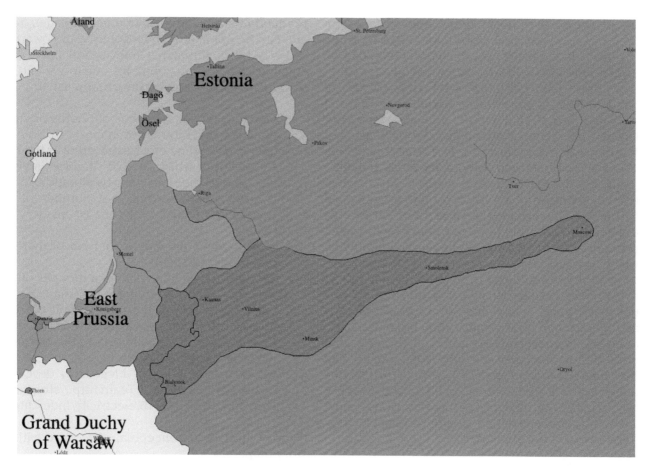

FIGURE 9.7 Napoleon's incursion into Russia (1812), derived from the *Centennia Historical Atlas* by Clockwork Mapping – http://historicalatlas.com (see also Figure 8.2)

capacity, for instance, to numerically compare data for other regions with that of the user's home region (or reference region), so that the user can find out in a standard way whether people in other areas are worse off or not. For physical planning atlas information systems, the scenarios should simulate the planning process, so that the user gets the feeling of participation (as provided in the *SimCity* computer games), instead of just being a bystander.

As the scenarios would allow for more or less user input, there are also differences in structures between these various types of atlas information system. National atlas information systems would focus on one country and only occasionally compare the situation in this one country with that in the wider world. The opposite is valid for reference atlases, which would strive to contain a specific level of detailed topographic/chorographic information for the whole area covered.

Web atlases are not the same as map machines. The latter are websites that allow the visitors to select any area they want to see and that consecutively provide maps (or satellite imagery) of that area. There is no particular objective involved in the provision of these maps or images other than reference, and that is a reason not to consider them as atlases. An example of a map machine is the National Geographic website (http://mapmaker.nationalgeographic.org). The best-known ones probably are Google Earth, Google Maps and Bing Maps.

9.4 MAP MACHINES

Several times in this book, Google Earth and Google Maps have been mentioned (Sections 1.5, 6.3, Figures 1.18, 1.19). These programs, just as well as their counterparts Bing Maps, Apple Maps and OpenStreetMap (OSM), have been described not as atlases but as intentional combinations of specially processed spatial data sets, together with the software to produce maps from them. They allow for detailed views of any area on Earth, either consisting of satellite imagery or consisting of maps, based on files for different areas that still might have different resolution and degrees of generalization, but that are constantly upgraded.

Important is that these maps and images can function as a background for data added by users (Figure 1.9). Via mash-ups, this information can be incorporated in the users' own websites.

Of these map machines, OSM, covered also in Section 6.6.2, is produced by volunteers and made available free of charge. The data are mainly collected from aerial photographs and GPS devices, but may also be supplied by government agencies. They are processed into topographic maps and route planners. The OSM initiative was started by volunteers in countries where the copyright dues charged by the National Mapping Agencies were so high as to be a serious barrier to private cartographic enterprise. In order to be a homogeneous global product, which does not change when passing a national border, OSMs are produced to a common legend and to standardized specifications, by local communities that engage on mapping campaigns. Nowadays, as topomaps are being made available online, free of charge in an increasing number of countries, the erstwhile need for free-of-charge topographic data becomes less urgent; on the other hand, the OSM initiative is now linked to other volunteer activities like the standards of the Open Geospatial Consortium, whereas commercial GIS programs usually have their own global coverages.

9.5 STORY MAP

The terms 'spatial narratives', 'visual storytelling', 'data-driven stories' and 'story maps' are not new; in the literature, they appear more and more frequent, and its elements were also used in the early day of multimedia and web cartography (see Section 5.9).

However, the present tools, especially ESRI's StoryMaps, have made this map type available for both expert and non-expert mapmakers.

The available templates together enable one to upload one's data or maps and present them in a coherent, professionally looking form. It forces one if there is a specific point to make regarding geographical or spatial phenomena, or a space-related message to get across, to analyse who the actors are, what is happening to them, and then devise how best to get the narrative across. The medium can be selected ('Do I want to use static or dynamic interactive maps?', 'Do I want to add multimedia experiences to the maps?', 'Do I want to tell the story only from my perspective, or can somebody else's perspective visualized as well?', 'Do I want to add sound effects or not?' – all questions that should be answered before the actual production starts). Your data (maps, images, texts, etc) can be uploaded, and – contrary to PowerPoint slides – these templates (blocks) all have different interactive functionalities such as (https://storymaps-classic.arcgis.com/en/app-list/):

→ Map tour (linear sequence of photos or videos linked to one interactive map);
→ Swipe (a timeline can be moved over the map, to the left of which the old situation is visualized and to the right the new situation);
→ Journal (data inserted in a number of sections, each with its own maps, imagery and texts);
→ Series (which presents a number of maps via tabs under headers);
→ Shortlist (users can click on a number of tabs and find their location on the map, or vice versa, i.e. click on the map to find out what kind of attractions are located there);
→ Cascade (the map changes, as the accompanying text is scrolled down), etc. An example would be the *Atlas of Electricity* developed by the ESRI StoryMaps Team http://storymaps.esri.com/stories/2016/electricity/index.html.

Figure 9.8 gives a conceptual view on creating a story map.

9.6 ATLASES AT WORK: MAP-USE FUNCTIONS

In the past chapters, examples have been provided that illustrate what maps can do in analysing and communicating geospatial data. By providing them, readers will have become familiar with many aspects of, e.g., the Maastricht municipality in the Netherlands. Without these maps, it would have been difficult to decide on answers to the questions or problems stated, or to decide upon a course of research, or to understand the spatial impact of environmental factors. Maps help one in deciding what to analyse, and later on, they support one in formulating decisions in issues with a spatial impact, and in communicating these decisions. In this section, the various functions of maps will be highlighted, which are as follows.

9.6.1 Explaining Patterns

By juxtaposing the map of a specific pattern with other maps of the same area, with different map themes, correlations might be found, and these

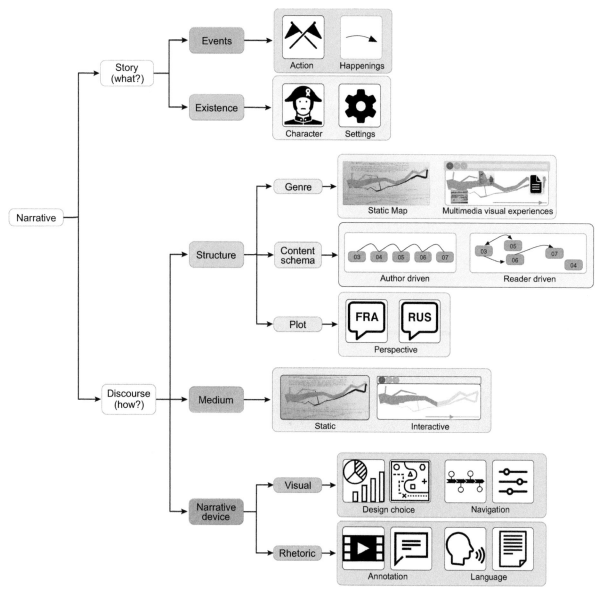

FIGURE 9.8 Creating a story map. Choices to be made before the actual incorporation of your data (maps, texts, photos) into story map blocks is effectuated. On the basis of this analysis, fitting story map blocks are selected to which your data can be downloaded. This particular scheme refers to the preparation of a story map presentation of Napoleon's 1812 incursion into Russia. For an example of such story map on the same topic, see https://1812.tass.ru/en

might help to find causal relationships. The relief map of the Kilimanjaro region in Figure 5.26 has been taken from the Digital Chart of the World (DCW). When this area is printed out with its highway infrastructure and settlements (Figure 9.9a), a discrepancy between the south-eastern and north-western parts of this mountain area emerges. This discrepancy can be explained by showing the extent of agricultural land, as, indeed, it is only on the south-eastern side of the mountain that an extensive area is used as arable land (Figure 9.9b). The extent of the arable land area can be explained, theoretically, by local soil patterns, slopes, heights and precipitation. As the first three factors are similar all around the mountain, it must be due to the influence of differences in precipitation, and indeed, when one compares these two maps with a precipitation map (Figure 9.9c), a strong correlation between the infrastructure and agricultural patterns and the rainfall pattern emerges.

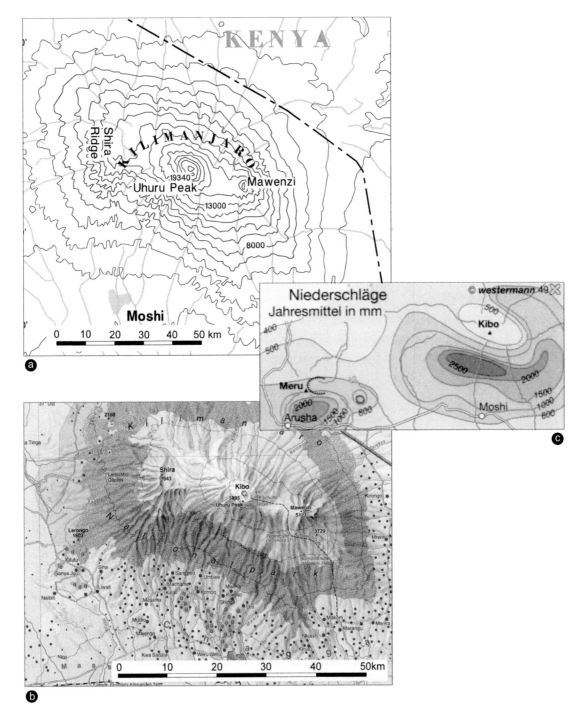

FIGURE 9.9 Understanding the Kilimanjaro region: (a) general topography (derived from the DCW; (b) land use map; (c) precipitation map (both courtesy Georg Westermann Verlag, Druckerei, and kartographische Anstalt GmbH & Co. KG, Diercke Weltatlas, 1989)

9.6.2 Comparison and Analysis

In similar ways, other patterns presented before might be explained: the differences in traffic accidents as shown in Figure 4.6 between the municipalities of Middelstum and Hoogezand-Sappemeer can be explained by juxtaposing them with the number of vehicles maps, population distribution maps (Figure 9.10a), road maps (Figure 9.10b), length of road maps or vehicle per km of roads per municipality maps of the same area. If all other conditions were similar, municipalities with more inhabitants would be likely to have more

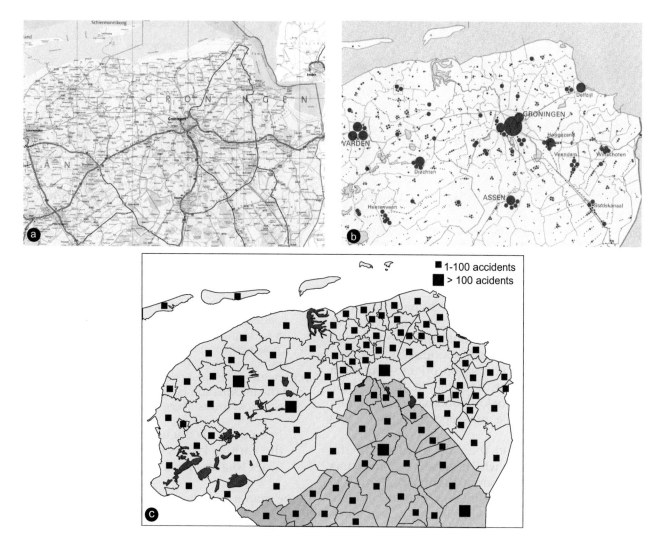

FIGURE 9.10 Northern Netherlands: (a) road network (derived from ANWB road map 2011); (b) population distribution (derived from National Atlas of the Netherlands, 2nd edition, volume I); (c) number of road accidents

traffic accidents; similarly, municipalities with motorway entries and exits and a larger overall road length, more cars or more cars per standard length of road would be liable to have more accidents. It would depend a bit on the local road safety conditions (as the Netherlands is known for its multitudes of undisciplined cyclists, separate bicycle paths would make a lot of difference), differences in alcohol consumption and general attitudes.

Differences in population numbers alone can never explain differences in accident numbers completely: Figure 2.13 shows that with only about 73 km², Barrow district has a population density of nearly 1000 persons per km² and is therefore pretty much urbanized. This would bring a higher chance of public transport, shorter distances to shops and malls, and therefore a lower percentage of car owners than in the other districts of Cumbria.

As soon as such hypotheses have been formulated, proof must be found, and the juxtaposition of maps is a prime way of ascertaining the factors that probably will explain the differences, and can be tested for their relevance for the issue statistically. Of course, there should be the facility to juxtapose the maps, and this calls for the possibility of splitting the monitor screen and showing at least two maps simultaneously.

9.6.3 Analysis and Decision-Making

Figure 9.11 shows the centre of Maastricht, with an overprint of newspaper distribution areas (newspapers delivered to subscribers in the Netherlands) and of the number of subscribers per area. When these numbers have been compared with the actual numbers of households per distribution area, it will

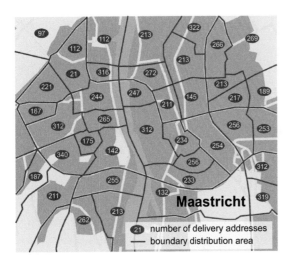

FIGURE 9.11 Newspaper sales in Maastricht

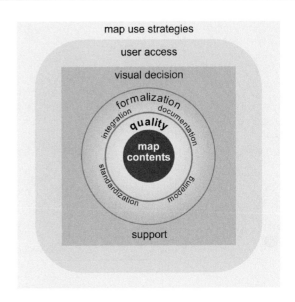

FIGURE 9.12 Visual decision support for spatiotemporal data handling. Keywords in the GIS cartography approach are map-use strategies (how people make decisions based on maps), public access (how people can work with the information), visual decision support (what the quality of the information is) and formalization (building expert systems) (based on Kraak et al., 1995)

be possible to assess the degree of penetration of the newspaper. Those areas with a relatively low percentage of subscribers could then be targeted by advertisements or special campaigns or offers for trial subscriptions at bargain prices – but such campaigns will only be sensible in those areas where the average socio-economic level would be compatible with the message brought by this newspaper.

The next step would then be to highlight the distribution areas for which positive action has been decided, where bargain subscriptions will be offered to the non-subscribers, and to circulate these maps to those working in the newspaper's marketing department.

9.6.4 Conditions for Proper Use of the Maps

When one wants to profit from this power of maps, certain conditions have to be met:

➤ One should be familiar with suitable map-use strategies. This can be defined as a conscious sequence of specific map-use tasks;

➤ One should have access to the relevant data sets;

➤ Preferably, meta-information on data quality should be available to assist in the decision-making process;

➤ It should be possible to integrate the various data sets, if necessary by modelling them (for instance, generalization); this integration should have been made possible through standardization of exchange formats and geometrical frames of reference. Both the relevant data set (2) and the relevant meta-information (3) can only be found if they have

been properly documented. Figure 9.12 shows the various issues one has to contend with in order to allow for a sensible bout of map use.

These map-use tasks consist of identifying the relationships between the mapped objects and their locations (after having identified the mapped area and the mapped topic, and the way the topic has been encoded in the legend). For this identification, different map-use tasks can be discerned, such as locate, identify, verify, establish (order, patterns), describe, detect, structure and, in general, obtain insight (Anson and Ormeling, 1995, Elzakker 2004). In order to execute these map-use tasks, specific map-use activities are necessary.

In the case (in Chapter 1) of the high-speed train in the Netherlands, the map-use task of identifying the optimal route consisted of the following map-use activities: *select* maps of suitable routes and restrictions, *identify* and *delineate* restricted areas, *compare* and *add up* or *score* restricted areas, *determine* or *identify* the route with the lowest score and *decide* on the least noxious route (map-use activities have been indicated in italics).

In the case of solving the conduit or circuit problem in the utility map example of Chapter 4 (Figure 4.4), the map-use tasks consist of the following activities: *monitor* the map, *detect* the problem, *decide* on direct action (initiate measures to remove or solve the problem), *determine* the consequences (*decide* on the nature of the effects and *delineate* the area being affected), *decide* on the course to mitigate or offset these effects (reroute traffic, for example), *determine* other utilities that might be affected, *inventorize* equipment in damaged section, etc.

So map-use strategies can be subdivided into the numbers of individual tasks, during which links or relationships are identified, data are combined, anomalies are signalled and alternatives are identified; these tasks in turn are solved by performing a number of map-use activities, elements are counted or their numbers estimated, sizes are compared, etc. In order to arrive at a specific answer, these individual tasks and activities are mostly performed in a specific sequence. It will vary from task to task. A map-use strategy sanctioned by experience is that of navigation: a typical map navigation problem would be solved with the following tasks: *search* and *locate* one's position on the map (provided the GPS contained in your device has not already done so); *orient* the map; *search*, *identify* and *locate* one's destination on the map; *determine* options for alternative routes; *select* one of the options; *set* a course; *determine* landmarks by which the course can be identified; *follow* the course on the map; *check* landmarks; *verify* the destination; and – if one wants the possibility to retrace one's steps – *store* the browser history.

These map-use tasks are perhaps a bit at odds with the traditional subdivisions of map use into map reading, map interpretation and map analysis. In map reading, the relations between the geographical features and the way they are rendered on the map are established, the map analysis activities consist of establishing the relationships between the various geographical features or phenomena that have been recognized in the previous step, and during map interpretation, logical conclusions are derived from those relationships.

9.7 WORKING WITH (WEB-BASED) ELECTRONIC ATLASES

In an electronic atlas environment, these map-use tasks and activities would be similar although at present, one is still restricted by the fact that two or more maps cannot yet be visually compared on one monitor screen in most commercially available products. A new element of map use in electronic atlases is the aspect of navigation through the atlas: this gives a new meaning to the '*select* map'

activity as the potential map titles will usually be provided in a thematic index. Other new map-use activities are the *clicking* of map objects in order to query them for their attributes and have these presented alphanumerically (such as name of the object, capacity and size).

The *aggregate* option in French electronic atlases refers to the possibility to view the thematic data at other levels of enumeration areas: instead of at the municipality level, these data might also be viewed at a district or canton level, a département level or that of an economic region (a grouping of a number of départements). Each new level of presentation would allow one to discover new geospatial patterns (see, e.g., Figure 3.4, where the high voting percentages for labour reflect the vicinity of factories or the location of housing blocks developed by socialist building societies at the voting ward level within a city). At a municipality level, the high percentage of labour votes would indicate urban rather than rural environments, while at a provincial level, religious denomination patterns would still govern the image.

With an '*add* layer' activity, names, e.g., can be added to the map studied, or a more detailed topographic background map. A *highlight* action would lead to the isolation of a specific category or class of objects from its surrounding values or categories, which will be left white or grey. This functionality is, e.g., contained in *Centennia* (see Section 9.3). In Figure 9.13, the Habsburg empire's many possessions in south-western Germany would pass unnoticed but for this highlighting function, which allows for their presentation in a customized colour, sharply contrasting with a white or grey background.

Other new map-use activities are *panning* and *zooming*, and these refer to changing our window on the world provided by the monitor screen. The fixed, isolating frames of paper maps are exchanged here for flexible boundaries, to be adjusted at will. Though *compare* is a frequently used map-use activity in a paper map environment, few electronic atlases allow for it, the Web atlas of Sweden (http://www.sna.se/e_index.html) and the Dutch National Public Health and Health Care portal (https://www.volksgezondheidenzorg.info/onderwerp/atlas-vzinfo/alle-kaarten for a list of all maps) being positive exceptions.

The multimedia environment of current electronic atlases brings with it frequent use of the *hop* command: hop to another presentation mode wherever icons or function buttons indicate that possibility. *Browse* is the sort of map-use activity hardly ever acknowledged, though it is still necessary to

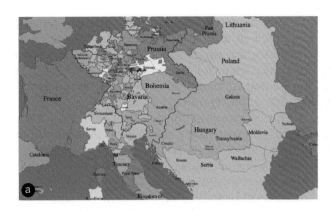 

FIGURE 9.13 Western Europe in 1789: (a) political map; (b) highlighting the Habsburg possessions (derived from the *Centennia Historical Atlas* by Clockwork Mapping – http://historicalatlas.com)

provide us with ideas. *Mark* or *annotate* allows one the possibility, as in *highlight*, to simulate the marking with colour pencils/crayons points or areas on the map. The *search* activity is usually directed at the electronic index, and this action is matched with the presentation, on the largest scale available, of a map of the object that goes under this name. By the *save, download, copy* and *export* map-use activities, the images in our files are transferred to other data carriers. *Print* will do the same in a visible way. The *time* and *coordinates* options are used to ask for the (local) time and geographical coordinates of the point at which the cursor is directed.

Toggle will let the map user alternate between different map types (such as the physical maps and the socio-economic maps of the same areas in post-war Austrian and German school atlases), between different areas or between maps showing the situation at different times. *Rank* is a map-use activity that will list a number of states or regions/cities on the basis of their score for specific topics (such as the number of golf courses and/or the length of the yearly period over 18°C). It can also be used for classifying data into quantiles (e.g. with the top 20% of the observations designated in the legend as 'highest', and the bottom 20% as 'lowest'). *Filter* is the map-use activity that will set minimum or maximum values to the data presented per enumeration area through proportional circles. Finally, *score* will result in the calculation operations undertaken on the data retained. And if one is interested in the maps visited and their sequence, the *browse history* option could be activated in order to reconstruct one's path through the atlas.

Regarding the map-use strategies or map-use task sequences, these will typically be executed by a series of clicks in order to get to a specific area displayed at a given scale, asking for thematic options to be displayed on the monitor, studying them, having them printed and comparing them, in order to assess their similarities and relationships, and taking account of them in providing the answers to the problems stated at the outset.

An approach akin to working with paper atlases would be to browse through electronic atlas maps, marking those maps one would be interested in or think helpful in solving a problem, and then asking again for the maps marked out, in a specific sequence. In a multimedia environment, this would go together with hopping to tables, imagery or alphanumeric explanations, whenever deemed necessary or interesting.

FURTHER READING

Aditya, T. 2007. The National Atlas as a Metaphor for Improved Use of a National Geospatial Data Infrastrcuture. Enschede: ITC.

Cron, Juliane (2006) *Kriterienkatalog. Umsetzung von Funktionen interaktiver Atlanten.* Zürich: ETH

Elzakker, C.P.J.M. van (2004) *The use of maps in the exploration of geographic data.* Netherlands Geographical Studies no 326. Utrecht: Utrecht University .

Kimerling, A. J., A.R. Buckley, P. C. Muehrcke & J. O. Muehrcke (2009). *Map use: Reading and analysis.* Redlands, California: ESRI Press Academic.

Ormeling, Ferjan (2015) Map use and map reading. Chapter 2 in Ormeling, F.J. and B.Rystedt *The World of Maps.* ICA. Downloadable for free from https://icaci.org/publications/the-world-of-maps/the-world-of-maps-english/

Sieber, René and Voženílek, Vít (eds.): *The Atlas Cookbook – Ten steps towards a successful atlas edition.* https://atlas.icaci.org/awards-and-publications/atlas-cookbook/ Zürich 2020 (can be downloaded for free)

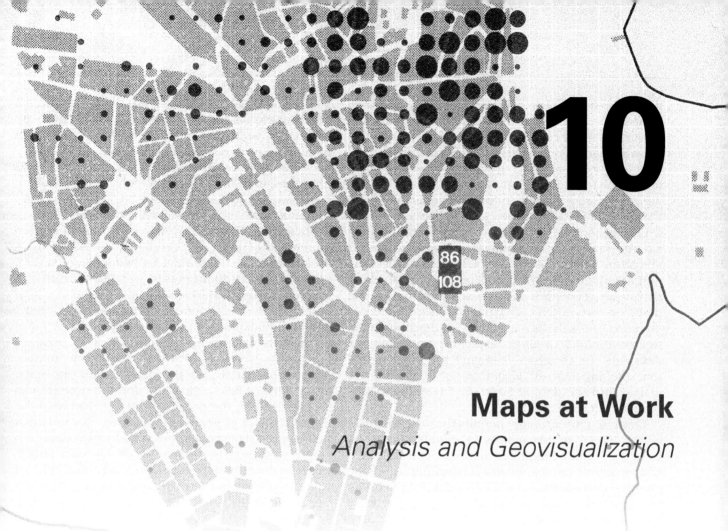

Maps at Work
Analysis and Geovisualization

The previous chapters have shown that the environment in which maps are produced and used is changing continuously. Although cartographers, geoscientists and computer scientists have witnessed highly dynamic and important developments in the fields of acquiring, managing, analysing, interacting with and visualizing large amounts of geospatial data since 1980, the basic map concepts as explained in Chapters 6–8 remain valid. Next to static maps, nowadays immersive and highly interactive virtual environments can be used to explore and present dynamic geospatial data.

Before the GIS era (the 1990s), paper maps and statistics were probably the most prominent tools for researchers to study their geospatial data. To work with those paper maps, analytical and map-use techniques were developed, which can still be found behind many GIS packages' commands. Today the same researchers have access to large and powerful sets of tools such as spreadsheets, databases and cloud computing and storage facilities, as well as graphic tools, to support their investigations of ever-increasing streams of data. There is also a clear need for this since the magnitude and complexity of the geospatial data sets available poses a challenge on how to transform the data into information and ultimately into knowledge. Comparing the on-screen approach with the traditional approach not only reveals a difference in processing effort and time (computers do in seconds what took weeks using manual methods), but now the user can interact with the map and the data behind it. This puts maps and graphics in a different perspective as they become interactive tools for exploring the nature of the geospatial data at hand. As stated in Section 1.1 (see Figures 1.1 and 1.7), the map should now be seen as an interface to geospatial data that can support productive information access and exploratory activities, while it retains its traditional role as a presentation device.

The above trend in cartography is strongly influenced by developments elsewhere, especially in scientific visualization, and has given the word 'visualization' an enhanced meaning. These developments have linked the word to more specific ways in which modern computer technology can facilitate the process of 'making visible' in real time to strengthen knowledge. Specific software toolboxes have been developed whose functionality is based on two keywords: interaction and dynamics. The relations between the fields of cartography and GIS on the one hand and scientific visualization on the other hand have been discussed in depth by Hearnshaw and Unwin (1994) and MacEachren and Taylor (1994) (see also Figure 3.1). Next to scientific visualization, which deals mainly with 3D medical imaging, process model visualization and molecular chemistry, another branch of visualization can be recognized, called 'information visualization' (Card et al., 1999; Ware, 2020; Spence, 2014), which focuses on the visualization of non-numerical information. In information, visualization maps are often used as a metaphor to access the non-numerical information. In this process, called 'spatialization', concepts such as scale, distance and direction as well as cartographic design concepts are used to create two- and three-dimensional map representations of often extensive and complex data sets. In all cases, it should be remembered that the purpose of visualization is to improve insight into the data at hand, for instance, to prepare for decision-making.

From a cartographic perspective, a synthesis of the above trends results in *geovisualization*. Geovisualization integrates approaches from scientific visualization, (exploratory) cartography, image analysis (a technique used in interpreting satellite imagery), information visualization, exploratory data analysis (EDA) and GIS to provide theory, methods and tools for visual exploration, analysis, synthesis and presentation of geospatial data (any data having geospatial referencing) (Dykes, MacEachren and Kraak, 2005a). In this context, it is required that cartographic design and research pay attention to human–computer interaction – the interfaces – and revive the attention for the usability of their products. In a geovisualization environment, maps are used to stimulate (visual) thinking about geospatial patterns, relationships and trends. This is realized by viewing geospatial data sets in a number of alternative ways, e.g. using multiple representations without constraints, the emphasis of this chapter. This is well described by Keller and Keller (1992), who in their approach to the visualization process suggest removing mental roadblocks and taking some distance from the discipline in order to reduce the effects of traditional constraints. Why not choose an alternative mapping method? For instance, show a video of the landscape next to a topographic map accompanied by a three-dimensional map. New, fresh, creative graphics could be the result; they might also offer different insights and would probably have more impact than traditional mapping methods.

Let us consider the functionality required in a geovisualization environment to explore the data at hand by creating alternative views of these data. It should allow the geoscientists to link different data sets together in any combination, at any scale, with the aim of seeing or finding geospatial patterns or relationships, and let them create maps and graphics for a single purpose and function as an expedient in the experts' attempt to solve a geoproblem. Here, two aspects are of importance: the view environment and its functionality and the graphic representation of the data. The view environment is often composed of a set of coordinated multiple views (Roberts, 2008). It means that one looks at several views (windows) that contain graphics of the same or related data, and which are linked together. If an object is selected in one window, the same object will be highlighted in all other windows. A basic example is derived from Figure 7.9 and shown in Figure 10.1. In each window, it shows one component of geospatial data: the location view (the map), the attribute view (a diagram) and the time view (a timeline). Here, the population development of the city of Maastricht per neighbourhood over time can be explored. This concept of visually exploring geospatial data was introduced by Monmonier (1989) when he described the term 'brushing'. Selection of values in the diagram will highlight the corresponding areas in the map. Conversely, one can also select areas in the map and see the corresponding elements highlighted in the diagram. Depending on the view in which one selects the object, one can call it 'geographical brushing' (clicking in the map), 'attribute brushing' (clicking in the diagram or table) and 'temporal brushing' (clicking on the timeline). Alternatively, other variables in other maps or diagrams, photos or videos could be shown on other linked views as well.

This requires functions such as (Figure 10.2):

→ *Basic display* (Figure 10.2a): Map displays need tools to allow the user to pan, zoom, scale, transform and rotate. These geometric tools should be available and independent of the dimensionality of the displayed geospatial data (see also Sections 5.1 and 5.2).

→ *Orientation and identification* (Figure 10.2b): This involves the keys to the map. At any time, the user should be able to know where the view is located and what the symbols mean. The orientation functions are particularly necessary for a 3D environment.

→ *Query data* (Figure 10.2c): During any phase of the visualization process, the user should have access to the spatial database to query the data. The questions should not necessarily be limited to simple 'What?', 'Where?' or 'When?' Access to the database is somehow

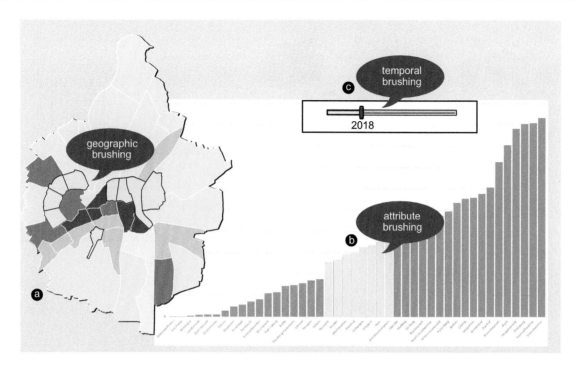

FIGURE 10.1 Coordinated multiple views: (a) the locational view (map); (b) attribute view (diagram); (c) time view (timeline) and the brushing technique. Since a link exists between the map and the diagram, selection of values in the diagram will highlight the corresponding areas in the map. Conversely, one can also select areas in the map and see the corresponding elements highlighted in the diagram

contradictory to the concepts of the digital landscape model (DLM: the 'database') and the digital cartographic model (DCM: the 'drawing code'), which clearly separates database and display. One example is animation, which, in most cases, consists of bitmaps derived from a database, without a link to it. In an exploratory environment, such a split is undesirable.

Since working in an exploratory environment often means dealing with a large amount of (partly unknown) data, one needs an exploration strategy. Shneiderman (1996) introduced his visual information-seeking mantra, which proved to be a very suitable approach. It consists of three major steps and is illustrated in Figure 10.3, which shows the trajectories of calving icebergs near Antarctica:

➡ *Overview first*: All relevant data is displayed. In Figure 10.3a, all observed positions of three icebergs during a particular time frame shown.

➡ *Zoom and filter*: One can zoom into a particular geographical area or moment in time. Filtering is done based on certain attribute values. In Figure 10.3b, the scientist has zoomed in on the south-east area and selected the 'red iceberg'.

➡ *Details on demand*: For certain objects, all information available is retrieved from the database. In Figure 10.3c, this is done for the red iceberg at one particular moment in time.

To stimulate visual thinking, an unorthodox approach to visualization was recommended earlier in this chapter. This requires options to manipulate data behind the map or to offer different visual representations to display the data. An example of data manipulation is the application of several classification systems (see Section 7.3), while the use of different advanced map types to view the same data represents display differences (see Figure 7.17). The principle is illustrated in Figure 10.4. It shows several alternative views on Minard's map of Napoleon's march on Moscow. The map above the 'original' in Figure 10.4b shows how this map is projected on a historical map from the mid-eighteen hundreds. The three maps on the left of the figure show different options to display the progress of the invasion. Figure 10.4b uses the coordinate multiple view approach to show the changes in the size of the army over time. This representation and the space–time cube (Figure 8.7) both reveal that Napoleon stayed for over a month in Moscow before he started the way back, something not clear from Minard's original map. Figure 10.4d

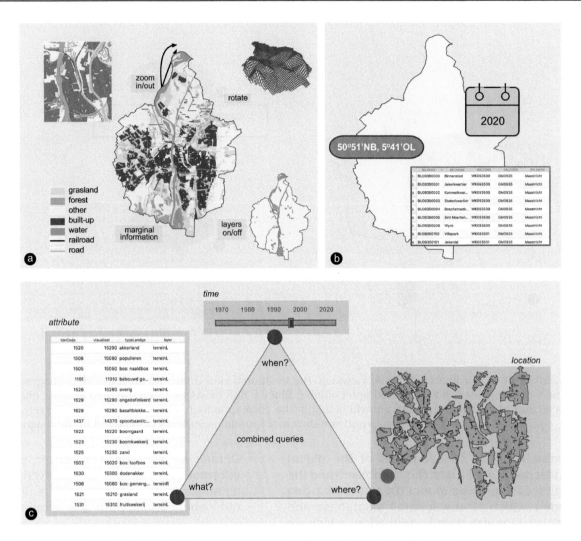

FIGURE 10.2 Basic functionality of an exploratory environment: (a) basic display with options to zoom, pan and rotate maps as well as switch layers on/off and display the legend, or use different visual representations; (b) navigation and orientation; (c) query, to ask elementary questions related to the what, where and when

gives a three-dimensional view of the size of the army. The height of the symbols represents, like those at the bottom of Figure 10.4b, the number of troops. Figure 10.4c gives different views on how the land 'behind' Napoleon might have been occupied by his troops. Figure 10.4e shows by smilies (from Napoleon's perspective) who won the battles. These could be seen as simplified Chernoff faces. These are used to visualize multivariate data. Faces are used because humans are able to perceive small differences in facial characteristics (Chernoff, 1973). For the variation in the face characteristics, one can use, among others, head eccentricity, eye eccentricity, eyebrow slope, nose size and mouth openness. Battle characteristics could be used as variables here. The objective of these alternative graphic representations is potentially to reveal patterns that would otherwise remain unnoticed.

However, from a map user's perception, the graphics or maps may look 'wrong' or unfamiliar.

Next to all these different maps and diagrams, all kinds of other graphic representations can be used. These could be scatter plots, line graphs or alternatives such as parallel coordinate plots (PCPs) or box plots, but also all kinds of multimedia expressions. In all situations, the graphics are linked to the map, something that is, for instance, relatively easily realized in a web environment. The PCPs should be mentioned separately. These popular diagrams have the ability to display *multidimensional* data in a single representation (Inselberg, 1985). Each geographical unit is represented by a line that intersects a number of vertical lines. Those vertical lines represent the variables. The variables are scaled (normalized) along the lines with the maximum value at the top and the

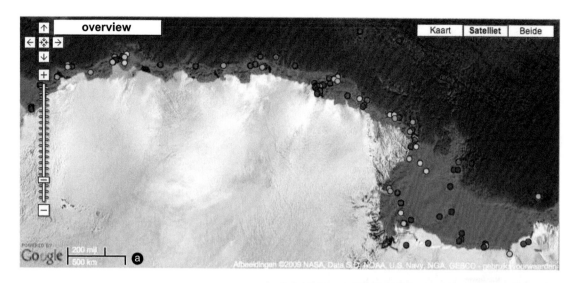

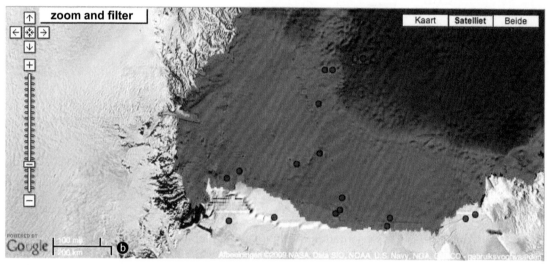

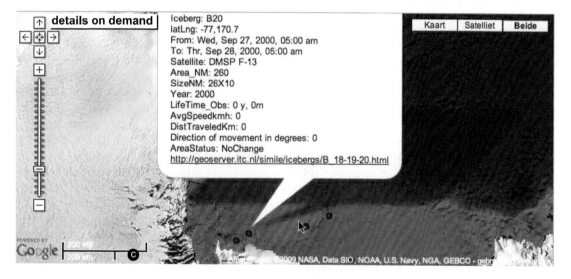

FIGURE 10.3 Shneiderman's visual information-seeking mantra illustrated: the movement of icebergs in the Antarctic region (a) overview; (b) zoom and/or filter; (c) details on demand (©2019 Google LLC, used with permission / data NASA / mashup: authors)

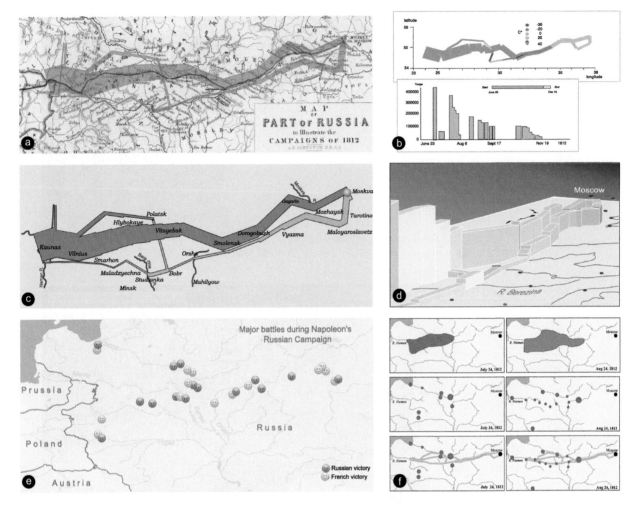

FIGURE 10.4 Alternative visual representation of Minard's 'Napoleon's march to Moscow' map: (a) Napoleon's path plotted on historical map; (b) path linked to diagram; (c) the path; (d) number of troops in 3D; (e) battle result from French perspective; (f) alternative small multiples to show campaign at two moments in time

minimum value at the bottom of the lines. The line of a particular geographical unit crosses a vertical line relative to the minimum and maximum values of the variable.

10.1 GEOVISUAL ANALYTICS

In the previous paragraph, it has been explained that a geovisualization environment offers interactive access to multiple alternative graphic representations that stimulate (visual) thinking about geospatial patterns, relationships and trends, and as such, it supports knowledge construction.

However, this is not sufficient to deal with the global challenges related to, for instance, climate change, health and energy. Something has to be done with the knowledge; e.g., visualization has to be combined with analytics to allow reasoning among the different disciplines involved. The National Visualization and Analytics Center

(NVAC) in the United States introduced the term 'visual analytics', which stands for 'the science of analytical reasoning facilitated by interactive visual interfaces' and originates from the research agenda: 'Illuminating the Path: The Research and Development Agenda for Visual Analytics' (Thomas and Cook, 2005), which has as motto 'detect the expected and discover the unexpected'. These interfaces could be the maps as described in a geovisualization context above and, when concentrating on the spatial aspect of the process, one can name it 'geovisual analytics'.

The geovisual analytics process allows one to find, assimilate and analyse continuously changing data about time-critical, evolving real-world situations. For instance, for coastal protection, one is interested in all kinds of parameters such as wind speed, direction and strength, water and wave heights as well as the current situation of the dykes protecting the land. The findings of the analysis have to be communicated to a range of interest groups, including keepers of sluices and river barriers, but also to shipping control and local authorities, who will have to take necessary action. It is obvious that it is a process where different experts need to work together. In other words, this requires reasoning techniques that enable one to gain insight into the situation, and apply available knowledge. Results, often hampered because data will be both incomplete (a weather station fails due to the storms) and uncertain (not enough active sensors for trustworthy interpolations in both space and time), have to be converted in appropriate information based on which decision can be made. This puts pressure on map design since one has to incorporate a message with a certain quality aiming at a particular kind of decision-makers involved.

Let us look at a simple example to illustrate the geovisual analytics process as such (see Figure 10.5). It is based on running data and includes locations and heart rate values. From the first variable, others such as speed and pace are derived. One has to realize that the accuracy of the measurements is reasonable but the device is 'sensitive' to noise, which should be filtered out before analysis. Figure 10.5a displays a map and diagrams for both speed and heart rate data for a short 10 km run. It can be observed that the heart rate values

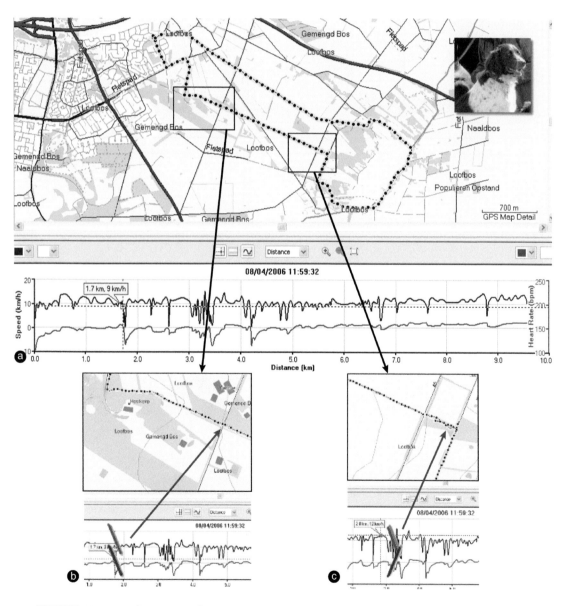

FIGURE 10.5. A simple geovisual analytics example: (a) the representation of a run in a map and graph (location, attribute and time); (b) detail with normal pattern (speed down, heart rate down); (c) detail of strange pattern (speed down, heart rate up)

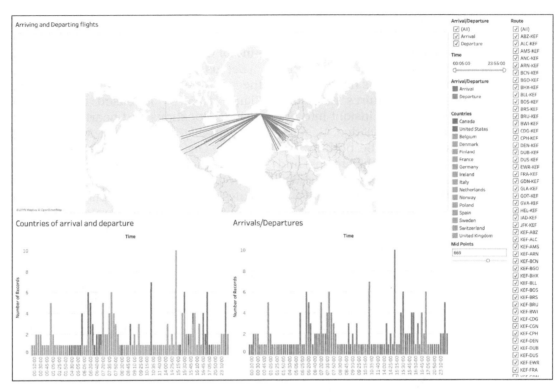

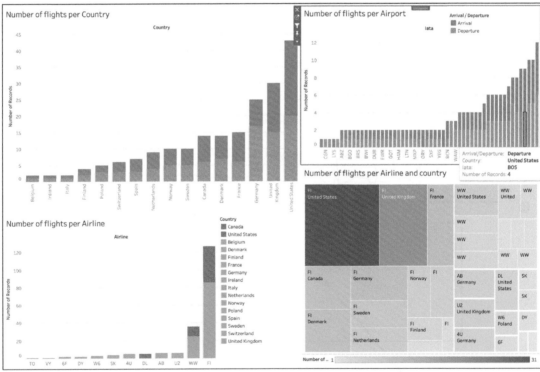

FIGURE 10.6. Example of an analytical dashboard using the Tableau software, with statistics of daily flights to and from Keflavik airport (Iceland) in multiple graphic representations

(the lower line in the graph) 'follow' the speed; e.g., running faster will soon result in a higher heart rate value. This is illustrated in Figure 10.5b where two trend lines are plotted at a point where the speed is reduced. These downward peaks can be recognized at several places in the overall graph. A question that will arise is: 'Is this a runner in bad shape who has to stop every so many metres, or is something else happening?' Without specific knowledge of the capabilities of this runner, the linked map provides the answer. The locations of the slowdown events seem to happen at crossings (Figure 10.5b) where the runner obviously watches out for traffic before crossing. This seems to be a plausible reasoning.

While studying the graph in more detail with the above in mind, some anomalies can be observed. Around 3.3 km, the graph shows a high density in changes in both speed and heart rate (Figure 10.5c). But if trend lines are plotted, it can be seen that while the speed goes down, the heart rate increases. This is contradictory compared to the earlier established trends. What goes on? Can the map assist? The map reveals no crossing around the 3.3 km point, and studying the track in more detail shows a forward and backward pattern along the road. With just common sense, it is not possible to find an explanation.

More information, not available in the data collected, like particular habits of this runner, is required. This is an example of the wide scope of geovisual analytics: one is often required to deal with incomplete data and the geo-expert often has to discuss/reason with other experts. In this particular example, the runner is accompanied by his dog. This might explain why at every crossing, the runner slows down, but does not explain the above contradictory pattern. However, if we know the dog is a hunting dog, and is running off-leash, and that at the location of the anomaly it observed and followed a rabbit, one will realize something different is going on. The runner slowed down, but not his heart rate because he was yelling at his dog to follow him instead of chasing a rabbit!

Another approach to data exploration is by using a dashboard. Few (2006) defined a dashboard as 'a visual display of the most important information needed to achieve one or more objectives; consolidated and arranged on a single screen so the information can be monitored at a glance'. One distinguishes between operational, strategic and analytical dashboards. The first is simple in design and should attract attention in the case of an emergency. It depends on the real-time data. The second provides an overview of the parameters decision-makers need to develop their strategies. It is based on existing data, and like the first, interactivity is not required. The third holds visual information on complex data, and interactivity is required to allow the identification of trends, patterns and anomalies in the data to get insight. An example for such a dashboard is given in Figure 10.6 and contains different views of the arrivals and departures on a single day at Iceland's Keflavik airport. It has been created using the tool Tableau. Compare the map with Figure 8.4. The dashboard has the default map in a Mercator projection, which is less appropriate here (see also discussion in Section 6.2).

FURTHER READING

Andrienko, N., G. Andrienko, and P. Gatalsky. 2006. *Exploratory Analysis of Spatial and Temporal Data - A Systematic Approach*. Berlin: Springer Verlag.

Card, S. K., J. D. Mackinlay, and B. Shneiderman. 1999. *Readings in Information Visualization: Using Vision to Think*. San Francisco, CA: Morgan Kaufmann.

Dykes, J., A. M. MacEachren, and M. J. Kraak, eds. 2005. *Exploring Geovisualization*. Amsterdam: Elsevier.

Few, S. 2006. *Information Dashboard Design: The Effective Visual Display of Data*. Sebastopol, CA: O'Reilly Media.

Hearnshaw, H. M., and D. J. Unwin, eds. 1994. *Visualization in Geographical Information Systems*. London: John Wiley & Sons.

Keller, P. R., and M. M. Keller. 1992. *Visual Cues, Practical Data Visualization*. Piscataway, NJ: IEEE Press.

MacEachren, A. M., and D. R. F. Taylor, eds. 1994. *Visualization in Modern Cartography*. London: Pergamon Press.

Shneiderman, B. 1996. "The eyes have it: A task by data type taxonomy for information visualization." *Proceedings IEEE Symposium on Visual Languages*, Boulder, Colorado.

Spence, R. 2014. *Information Visualization*. 3rd ed. Berlin: Springer.

Thomas, J. J., and C. A. Cook, eds. 2005. *Illuminating the Path: The Research and Development Agenda for Visual Analytics*. Washington, DC: IEEE Press.

Ware, C. 2020. *Information Visualization: Perception for Design*. 4th ed. Burlington, MA: Morgan Kaufmann Publishers.

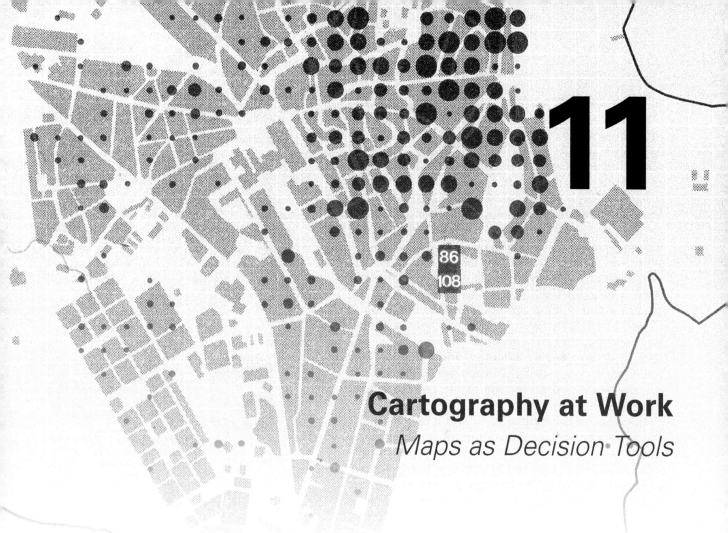

11

Cartography at Work
Maps as Decision Tools

11.1 AGAIN: WHY MAPS?

In Chapters 1-10, examples have been provided that illustrate what maps can do in analysing and communicating geospatial data. By providing them, readers will have become familiar with many aspects of, e.g., Maastricht municipality in the Netherlands or the Lake District in Britain. Without these maps, it would have been difficult to decide on answers to the questions or problems stated, or to decide upon a course of research, or to understand the spatial impact of environmental factors. Maps help one in deciding what to analyse, and later on, they support one in formulating decisions in issues with a spatial impact and in communicating these decisions. Maps can help in explaining patterns (see Section 9.4.1), in comparing them (see Section 9.4.2), in analysing them (Section 9.4.3 and Chapter 10), in using them as an interface with databases (see Section 9.4.4) and in letting them act as a stimulus for visual thinking (Chapter 10). But in order to do so, maps have to answer a number of requirements, relating to their quality (see Section 1.3), which includes the aspect whether the data are up to date, whether they can be reformatted so that they can be exchanged according to widely accepted

standards (see Section 1.5) or whether information on the data files is well documented. That caution is needed here as demonstrated in Section 11.3, with the example of the DCW (Digital Chart of the World). But even if the data are up to date, they are not always accessible, and that is where nowadays spatial information policies come in (see Section 11.4). And even if the information can be made available everywhere, anytime, it might come at a price one cannot afford. That is why copyright is such an important aspect of dealing with spatial data (Section 11.5). Finally, if all conditions for technical and organizational exchange of data have been met, there still remains the usability aspect: can the data actually be applied for one's objectives (Section 11.6)? In Section 11.7 using the example of Minard, a synthesis of these issues is presented.

11.2 MANAGEMENT AND DOCUMENTATION OF SPATIAL INFORMATION

Geospatial data that have been generated must be stored in such a way that they can be retrieved and used easily. For the storage of geodata in paper form – on a map – a number of auxiliary comments in the map margin have been developed that together are termed 'marginal information'. These comprise not only the title and legend of the map and other items that are immediately necessary for the map's use but also items ultimately aimed at enabling the user to find the map in a catalogue. The map user needs the title in order to get an idea of the area and theme depicted (external identification), and the legend in order to see what characteristics or attributes of the mapped objects are rendered (internal identification). But she will also need information that will enable her to judge the quality of the geodata, and their fitness for use, in order to evaluate the conclusions she will be able to draw from the map. In Section 2.5, the various aspects of data quality have already been mentioned. The date at which the data have been collected and the way they have been processed are important aspects. So are the scale and projection of the map (together called 'mathematical data').

In order for the map user to find a specific map, the producer has to enter items of information in the map margin, which will be found by a documentalist or map curator and be entered in catalogues or inventories from which the map can be retrieved. Apart from the map title, the bibliographical data consist of the map's imprint statement (mention of the publication place, publisher and year of publication, in this sequence), an annotation (other relevant notes regarding the sources used, authors or institutions responsible for the production) and the collation (a reference to the map's dimensions).

For data files, the same map-use and bibliographic information is necessary as for maps. The nature of the data will be different (instead of scales, spatial resolution will be mentioned). The term used to indicate the marginal information in a digital environment is metadata (see Section 1.5). Figure 11.1 shows the location of marginal information on a paper map and the additional metadata that would be combined with a geodata file.

The metadata with which spatial data are documented consists of information that identifies the data set, that indicates its quality (fitness for use), that shows how the data files are organized, what spatial reference system is used, what kind of attribute information is contained, how the database can be accessed (for instance, by showing an URL by which it can be accessed on the Internet) and how the metadata itself is referenced and organized. Information on those that provided the metadata, and on the time period the data are valid for and contact information on those that collected the data are part of the documentation as well.

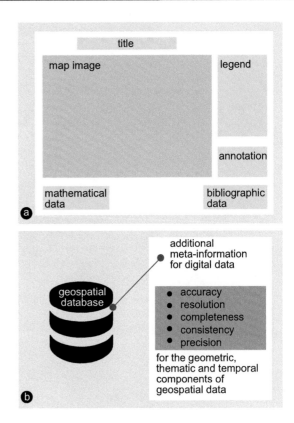

FIGURE 11.1 Marginal information (a) and meta-information (b)

11.2.1 Retrieving Geodata

A typical question to a map curator would be: 'Is there in your collection a geological map on a scale between 1:50 000 and 1:250 000 of the Cape Leeuwin area in Western Australia produced after 1960?' If the maps were described on index cards, the curator would have to manually check all index cards on Australia from 1960 onwards (in a temporally ordered index-card catalogue) or check all the maps on Western Australia (in a geographically ordered index-card catalogue). If the catalogue was automated, the curator would just have to set all the criteria (area, time, topic, scale), and the system would check automatically whether a map answering the requirements was available in the collection.

There are international rules for automated map descriptions, the ISBD (International Standard Bibliographic Description) Consolidated Edition rules. These rules define the structure of bibliographic records (with specific fields for title, author, edition, imprint statement, etc) and provide a framework for the manner and sequence in which the various bibliographic elements should be entered. They are now being extended to enable coverage and description of digital geodata files. This is a prime and necessary step in order to be able to organize clearing houses for geodata. The descriptive elements themselves are modelled on the DCRM (C) standard (Descriptive Cataloguing of Rare Materials – Cartographic, 2016). Originally based on AACR2 (Anglo-American Cataloguing Rules, 2nd edition), DCRM (C) will be gradually adopted to comply with the successor to AACR2, the internationally adopted RDA (Resource Description and Access) standard.

An elementary form of a geodata clearing house (see Section 1.6) can be described as a service hatch through which a user can find out digitally whether any geodata files of a specific area, answering specific requirements (such as topic, collection date, resolution) exist, and where these can be obtained on what conditions. In a more progressive form, it would be the central counter where not only the metadata but also the data themselves could be procured digitally. Such a clearing house or data warehouse is a part of the National Spatial Data Infrastructure (NSDI). Such an infrastructure would be economical, would prevent the overlap of geodata collection endeavours, would inform users of the available geo-data sets and thus would lead to a better planning of the available resources. In an international context, a number of organizations are working to promote the GSDI (Global Spatial Data Infrastructure) initiative, by implementing standards and other technical measures that would enable and speed up the exchange and integration of geodata. They are now represented by the GGIM (Global Geospatial Information Management), a group of experts that on behalf of the United Nations support the Global Spatial Data Infrastructure.

A number of the institutions that are busy incorporating the spatial data sets they collect and maintain into these national clearing houses are also making them available – whether for free or otherwise – on the Internet. The spatial information available on the Web is increasing at a very strong pace, and it is impossible to keep track of it unless helped by geoportals – websites specializing in conveying their visitors to the spatial information required.

11.3 OUTDATED DATA: AT WORK WITH THE DIGITAL CHART OF THE WORLD

There are a large variety of GIS data sets available today. For small-scale applications, the Digital Chart of the World (DCW) at scale 1:1 million used to be one of the more popular data sources. Many vendors have their own version (= format) of the DCW (see ESRI's Living Atlas – Figure 6.45), which was originally published by the Defense Mapping Agency in 1992. How useful are these data in GIS and cartographic applications? This section will illustrate the potential use of DCW data for two simple spatial analysis operations. If no other data are available, such data sets still prove to be a valuable data source. However, some care is needed. Especially, the age of the data can cause problems. Although the CD-ROMs on which the DCW was distributed bear the 1992 imprint, most of the data themselves were even older than 1970. This is especially the case for those areas where the DCW is likely to be the only (digital) source available. Today, DCW type of data is often replaced by OSM (OpenStreetMap) data, which partly suffers the same problem although the time intervals are much smaller. Moreover, height information is lacking in OSM data.

11.3.1 Case 1: The Netherlands' Railroads

How suitable is the DCW when one intends to use its data for a network analysis? The railroad layer is one of the many data layers available in the DCW. Data for this layer were digitized from existing maps 1:1 million. For the Netherlands, ONC (Operational Navigation Chart) E-2, last revised in 1985, was used. The map image contains two line symbols for railroads, one for single and one for double tracks (Figure 11.2a). Those tracks visible on the map have been digitized. Because of the high symbol density on the map, it is sometimes difficult to establish if one is dealing with single or double track. From the map, it can also be seen that the tracks stop at the edge of built-up areas. To provide a 'connected' network, the gaps in the urban areas have been closed. For several important cities, this was done by using other data sources (see Leeuwarden – Figure 11.2b).

Next to these variable rail types, the layer also contains information on the variable status of the rail lines, i.e. information on the nature of their use. Were they functioning or abandoned, or was their functioning doubtful? Were they added to fill gaps in the network? Actually, all railroads north of the east–west line on the map were abandoned long before 1985.

From this experiment with DCW data in a well-mapped area, it can be concluded that although metadata are available, one should be careful in using the data. Visualizing the data itself as well as its meta-information can help the user decide if it is fit for use, and how much work has to be done to edit the data. It should also be established as to what purpose was in mind when the original map was compiled (e.g. navigation charts for fighter pilots).

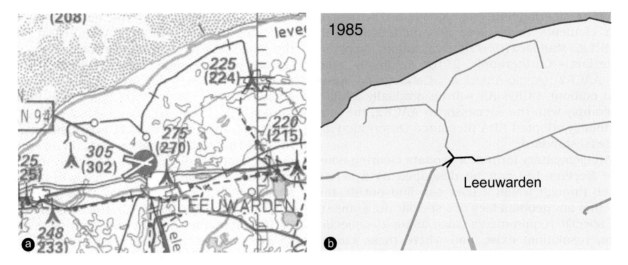

FIGURE 11.2 Netherlands' railroads: (a) part of the northern Netherlands from the ONC sheet E-2 1985; (b) the same area from the DCW's railroad layer. (https://commons.wikimedia.org/wiki/Category:Operational_Navigation_Chart)

11.3.2 Case 2: East African Highlands

For some applications, the DCW data are not available. Surface analysis is such an example. Relief data for the Earth's more remote areas could not be retrieved from the original aeronautical charts. However, as can be seen on the left in Figure 11.3, the paper map provides an overprint warning that the heights are believed not to exceed a certain number of feet. In the DCW database, these areas come without contour lines or height points.

11.4 ACCESSIBILITY: CARTOGRAPHY, GIS AND SPATIAL INFORMATION POLICY

Geographical information, just as geodetic data, does not just happen. It has to be collected, usually at heavy costs to the taxpayer. Sometimes it will be made available to the taxpayers again, at a price. Society decides about the data categories that will be needed for government, and should be collected therefore. Governments need topographical maps for defence, census maps for implementing

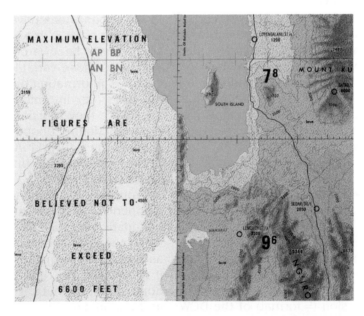

FIGURE 11.3 East African Highlands: detail from ONC sheet M-5 1981 at left a gap in the contour layer (https://commons.wikimedia.org/wiki/Category:Operational_Navigation_Chart)

socio-economic policies, resource mapping for physical planning, highway mapping for maintaining that crucial infrastructure and environmental mapping for offsetting the effects of physical planning and highway building.

Nowadays, it is considered wasteful that all the government agencies that engage in collecting geospatial data, processing them, putting them in databases and producing maps from these do so only on their own behalf and do not consult each other. The trends towards database exchange could be offset by uncontrolled non-standardized construction of databases. As this would force society to pay enormous extra sums because of duplicating existing work, the strategy has been chosen in a large number of countries to formulate geospatial information policies. In the United States, the US Federal Geographical Data Committee (FGDC) and in Europe the INSPIRE initiative have defined such policies (see also Section 1.5). Implementing these geospatial information policies would allow for the integration of different databases, developed and kept up to date by different government agencies that could all be used for solving specific problems. By preparing maps only from databases kept up to date, a number of problems of paper maps could be offset: the maps would not age, show only a limited amount of detail, would not bear street names when not required or would not be intersected by a map frame.

If such a national spatial data infrastructure, linked to global standards, were implemented globally, conditions for the use of maps as tools for analysis in GISs and for communication would be boosted even more, as it would entail continuously available up-to-date geospatial information. In the Netherlands, the GDI (Geospatial Data Infrastructure) is 'extended' with the National Atlas as an alternative entry to the GDI. Why use a National Atlas at all when searching for geodata? All users of geospatial information had atlases when they were first confronted with this kind of data. At school, they were taught how to deal with them through the concepts that the school atlases were based on: areal and thematic subdivisions, map comparison, georeferencing, datums, etc. Of course, it is not just the atlas information it would give access to, but also the underlying data sets and – when functioning as a geo-portal – also to all other related data sets made available by the national geospatial data providers. Thus, it offers these data providers also a 'presentation outlet'. The main benefits of having the National Atlas as the portal towards the nation's geospatial information are ease of use because of familiar concepts

and ease of access because of the topical atlas structure.

Figure 11.4 summarizes the main atlas components and their link to the GDI. Of course, displaying maps is the main objective of the atlas. If a user selects a topic, for instance, the number of inhabitants per municipality (A-I), the request goes to a geoservice in the GDI which returns the necessary data that allows for the creation of an interactive map (A-II). Alternatively, it is possible to search for a topic or for a geographical name (B-I). In this last case, the geoservice will return all names with the text string entered and map topic on the base map (B-II).

All maps are interactive and allow for the display of the data behind the symbols. It is also possible to search the GDI for alternative data sets (C-I). Through the atlas maps, which will display the footprint of the available data sets, the metadata of those data sets can be evaluated (C-II). One of the characteristics of an atlas is that one can compare different themes, for instance, the distribution of the young (population under 18) or the elderly (population over 65) (D-I). Such request will result in two maps that allow for the comparison of spatial patterns. For each topic, the atlas provides a narrative, the story behind the map in a wider context (E-I). This will also result in access through web links to other related information accessible via the GDI (E-II). Finally, it is possible to export (F-I) atlas maps to an external environment where users might combine the particular maps with their own data (F-II). However, it will also be possible to import (G-I) user data to be combined with the atlas maps (G-II).

Among the international organizations engaged in promoting this ideal situation, the former GSDI, the Global Map initiative (Sections 4.4, and 6.6.2), Digital Earth (a global project aimed at virtual representation of the Earth that enables a person to explore and interact with the geospatial data sets gathered about the Earth) and the United Nations GGIM can be named, as well as the ICA (International Cartographic Association). The latter, ICA, founded in 1958 in order to overcome national and international barriers in information exchange, functions as the international forum at which new developments regarding the issues discussed in this textbook are presented. It has its own definition of cartography and GIS (the flexible and versatile facility for the examination, analysis and presentation of geospace through maps), and the scientific structure (journal, biannual conferences, website, international commissions and working groups) that allows it to promote and develop the discipline of cartography and geographical information

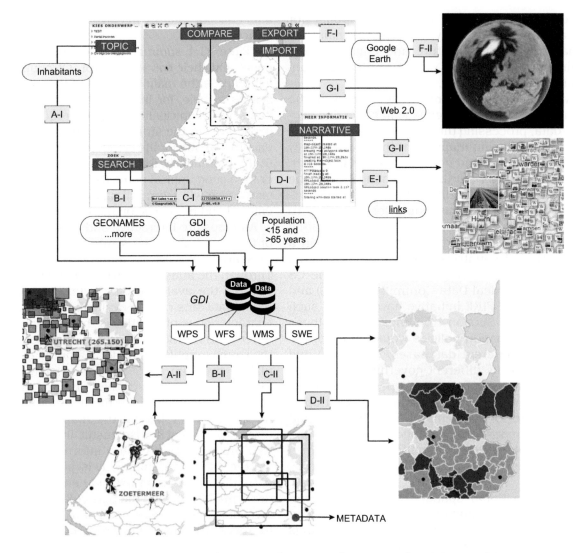

FIGURE 11.4 The Dutch National Atlas integrated into the National Geospatial Data Infrastructure (NGDI). The specific atlas functions have been integrated with geo-web services available via the geodata infrastructure (Kraak et al., 2009)

science in an international context; and the potential to apply its expertise for the benefit of a sustainable development of the Earth.

11.5 COPYRIGHT AND LIABILITY

The maps one produces or uses are usually based on both topographical (boundary) files and attribute data files. In some countries, the use of these source data is free as they have been collected by government agencies. In other countries, governments would like to share in the profits made with the data they collected, and therefore apply high copyright fees to their data. Apart from that, in a commercial environment, those that spent much time and creativity in visualizing the data would like to see a return on their achievements as well.

One of the means to help them there are copyright acts, but there are other means as well.

Part of the data to be used for maps can be found in the public domain and is, for instance, downloadable from the Internet. This does not mean that in such a situation, the sources need not be referred to, if only for one's own documentation. If the map has to be updated or extended, it surely helps if the sources of the original data have been documented. Copyright only protects original creative achievements, and not the application of specific techniques. In order for such a technique (like producing plastic relief models with a pantograph-linked milling machine, or like photographing plastic relief models in order to produce hill-shading images) to be protected, one needs to have it patented.

The protection of a cartographer's creativity (or of the results of her achievements) has a downside as well: it may impede the free flow of information. The more barriers there are constructed against freely copying spatial data or map images from others, the more difficult it becomes to get access to spatial information. This can be because costs get too high, because the data are embargoed, because they are only to be used by civil servants, or – and this certainly happens to maps or cartographic data files – just because they are kept secret.

11.5.1 Copyright

Copyright can be defined as the exclusive right of the author or producer of a literary, scientific or artistic work to publish and reproduce it. Among these literary, scientific and artistic works, maps, charts and plans are also included. Although there might be some differences from country to country in copyright laws, the following provides a general gist of the law and what it entails. In cartography, the author can be defined as the person, group of persons, institution or firm that conceives of the idea for a map and develops it, takes part in the production and in the decisions regarding the final design and has the primary responsibility for the selection of the information rendered.

When a person produces something original and creative, copyright sees to it that this person has and keeps all the rights to reproduction. This is termed the 'exclusive exploitation right'. This person would be able to transfer these rights on his or her own conditions, in writing. The conditions that apply can be, for instance, transferring copyright against certain remuneration, on condition of the author's name being cited or on the condition of adhering to specific reproduction modes. One of the conditions can be, for instance, that the author gets the right to check the galley proofs before final printing.

Instead of a transfer of copyright, a licence agreement can be signed, under which the author keeps the copyright, but the licensee gets a limited exploitation right. This right can be restricted to a specific agreed use so that, if the licensees want to reproduce the work for other uses as well, they have to renegotiate with the copyright holder. In all such agreements, it should be codified who holds the copyright, who has the ownership of the reproduction originals and who has the right to access and use these originals. In a pre-digital environment, these originals will be the films; in a digital environment, these originals will be the data sets.

For a (cartographic) work to fall under copyright law, it has to be original. In a cartographic context, this condition implies that only the original visualization of the contents is protected, not the contents itself. This originality is implicit in the design and in the selection that has been made: of all the possible data that could have been rendered, only a limited number have been selected. The design can be original in its colour combination and contrast, in the map scale and especially in its generalization. The person that reproduces (and scanning and digitizing are also regarded as reproduction) this original work without an express agreement with the author is punishable or actionable. The penalties can consist of payment of any profits, damages, court costs and legal fees.

A number of lawsuits may be of interest here: in a 1990 case of a cartographic firm against an advertising firm in the Netherlands, the latter was accused by the former of having copied a town plan produced by the former, and having reproduced it with modified scale, lettering and colour combination on advertising pillars. The plaintiff could prove to the judge that his map had been copied by highlighting a large number of completely identical ways of generalizing on his town plan and on the challenged map. On the basis of this evidence (identical generalization instances on both maps), the judge ruled against the defendant. In a similar case of a municipality against a publisher, this was made even more clear by another judge: even though the defendant had used another scale, another letter font and other colours on a town plan, and had widened all the streets in order to accommodate a larger letter font for the street names, the judge held that it had been proved by the plaintiff that the defendant had kept the same generalization. As it is impossible that in an independent generalization, the same decisions would be taken all the time, this proved that the map had been copied by the defendant, and judgement was passed accordingly (see also Figure 11.5).

In order to ease the winning of potential cases, some map producers introduce errors in town plans on purpose. When someone else copies such a map or file, errors included, proving a copyright infringement becomes simple. Of course, such errors may only pertain to subordinate, non-essential points, because the producer would otherwise expose itself to liability cases, as map users could sustain damages on the basis of purposely erroneous maps.

Copyright protects not only the results of creative, original achievements but also the link between author and product. This is called the 'moral right'.

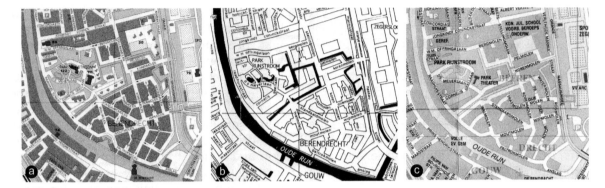

FIGURE 11.5 Copyright issues: (a) base map produced by a municipality in the Netherlands; (b) derived map produced by firm that was assumed to have copied base map (a); (c) derived map produced by another firm, with independent generalization

Even though a work may have been commissioned, or has been sold, the author has the right to have his name attributed and linked exclusively to the work (attribution right) and the right to prevent any distortion, mutilation or other modification of the work (right of integrity). This would imply for maps that without the author's consent, changes in colour, reductions or enlargements and additions would be forbidden.

Being entitled to something is not the same, however, as seeing justice done to oneself. Often, the latter implies conducting a laborious, expensive and lengthy lawsuit not everyone has the money and time to indulge in. That is why in many Western countries, associations of graphic artists have been founded that – against payment of membership dues and administrative fees for inventorying one's products – take action on behalf of their members against copyright infringements.

11.5.2 Exceptions to the Copyright Law

A number of exceptions to the copyright law exist. We think the laws and jurisprudence from several Western countries allow us to make the following assumptions: a product may be reproduced without any problems if:

- The producer died over 70 years ago. When the producer is not a person but an institution or firm, the copyright expires 70 years after the work has been produced;
- It is used by the press for news reporting purposes;
- It is used for scientific or information purposes, such as for commenting, criticizing or research, as, for instance, in a review;

- The work concerned is produced by the government and lacks a specific copyright notice. For such a specific copyright notice on a government-produced map (see Figure 11.8). On works published by non-governmental agencies or firms, a specific copyright notice (such as '© Publisher X' or '© copyright publisher Y') is not required. The mere publication already entails copyright protection;
- Publication is aimed at educational use. Map details may, for instance, be copied in textbooks if a reasonable remuneration is paid. As what constitutes a reasonable remuneration is subject to interpretation, prior contacts with the producer are advisable. In assessing whether the use made of a work is fair, the purpose and character of the use (is it for commercial or non-profit educational use) and the size and substantiality of the portion of the work used will be considered.

All these rights refer to a work that has been produced independently. It is different when a cartographer produces a map on behalf of her employer: in that case, she would have no rights to the work, with exception perhaps of the moral rights – this would prevent the employer assigning the work to someone other than the actual producer. Otherwise, all the rights by virtue of the employment contract have been transferred to the employer. When someone freelancing produces a work commissioned to him, he will have to settle with the principal as to who will have what rights to the work. In freelancing, name recognition and thus citation of the author's name seems to be an important issue in the settlement or contract.

11.5.3 Doubtful Copyright Protection of Geographical Information

Does a copyright notice as shown in Figure 11.8 actually protect the institution that collected the data, processed it and published it in map form against infringements? On the basis of the jurisprudence shown above, one would agree. Didn't the cartographers of the survey take a number of original decisions regarding the many generalization and selection options they were faced with? And were their rights by virtue of their employment not transferred to the survey? European experts, however, claim that the copyright laws in their present form do not provide suitable arguments for legal protection of spatial data. They base their verdict on their assumption that the procedures under which geo-information is created answer insufficiently the criterion of originality, based as they are on technical rules, regulations and standards. See also the scheme in Figure 11.6. Individual data and coordinates at least are never protected. A collection of individual data is protected as this collection would have been realized as a selection according to specific criteria, expressed in the file structure. The sheer application of rules, regulations and standards alone, or the sheer result of technical operations like the production of aerial

photographs, cannot be deemed original and cannot therefore be protected.

But an increasingly strong case is being made for the protection of databases. In the legal context, a cartographic database is a collection of independent data, arranged systematically or methodically and individually accessible. The relevant database directive of the European Union, which was brought into law by the EU member nations in January 1998, states that the author of a database has the exclusive right to authorize any form of distribution to the public of the database or of copies thereof for a period of 15 years (this protection period is extended each time substantial changes have been made to the database). As it is based on originality considerations, this is in fact an extension of the copyright act. But in order to protect the investments made in the production, verification and presentation of databases, the database directive also states that the author has the right to prevent extraction or re-utilization (for commercial purposes) of the whole or of substantial parts of the contents of his database, be it in analogue or digital form. But this right applies only if the database author can prove that substantial investments had been made in its production. For geodata, it is not so much the originality or creativity of their collection that has to be protected but the amount of labour invested

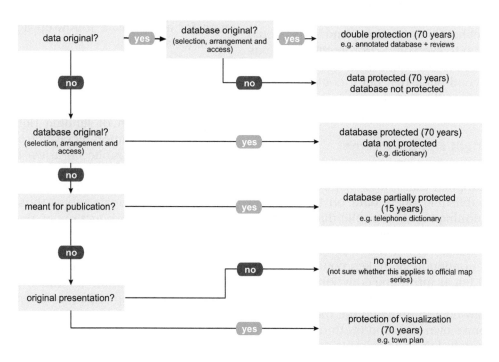

FIGURE 11.6 Scheme for copyright protection of geodata, inspired by an earlier scheme by RAVI (1994). There is a difference of opinion as to whether original representation alone leads to protection under the copyright law for serial map products. Combination with original data collection should always provide a sufficient case for protection

and therefore their monetary value. The producer has the right to set the conditions for the use of the resulting databases in the form of supply contracts. For the Ordnance Survey in Great Britain for instance, the licence fee for the use of its digital geodata would depend on the amount of data, the duration of their use, the number of terminals on which they would be used and the use objectives.

11.5.4 Freedom of Information Act

Even if the government or its subsidiaries retain the copyright to the maps and geodata files it produced, this does not mean that it can keep these files only for its own use, or only sell it to some firm of its choice and not to other firms. Under the Freedom of Information Act or similar acts, the government is required to communicate the information it holds to its citizens. Unless the public harm caused by disclosure outweighs the public benefit of having the information made available, the government and its subsidiaries are required to do so, but not necessarily free of charge.

Take the case of a Dutch municipality, Dordrecht, which used to make its digital town plan files available to a private publisher on the condition that this publisher would produce an updated town plan every year. On the basis of this exclusive arrangement, another publisher was initially refused access to these basic geodata files. By appealing in court to the Freedom of Information Act – which provides access to government-held information (and this would include geodata) to everyone – Dordrecht municipality was forced to provide these basic geodata files as well to the second publisher. But it could of course charge for them as the Freedom of Information Act does not claim that this should be for free.

11.5.5 Copyright and the Internet

The use of the Internet seems to depart implicitly from the assumption that it implies public use; i.e. to say that publishing one's data on the Internet implies abandoning them to public disclosure. As soon as something is on the Web, everyone has access to it. While surfing the Web, a computer, by definition, continuously makes copies, but this copying is not felt as infringing copyright, similar to xeroxing something for one's own use. But one may go beyond viewing and applying the material from the Web for one's own use, and try to market it. As this is not generally felt as a desirable development, web page builders increasingly claim copyright. And illustrations on websites are increasingly protected against copying by watermarking them digitally, thus modifying and disturbing the imagery.

11.5.6 Creative Commons Licences

In today's world, increasing availability of open data copyright is seen differently. Open data can be freely used, reused and redistributed by anyone, but is often subject to a Creative Commons (CC) licence. It can ask for Attribution (BY), Share-alike (SA), Non-commercial (NC) or No Derivative works (ND) – https://creativecommons.org. OSM used the CC-BY version.

11.5.7 Right of Possession

Maps that were produced over 70 years ago are no longer subject to copyright. In that case, anyone could produce facsimiles of them without having to pay any dues. But this is only a valid assumption for someone who actually holds possession of the map in question. But when it refers to rare maps that are part of a map library or are owned by a collector, rights of possession come into play. This right would protect the owner against any depreciation in value by reproduction: the value of a historical map could well decrease by circulating its reproductions. This implies that prior to facsimiling them, one should come to an agreement with the map owner. When commissioning photographs or scans of old maps from libraries, the libraries would set the conditions for the use of these negatives or scans, e.g. for single reproduction in a book. This would mean that even for a reprint, a follow-up agreement should be negotiated.

11.5.8 Public Lending Right

In most countries of the European Union, there exists some form of lending right remuneration. This is a remuneration received from public lending establishments (public libraries) by a registrar on the basis of the number of times books have been lent. According to some fixed ratio, these remunerations are then allocated to all those that have been involved in the production of the book: authors, translators, photographers, illustrators, designers, graphic artists, cartographers and publishers.

The exclusive right to authorize or forbid lending belongs to the author (or those to whom he has transferred his copyright), as far as the original and copies of his work are concerned; once given, all those involved in the production would share in the proceeds. The remuneration does not take place automatically; it has to be applied for by the author or the other contributors, who have to submit a list of the works or the contributions to the works for which the remuneration is asked.

11.5.9 International Differences

In the publication Legal Protection of Geographical Information (Eechoud, 1996) commissioned by EUROGI (European Umbrella Organisation for Geographic Information), the differences in the present copyright legislation between the various European countries are discussed. In most countries, originality is a requirement for cartographic products to fall under the copyright act; in Great Britain and Ireland, it is sufficient for a map to be a 'work of skill and labour' in order to be protected under the copyright act. Apart from the rules relative to maps and geodata, different standpoints are valid among the EU countries regarding the protection of photographs and catalogues. Regarding photographs, protection under the act ranges from exclusive protection of reproduction and distribution rights to a meagre protection of the moral right if the photograph is a personal artistic creation, or no protection at all where it is regarded as a purely technical achievement, as in the case of aerial photographs. In Germany, Sweden and Portugal, one needs special consent for duplicating information resulting from large-scale surveying or mapping operations, and there are differences in the criteria regarding the interpretation of the originality concept. In countries where geographical information is not protected as yet by copyright, it is often possible to appeal to the unfair competition act in order to safeguard one's interests.

11.5.10 Liability

Apart from rights, there are duties as well. In Europe, the latter are not so clearly defined as in the United States, where the liability of a database provider for damages is a well-known concept. To what extent is the producer of a map or database answerable for any errors in his product? He is not answerable when the errors are due to causes beyond his control. But he may well be deemed answerable when he has been demonstrably making mistakes, has been careless or has checked the information insufficiently.

Is a go (Figure 11.8) vernment department liable when it distributes incorrect data in its GIS files? In the United States, the verdict is no when the department only distributes the files because of a legal obligation to do so, like under the Freedom of Information Act. But when this department would actively market these files, it would most certainly be liable for damages. When the general public trusts the quality and accuracy of files that are incorrect because of the carelessness of government departments, government is answerable for the damages that occurred because of data misrepresentation. It is not quite clear whether the incorporation of disclaimers or waivers (see Figure 11.7) on the products makes much difference here. Their major role would lie in the fact that they would point out to the users

The information on this map is based upon or drawn from various authoritative sources and whilst all reasonable care has been taken in the preparation of this map no warranties can be given as to its accuracy and/or no reliance should be placed upon the same without further detailed inspection and survey. Therefore the publishers cannot accept any liability or responsability for any loss or damage and indeed would be grateful to receive notification of any errors or inconsistencies.
No reproduction of this map or any part thereof is permitted without prior consent of the copyright owners. The international and other boundaries in this map are taken from authoritative sources and are believed to be accurate as at the date of publication of this map.
© Rabobank International

Kartografie: ROVU, afd. Vastgoedinformatie, gem. Utrecht / DPA buro voor kartografie, Den Haag.
Auteursrechten voorbehouden aan Gemeente Utrecht, Dienst Ruimtelijke Ordening, Afdeling Vastgoedinformatie 1992.
Uitgever/maker is niet aansprakelijk voor schade die het gevolg zou zijn van verkeerde vermelding van enig gegeven op deze kaart.

© SUURLAND - FALKPLAN BV - EINDHOVEN
Hoewel aan de vervaardiging van deze kaartde grootst mogelijke zorg is besteed, is de uitgever/maker niet aansprakelijk voor schade die het gevolg zou zijn van verkeerde vermelding van enig gegeven op deze kaart.
Heeft u op- / aanmerkingen, schrijf ons dan: suurland - falkplan - postbus 9510 - 5602 LM Eindhoven.

FIGURE 11.7 Examples of disclaimers or wavers, which state that the producer considers itself not answerable for any damages that could occur because of incorrect data rendering in the map

Copyright © 1997 Topografische Dienst, Emmen
Alle rechten zijn voorbehouden aan de Staat der Nederlanden.
Niets uit deze publicatie mag worden vermenigvuldigd, opgeslagen in een geautomatiseerd gegevensbestand of openbaar gemaakt door middel van druk, fotokopie, fotografie, digitale registratie of enige andere methode, zonder voorafgaande schriftelijke toestemming van de Direkteur van de Topografische Dienst.
All rights reserved to the State of the Netherlands.
No part of this publication may be reproduced, stored in a retrieval system, or published by printing, photocopying, photography, digital recording or any other means without the prior written permission of the Director of the Topografische Dienst.

FIGURE 11.8 Example of a copyright notice

that they should not completely trust the information rendered.

In the United States, there seem to be two reasons for lawsuits in the field of liability regarding geodata (Epstein, 1987): (a) errors and omissions and (b) unintentional use. Examples of errors and omissions are an incorrect representation of a TV aerial on an official aeronautical chart and the representation of a lighthouse on a chart that was not in service any more and consequently did not radiate light – which caused shipwreck. Another example is the commercial aeronautical chart produced by Jeppesen Inc., which showed the vertical and horizontal projection of the start and landing procedures at different scales, which confused the pilots because the graphic presentation suggested that the scales were the same.

Another example is building a house in the floodplain (the area that statistically is inundated by rivers once every 100 years). The limits of the floodplain had not been rendered on the basis of scientific data, but had been misrepresented because of political considerations. This led to damages being awarded to the claimant by the map-producing institution when the area was flooded.

Examples of unintentional use refer to taking the wrong conclusions from maps that are inherently correct, but have been used for the wrong objectives. The only way to prevent such actions is to state explicitly on the document for what kinds of use the map has been produced.

11.6 MAP USE AND USABILITY

When producing a map, one will be concerned with selecting the contents of the map adapting so as to reach a given communication aim, taking account of the target audience. To this end, map-use research has developed, investigating use and user requirements, or generally all aspects and elements of the communication process.

In the communication chain, many steps may go wrong: have the proper data been selected for the map, have these data been visualized in the correct way, did the map user perceive the mapped information in the intended way and did she derive the correct conclusions from the mapped image (see the communication model in Figure 3.9). In a computer environment, the ease how to deal with the digital equipment is added, and here we come across the usability concept (defined in Section 3.3):

It is the capability of the map to be attractive, understood and used under specified conditions. Usually, it is measured in three aspects (Nielsen, 1996, Faulkner, 2000):

- Effectiveness – the ability of a given system (here a map) to accomplish user's task correctly);
- Efficiency – how quickly the task can be accomplished in a given time period;
- Satisfaction – the degree of comfort felt by users when executing the task.

Usability in its strict sense was developed as a term to indicate how easy user interfaces are to use. It referred also to methods for improving ease-of-use during the software design process. Aspects of usability are the ease to learn to handle the software, the speed with which one may learn it, how easy it will be to remember how to deal with it, how many errors one is bound to make and finally whether one will feel good to use the software. Usability will be tested, and through this test process, one gains insight into the interaction between the user and the software product.

But the usability concept has moved beyond the computer interface in cartographic research and now denotes generally the research into the effectiveness and efficiency with which map users may reach a specific correct conclusion under specific (map-use) circumstances. Examples of evaluation methods are the think-aloud method (see below), heuristic evaluation (here a few evaluators examine the software or digital product systematically on the basis of its conformity with accepted usability principles), focus groups (interactive evaluation), questionnaires and eye-tracking (Figure 11.9).

An example of the think-aloud method is shown in Figure 11.9b. Here, a test person is working with a GIS program, and voicing her comments (why she is performing the task in a given manner and taking specific decisions, generally speaking out what she thinks during the execution of the task). Her comments are recorded, as are the images the video camera will take of her and simultaneously the images generated on her computer screen and the paper maps she has taken out to use as well. Additionally, her eye movements are tracked. By analysing all these recorded data simultaneously,

FIGURE 11.9 Think-aloud method in usability research: (a) mobile eye-tracking; (b) fix laboratory set-up with eye-tracker, think-aloud set-up (courtesy ITC)

more insight is gained in the spatial data communication process and procedures and in the usability of the graphical user interface. Figure 11.9a shows an usability experiment with eye-tracking and think-aloud methods using a mobile mapping application.

Still too little is known about the way in which one derives information from maps and how this information is inserted into the existing spatial consciousness or cognitive map.

11.7 MAPS AND GISCIENCE REVISITED

In Chapter 1, it was worded that GIScience either addresses the fundamental research principles on which GISs are based or refers simply to the use of GIS in scientific applications (e.g. research with GIS). Since geovisualisation studies promote all kinds of (map) graphics that stimulate visual thinking, the approach to GIScience as research on GIS would fit best. Cartographers try to develop the visual methods and techniques (tools) to present, analyse, synthesize and explore geospatial data, but one is also interested in their effect on problem-solving (efficiency, effectiveness). In his book chapter, entitled

'Beyond tools: visual support for the entire process of GIScience', Gahegan (2005) addresses this particular problem. He described the GIScience process and projected possible maps and graphics, as well as computational methods on each of the process steps. He writes: 'To better support the entire science process, we must provide mechanisms that can visualize the connections between the various stages of analysis, and show how concepts relate to data, how models relate to concepts, and so forth' (Gahegan, 2005, p. 85).

Figure 11.10, based on Gahegan's ideas, is a simplified version of the process illustrated with the data behind Minard's map (see also Figure 10.4). The process contains the steps exploration, synthesis, analysis, evaluation and presentation, but not necessarily in this sequential order. Let us look at Napoleon's campaign and follow the process. In this example, the objective is to understand what happened during the campaign. The data available consist of statistics and maps derived from all kinds of sources such as official French army documents, diaries of individuals, Russian reports and other references. To get insight into these data, one has to explore the data, for instance, via a parallel coordinate plot in which all variables, such as number of troops, temperature, battles, number of deaths and wounded, are given. The geographical locations for which these facts are known are represented by 'horizontal' lines. At such a stage, one might formulate a hypothesis like: 'Since Napoleon lost the campaign, he probably lost most battles'. This could be checked via an iconic display to show who won the battles. Since he won most battles, one will wonder why he lost the campaign and might be interested to view the relation between the loss of troops and factors such as weather and geography over time. Interactive visualization tools could reveal interesting patterns. Alternatively, one could create time cartograms to better understand the influence of geography on time. As a result of the analysis, one can create time series to compare individual events or create a visual overlay of the layers of information and can display the relations between the different thematic layers. Before the final maps are drawn, one should evaluate the result via uncertainty visualization. This graphic could indicate the accuracy of the location of the campaign path, as well as of the numbers of troops. In a last phase, the findings are put in well-designed maps and diagrams. Minard's map could be an example, but it could equally well be an interactive space–time cube. For each of the individual steps, particular visualization tools exist, often created with a specific domain-dependent task in

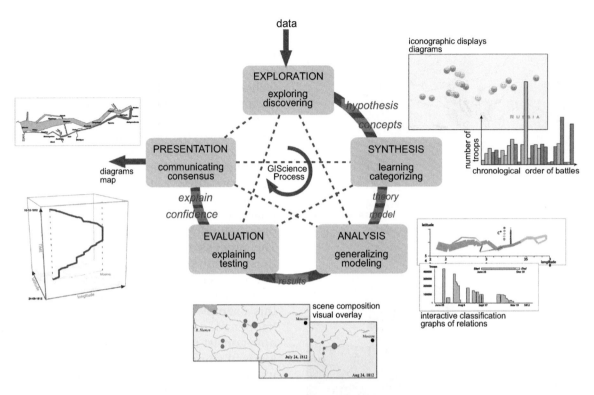

FIGURE 11.10 Map and the GIScience process illustrated by data derived from Napoleon's March on Moscow in 1812

mind. However, there is no single software environment that can handle all. In a fully fledged geovisualization environment, easy access to all required functionalities should be available, for instance, via geo-web services.

Although Figure 11.10 somehow includes all cartographic knowledge discussed in this book, it is not the final story. In Minard's case, it is obvious that the historic data, such as the number of soldiers involved and the losses during battle, is not that accurate. In the above process, the data quality issue is just one example of the gap faced between cartographic theory and technological possibilities. Today almost all seems possible, every time

of data can be converted into another type without realizing the consequences. Advanced exploratory software is available to 'play' with the data and to create all kinds of graphics. However, theory has not been able to keep pace with the developments in data processing techniques offered in GIScience, being it a 'simple' overlay or the use of advanced multiple coordinated view environment with complex models behind the views. It is this gap between theory and possibilities that needs being addressed (see Figure 11.11).

FURTHER READING

Elzakker, C. P. J. M. van. 2004. *The Use of Maps in the Exploration of Geographic Data*. Utrecht: Utrechtse Geografische Studies, 326.

Faulkner, X. 2000. *Usability Engineering*. New York: Palgrave.

Nielsen, J. 1994. *Usability Engineering*. San Francisco, CA: Morgan Kaufmann.

Ormeling, F.J. and M-J. Kraak (2008) Maps as predictive tools - mind the gap. Cartographica, 43 (3), 125–130.

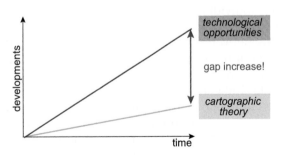

FIGURE 11.11 The increasing gap between the development of cartographic theory and technological opportunities (Ormeling and Kraak 2008)

References

Aditya, T. 2007. *The National Atlas as a Metaphor for Improved Use of a National Geospatial Data Infrastrcuture*. Enschede: ITC.

Andrienko, N., G. Andrienko, and P. Gatalsky. 2006. *Exploratory Analysis of Spatial and Temporal Data - A Systematic Approach*. Berlin: Springer Verlag.

Anson, R. W., and F. J. Ormeling, eds. 1995. *Basic Cartography*. Vol. 1–3. London: Elseviers - Butterworth.

Anson, R. W., and F. J. Ormeling, eds. 2002. *Basic Cartography for Students and Technicians*. 2nd ed. Vol. 2. Oxford: Butterworth-Heinemann.

Bartholomew, C. 2018. *The Times Comprehensive Atlas of the World*. London: Times Books/HarperCollins.

Berry, B. J. L., and D. F. Marble. 1968. *Spatial Analysis*. Englewood cliffs, NJ: Prentice Hall.

Bertin, J. 1967. *Semiology Graphique*. Den Haag: Mouton.

Bertin, J. 1983, 2011. *Semiology of Graphics*. Redlands: ESRI Press.

Blok, C. 2005. "Dynamic visualization variables in animation to support monitoring." *Proceedings 22nd International Cartographic Conference*, A Coruna Spain.

Brassel, K. E., and R. Weibel. 1988. "A review and conceptual framework of automated map generalization." *International Journal of GIS* 2 (4):229–244.

Brewer, C. A. 1994. "Color use guidelines for mapping and visualization." In *Visualization in Modern Cartography*, edited by A. M. MacEachren and D. R. F. Taylor, 123–147. Oxford/New York: Pergamon.

Brewer, C. A. 2015. *Designing Better Maps: A Guide for GIS Users*. 2nd ed. Redlands: ESRI Press.

Brown, A. 1982. "A new ITC colour chart based on the Ostwald colour system." *ITC Journal* 2:109–118.

Brown, A., and W. Feringa. 2003. *Colour Basics for GIS Users*. Upper Saddle River, NJ: Prentice Hall.

Brunet, R. 1980. "La composition des modèles dans l'analyse spatiale." *L'Espace Géographique* 8 (4):253–265.

Brunet, R. 1989. *Les Villes européennes*. Paris: Datar.

Buckley, A., and B. Rystedt. 2015. "Geographic information, access and availability (https://icaci.org/files/documents/wom/00a_IMY_WoM_en.pdf)." In *The World of Maps*, edited by F. J. Ormeling and B. Rystedt. International Cartographic Association.

Burrough, P. A., and R. MacDonell. 1998. *Principles of Geographical Information Systems*. Oxford: Oxford University Press.

Buttenfield, B. P., and R. B. McMaster, eds. 1991. *Map Generalization. Making Decisions for Knowledge Representation*. London: Longman.

Campbell, C. S., and S. L. Egbert. 1990. "Animated cartography/thirty years of scratching the surface." *Cartographica* 27 (2):24–46.

Canters, F., and H. Decleir. 1989. *The World in Perspective: A Directory of World Map Projections*. Chichester: Wiley.

Card, S. K., J. D. MacKinlay, and B. Shneiderman. 1999. *Readings in Information Visualization: Using Vision to Think*. San Francisco, CA: Morgan Kaufmann.

Cartwright, W., M. Peterson, and G. Gartner, eds. 2007. *Multimedia Cartography*. Berlin: Springer.

CensusResearchUnit. 1980. *People in Britain, a Census Atlas*. London: HMSO.

Chernoff, H. 1973. "The use of faces to represent points in k-dimensional space graphically." *Journal of the American Statistical Association* 68:361–367.

Cole, J. P., and C. A. M. King. 1968. *Quantitative Geography: Techniques and Theories in Geography*. London: Wiley.

Cowen, D. J. 1988. "GIS versus CAD versus DBMS: What are the differences?" *Photogrammetric Engineering & Remote Sensing* 54 (11):1441–1455.

Crampton, J. W. 2010. *Mapping: A Critical Introduction to Cartography and GIS*. Hoboken, NJ: Wiley-Blackwell.

Cron, J. 2006. *Kriterienkatalog. Umsetzung von Funktionen interaktiver Atlanten*. Zürich: ETH.

Dahlberg, R. E. 1967. "Towards the improvement of the dot map." *International Yearbook of Cartography* VII:157–167.

Dale, P. F., and J. McLaughlin. 2000. *Land Administration*. Oxford: Oxford University Press.

Dent, B. D., J. Torguson, and T. W. Hodler. 2008. *Cartography: Thematic Map Design*. 6th ed. Boston, MA: McGraw-Hill.

Depuydt, F. 1989. *Elkab IV Topographie. Archaeological Topographical Surveying of Elkab and Surroundings*. Brussels: Uitgaven van het Comité voor Belgische opgravingen in Egypte.

DiBiase, D. 1990. "Visualization in earth sciences." *Earth & Mineral Sciences, Bulletin of the College of Earth and Mineral Sciences* 59 (2):13–18.

DiBiase, D. 2015. "The Nature of Geographic Information. An Open Geospatial textbook (E-education.psu.edu/natureofgeoinfo/)." In State College: Department of Geography, Pennsylvania State University.

DiBiase, D., A. M. MacEachren, J. B. Krygier, and C. Reeves. 1992. "Animation and the role of map design in scientific visualization." *Cartography and Geographic Information Systems* 19 (4):201–214.

Dorling, D. 1995. *A New Social Atlas of Britain*. Chichester: John Wiley and Sons.

Douglas, D. M., and T. L. Peucker. 1973. "Algorithms for the reduction of the number of points required to represent a digitized line or its caricature." *Canadian Cartographer* 10 (3):112–122.

Dransch, D. 1997. *Computer Animation in der Kartographie: Theorie und Praxis*. Berlin: Springer Verlag.

Dykes, J., A. M. MacEachren, and M. J. Kraak. 2005a. "Advancing Geovisualization." In *Exploring Geovisualization*, edited by J. Dykes, A. M. MacEachren and M. J. Kraak, 693–703. Amsterdam: Elsevier.

Dykes, J., A. M. MacEachren, and M. J. Kraak, eds. 2005b. *Exploring Geovisualization*. Amsterdam: Elsevier.

Eechoud, M. van. 1996. *Legal Protection to Geographical Information. Copyright Inventory EU Countries*. Amersfoort: RAVI.

Elzakker, C. P. J. M. van. 2004. *The Use of Maps in the Exploration of Geographic Data*, 326. Utrecht: Utrechtse Geografische Studies.

Epstein, E. F. 1987. "Litigation over information: the use and misuse of maps." *Proceedings International Geographic Information Systems Symposium*, Arlington.

Fairbairn, D. 2018. "Creating a body of knowledge for cartography." *Proceedings of International Cartographic Association* 1:35. doi: 10.5194/ica-proc-1-35-2018.

Faulkner, X. 2000. *Usability Engineering*. New York: Palgrave.

Federal Office of Topography. 2002. Atlas of Switzerland 2.0. Berne: Federal Office of Topography.

Few, S. 2006. *Information Dashboard Design: The Effective Visual Display of Data*. Sebastopol, CA: O'Reilly Media.

Field, K. 2018. *Cartography*. Redlands: ESRI Press.

Fisher, P. 1994. "Animation and sound for the visualization of uncertain spatial information." In *Visualization in Geographical Information Systems*, edited by H. M. Hearnshaw and D. J. Unwin, 181–185. London: J. Wiley.

Freitag, U. 1992. "Cartographic conceptions: Contributions to theoretical and practical cartography 1961–1991." *Berliner Geowissenschaftliche Abhandlungen, Reihe C, Band* 13.

Gächter, E. 1969. *Die Weltindustrieproduktion 1964. Eine statistisch-kartographische Untersuchung des sekundären Sektors*. Zürich.

Gahegan, M. 2005. "Beyond tools: Visual support for the entire process of GIScience." In *Exploring Geovisualization*, edited by J. Dykes, A. M. MacEachren and M. J. Kraak, 83–99. Amsterdam: Elsevier.

Groot, D. and J. McLaughlin, eds. 2000. *Geospatial Data Infrastructure – Concepts, Cases and Good Practice*. Oxford: Oxford University Press.

Gruenreich, D. 1992. "ATKIS - a topographic information system as basis for GIS and digital cartography in Germany. From digital map series to geo-information systems." *Geologisches Jahrbuch A Heft* 122:207–216.

Guptill, S. C., and J. L. Morrison. 1995. *Elements of Spatial Data Quality*. Oxford: Pergamon.

Guptill, S. C., and L. E. Starr. 1984. "The future of cartography in the information age." In *Computer Assisted Cartography Research and Development Report. ICA Commission C*, edited by L. E. Starr, 1–15. Washington, DC: ICA.

Hägerstrand, T. 1967. *Innovation Diffusion as a Spatial Process*. Chicago, IL: University of Chicago Press.

Harley, J. B. 2001. *The New Nature of Maps. Essays in the History of Cartography*. Baltimore, MD: John Hopkins University Press.

Hearnshaw, H. M., and D. J. Unwin, eds. 1994. *Visualization in Geographical Information Systems*. London: John Wiley and Sons.

Heuvelink, G. B. M. 1998. *Error Propagation in Environmental Modelling with GIS*. London: Taylor & Francis.

Heuvelink, G. B. M. 2019. "Uncertainty and uncertainty propagation in soil mapping and modelling." In *Pedometrics*, edited by A. B. McBratney, B. Minasny and U. Stockmann, 439–461. New York: Springer.

Heywood, I., S. Cornelius, and S. Carver. 1998. *An Introduction to Geographical Information Systems*. Harlow: Longman.

Hootsmans, R. M., and F. J. M. van der Wel. 1992. "Kwaliteitsinformatie ter ondersteuning van de integratie van ruimtelijke gegevens." *Kartografisch Tijdschrift* 18 (2):51–55.

Hootsmans, R. M., and F. J. M. van der Wel. 1993. "Detection and visualization of ambiguity and fuzziness in composite geospatial datasets." *Proceedings of the Fourth European Conference on Geographical Information Systems*.

ICA. 2011a. ICA strategic plan 2011–2019. (https://icaci.org/files/documents/reference_docs/ICA_Strategic_Plan_2011-2019.pdf). International Cartographic Association.

ICA. 2011b. Strategic plan 2003–2011. https://icaci.org/strategic-plan/.

ICA, Commission II. 1973. *Multilingual Dictionary of Technical Terms in Cartography*. Wiesbaden: Steiner.

Imhof, E. 1963/1982/2007. *Kartographische Geländedarstellung [Cartographic Relief Representation]*. New York/Redlands: Walter de Gruyter / ESRI Press.

Imhof, E. 1972. *Thematische Kartographie*. Berlin: Walter de Gruyter.

Inselberg, A. 1985. "The plane with parallel coordinates." *The Visual Computer* 1:69–91.

Jenks, G. F., and M. R. C. Coulson. 1963. "Class intervals for statistical maps." *International Cartographic Yearbook* 3:119–113.

Kadmon, N. 2000. *Toponymy. The Lore, Laws and Languages of Geographical Names*. New York: Vantage Press.

Kadmon, N. 2002. *Glossary of Terms for the Standardization of Geographical Names*. New York: UNGEGN/UN.

Keates, J. S. 1993. *Cartographic Design and Production*. 2nd ed. New York: Wiley.

Keim, D., J. Kohlhammer, G. Ellis, and F. Mansmann, eds. 2010. *Mastering the Information Age Solving problems with Visual Analytics*. Goslar: Eurographics Association.

Keller, P. R., and M. M. Keller. 1992. *Visual Cues, Practical Data Visualization*. Piscataway: IEEE Press.

Kerfoot, H., F. J. Ormeling, and P. -G. Zaccheddu. 2017. *Toponymy Training Manual*. New York: UNGEGN/UN.

Kessler, F., and S. Battersby. 2019. *Working with Map Projections A Guide to Their Selection*. Boca Raton, FL: CRC Press.

Kimerling, J. A., A. R. Buckley, P. C. Muehrcke, and J. O. Muehrcke. 2016. *Map Use: Reading, Analysis, Interpretation*. 8th ed. Redlands: ESRI Press.

Koussoulakou, A. 1990. *Computer-Assisted Cartography for Monitoring Spatio-Temporal Aspects of Urban Air Pollution, Delft*. Delft: Delft University Press.

Koussoulakou, A., and M. J. Kraak. 1992. "The spatio-temporal map and cartographic communication." *Cartographic Journal* 29 (2):101–108.

Kraak, M. J. 1988. *Computer-Assisted Cartographical Three-Dimensional Imaging Techniques* (PhD-thesis). Delft: Delft University Press, 0.

Kraak, M. J. 2014. *Mapping Time: Illustrated by Minard's Map of Napoleon's Russian Campaign of 1812*. Redlands: ESRI Press.

Kraak, M-J., J. C. Müller and F. J. Ormeling. 1995. "GIS Cartography: visual decision support for spatio-tempotal data handling." *International Journal of Geographic Information Systems* 9 (6), 637–645.

Kraak, M. J., and A. Brown, eds. 2000. *Web Cartography - Developments and Prospects*. London: Taylor & Francis.

Kraak, M. J., and S. I. Fabrikant. 2017. "Of maps, cartography and the geography of the International Cartographic Association." *International Journal of Cartography* 2 (S1: Research Special Issue):9–31. doi: https://doi.org/10.1080/23729333.2017.1288535.

Kraak, M. J., F. J. Ormeling, B. Köbben, and T. Aditya. 2009. "The potential of a national atlas as integral part of the spatial data infrastructure as exemplified by the new Dutch National Atlas." In *SDI Convergance, Research. Emerging Trends and Critical Assessment*, edited by B. Loenen, 9–21. Delft: VGI.

Kramers, R. E. 2020. "Chapter 10. prototyping." In *Atlas Cook Book*, edited by R. Sieber and V. Vozenilek. International Cartographic Association.

Krygier, J. 1994. "Sound and cartographic visualization." In *Visualization in Modern Cartography*, edited by A. M. MacEachran and D. R. F. Taylor, 149–164. Oxford: Pergamon.

Krygier, J., and D. Wood. 2016. *Making Maps: A Visual Guide to Effective Map Design for GIS*. 3rd ed. New York: Guildford Press.

Langran, G. 1992. *Time in Geographic Information Systems*. London: Taylor & Francis.

Lapaine, M., and L. Usery 2017. *Choosing a Map Projection*. Berlin: Springer/ICA.

Laurini, R., and D. Thompson. 1992. *Fundamentals of Spatial Information Systems*. Vol. 37, APIC Series. London: Academic Press.

Lillesand, T. M., R. W. Kiefer, and J. Chipman. 1999. *Remote Sensing and Image Interpretation*. 7th ed. New York: Wiley.

Longley, P. A., M. F. Goodchild, D. J. Maguire, and D. W. Rhind. 2015. *Geographic Information Systems and Science*. 4th ed. Chichester: Wiley.

MacEachren, A. M. 1994. "Visualization in modern cartography: Setting the agenda." In *Visualization in Modern Cartography*, edited by A. M. MacEachren and D. R. F. Taylor, 1–12. London: Pergamon Press.

MacEachren, A. M. 1995. *How Maps Work: Representation, Visualization, and Design*. New York: Guilford Press.

MacEachren, A. M., and D. R. F. Taylor, eds. 1994. *Visualization in Modern Cartography*. London: Pergamon Press.

Maling, D. H. 1992. *Coordinate Systems and Map Projections*. Oxford: Pergamon Press.

McMaster, R. B., and K. S. Shea. 1992. *Generalization in Digital Cartography*. Washington, DC: Association of American Geographers.

Moellering, H. 1983. "Designing interactive cartographic systems using the concepts of real and virtual maps." *Proceedings Autocarto* 6.

Monmonier, M. 1989. "Geographic brushing: Enhancing exploratory analysis of the scatterplot matrix." *Geographical Analysis* 21 (1):81–84.

Monmonier, M. 2018. *How to Lie with Maps*. 3rd ed. Chicago, IL: Chicago University Press.

Muehlenhaus, I. 2013. *Web Cartography: Map Design for Interactive and Mobile Devices*. Boca Raton, FL: CRC Press.

Müller, J. C. 1991. "Building knowledge tanks for rule based generalization." *Proceedings ICA*, Bournemouth.

Neurath, O. 1930. *Atlas Gesellschaft und Wirtschaft*. Leipzig: Bibliographisches Institut.

Nielsen, J. 1994. *Usability Engineering*. San Francisco, CA: Morgan Kaufmann.

Ormeling, F. J. 1996. "Teaching animated cartography." *Proceedings of the Seminar on Teaching Animated Cartography*, Madrid 1995.

Ormeling, F. J. 2015. "Map use and map reading." In *The World of Maps*, edited by F. J. Ormeling, and B. Rystedt, https://icaci.org/publications/the-world-of-maps/the-world-of-maps-english/. International Cartographic Association.

Ormeling, F. J., and M. J. Kraak. 2008. "Maps as predictive tools – mind the gap." *Cartographica* 43 (3):125–130.

Parkes, D., and N. Thrift. 1980. *Times, Spaces, and Places*. Chichester: Wiley & sons.

Peterson, G. N. 2014a. *GIS Cartography: A Guide to Effective Map Design*. 2nd ed. Vol. Boca Raton, FL: CRC Press.

Peterson, M. P. 1995. *Interactive and Animated Cartography*. Englewood Cliffs, NJ: Prentice Hall.

Peterson, M. P. 2014b. *Mapping in the Cloud*. New York: Guildford Press.

Peuquet, D. J. 1984. "A conceptual framework and comparison of spatial data models." *Cartographica* 2 (4):66–113.

Peuquet, D. J. 2002. *Representations of Space and Time*. New York: The Guilford Press.

Pillewizer, W., and F. Töpfer. 1964. "Das Auswahlgezetz, ein Mittel zur kartographischen Generalisierung." *Kartographische Nachrichten* 14 (4):117–121.

Rendgen, S. 2018. *The Minard System: The Complete Statistical Graphics of Charles-Joseph Minard*. Princeton, NJ: Princeton Architectural Press.

Rhind, D. 1998. *Framework for the World*. Cambridge: Geoinformation International.

Roberts, J. C. 2008. "Coordinated multiple views for exploratory GeoVisualization." In *Geographic Visualization: Concepts, Tools and Applications*, edited by M. Dodge, M. McDerby and M. Turner, 25–48. Chichester: John Wiley & Sons Inc.

Robinson, A. H. 1967. "The thematic maps of Charles Joseph Minard." *Imago Mundi* 21:95–108.

Robinson, A. H., and R. E. Bryson. 1957. "A method for describing quantitatively the correspondence of geographical distributions." *Annals of the Association of American Geographers* 47 (4):379–391.

Robinson, A. H., J. L. Morrison, P. C. Muehrcke, A. J. Kimerling, and S. C. Guptill. 1995. *Elements of Cartography*. New York: John Wiley and Sons.

Roelfsema, C. M., A. van Voorden, and F. van Tatenhove. 1995. "Vormverandering van gletsjers." *Geodesia* 37 (4):193–198.

Samet, H. 1989. *The Design and Analysis of Spatial Data Structures*. Reading, MA: Addison-Wesley.

Šavrič, B., T. Patterson, and B. Jenny. 2016. "The Natural Earth II world map projection." *International Journal of Cartography* 1 (2):123–133.

Schiewe, J. 2014. "Physiological and cognitive aspects of sound maps for representing quantitative data and changes data." In *Modern Trends in Cartography - Lecture Notes in Cartography and Geoinformation*, edited by J. Brus, A. Vondrakova and V. Vozenilek, 315–320. Olomouc: Springer International.

Shneiderman, B. 1996. "The eyes have it: A task by data type taxonomy for information visualization." *Proceedings IEEE Symposium on Visual Languages*, Boulder, Colorado.

Sieber, R., and V. Vozenilek. 2020. *The Atlas Cookbook – Ten Steps Towards a Successful Atlas Edition*. Zürich: International Cartographic Association.

Slocum, T. A., R. B. McMaster, F. C. Kessler, and H. H. Howard. 2008. *Thematic Cartography and Geovisualization*. 3rd ed. Upper Saddle River, NJ: Pearson.

Snyder, J. P. 1987. *Map Projections: A Working Manual*. Vol. 1395, *U.S.G.S. Professional Paper*. Washington, DC: U.S. Government Printing Office.

Snyder, J. P., and H. Stewart. 1997. *Bibliography of Map Projections*. 2nd ed. Vol. 1856, *U.S.G.S. Professional Paper*. Washington, DC: U.S. Government Printing Office.

Snyder, J. P., and P. M. Voxland. 1989. *An Album of Map Projections*. Vol. 1453, *U.S.G.S. Professional Paper*. Washington, DC: U.S. Government Printing Office.

Spence, R. 2014. *Information Visualization*. 3rd ed. Berlin: Springer.

Stoter, J., M. Post, V. van Altena, R. Nijhuis, and B. Bruns. 2014. "Fully automated generalization of a 1:50k map from 1:10k data." *Cartography and Geographic Information Science*, 41 (1):1–13.

Taylor, D. R. F. 1991. "Geographic information systems: The microcomputer and modern cartography." In *Geographic Information Systems*, edited by D. R. F. Taylor. Oxford/New York: Pergamon.

Thomas, J. J., and C. A. Cook, eds. 2005. *Illuminating the Path: The Research and Development Agenda for Visual Analytics*. Washington, DC: IEEE Press.

Tobler, W. R. 1970. "A computer movie: Simulation of population change in the Detroit region." *Economic Geography* 46 (20):234–240.

Tobler, W. R. 1973. "Choropleth maps without class intervals?" *Geographic Analysis* 5:262–265.

Tomlin, C. D. 1990. *Geographic Information Systems and Cartographic Modelling*. Englewood Cliffs, NJ: Prentice Hall.

Tufte, E. R. 1983. *The Visual Display of Quantitative Information*. Cheshire, CT: Graphics Press.

Tufte, E. R. 2006. *Beautiful Evidence*. Cheshire, CT: Graphics Press.

Turner, A. 2006. *Introduction to Neocartography*. Boston, MA: O'Reilly.

Tyner, J. A. 2014. *Principles of Map Design*. New York: Guildford Press.

Van der Krogt, P. C. J. 2015. "The origin of the word 'cartography'." *e-Perimetron* 10 (3):124–142.

Ware, C. 2020. *Information Visualization: Perception for Design*. 4th ed. Burlington, MA: Morgan Kaufmann Publishers.

Williamson, I., S. Enemark, and J. Wallace. 2010. *Land Administration for Sustainable Development*. Redlands: ESRI Press.

Index

A

Absolute proportional method, 176
Accuracy, 38
    attribute, 39, 116
    geometric/planimetric/positional, 11,
      24, 39
    height, 125
Admiralty Raster Chart Service
      (ARCS), 28
Aerial photograph, 51, 98
Affine transformation, 110–11
Aggregation, 12, 34, 46–7, 148, 200
Allonym, 143
Animation, 18, 71, 98, 193–6, 204
    non-temporal, 194
    temporal, 194
Apple Maps, 112, 207
Area diagram, 178
Areal diagram, 178
Arithmetic series, 160
Aspect, 126
Atlas, 3, 18, 199–207
    information system, 204–7
    narrative, 199–203, 205, 229
    scenario, 199
    structure, 200
    types, 200, 202
Attribute data, 4, 34, 53, 87, 119,
      122, 147
Attribution right, 232
Augmented reality (AR), 130–1
Averages, 149–51, 160

B

BeiDou, 25, 112
Base map, 25–7, 48–9, 173, 200, 232
Bertin, Jacques, 75, 79–80, 95, 196
Big data, 4
Binary maps, 81–3
Bing Maps, 27, 112, 207
Block diagram, 203
Boundary files, 25–6, 44, 49, 230
Break points, 158–9
Brushing, 216–17

C

Cadastral map, 61–3
Cadastre, 61–2
Cartogram, 169, 182–3, 193
Cartographic body of knowledge
      (BOK), 20
Cartographic communication, 49–50,
    54
Cartographic data analysis, 163
Cartographic education, 20
Cartographic method, 16, 43–4, 52, 194
Cartography, definitions, 20, 42, 44–5,
    49–50, 199, 225, 228–9
Cathode ray tube (CRT), 82
Census, 24–5, 51, 65, 71, 148, 156

Certainty factor, 41
Change, mapping, 192
Charge-coupled device (CCD), 31
Chernoff faces, 218
Chorèmes, 179–81
Chorochromatic map, 168–70
Choropleth (map), 167–8
Choropleth map, unclassed, 170–3
Classification, 155
    equal steps, 159
    nested means, 160
Clearing house, 19, 227
Clipart, 89–90
Cloud, 30, 38, 54, 215
Cognitive map, 4, 6, 49–50, 87, 237
Cognitive overload, 196
Collation, 226
Colours, 81
    additive, 82, 94
    chart, 84
    hue, 78–82
    saturation, 80–2
    separation, 88, 93
    subtractive, 82, 94
    use on maps, 81
Comparison, 67
    geospatial, 67, 69
    temporal, 67, 72
    thematic, 67, 71
    topological, 70
Comparison, geographical, *see*
      Comparison, geospatial
Confusion index, 39
Contour line, 58, 72, 123–5, 127
Coordinate system, 87, 103
    geographical, 103–4
    local, 104–5
    national, 103–5
Coordinated multiple views, 216–17
Copyright, 230–6
Creative Commons Licences, 234
Cumulative frequency diagram,
    158–60
Curvilinear transformation, 110

D

Dashboard, 222–3
Dasymetric map, 171
Data acquisition, 23–4
Data adjustment, 155
Data analysis, 149, 163–5
Data components, 5, 166
Data correlation, 51
Data manipulation, 10, 87, 195, 217
Data quality, 10, 15, 39
Data structure, 32
    georelational, 34–5
    quadtree, 36
    raster, 32–4, 36
    spaghetti, 34–5
    topological, 34
    vector, 32, 34–5, 37

Database management systems
      (DBMS), 37
Database structure
    hierarchical, 37
    network, 37
    relational, 37–8
Decision-making, 12, 18, 211, 216
Delaunay triangle, 126
Density, 147
Density map, 171
Descriptive Cataloguing of Rare
      Materials-Cartographic
      (DCRM-C), 227
Desk top publishing (DTP), 89
Diacritics, 142, 144
Diagram maps, 167, 178–9
Digital cartographic model (DCM), 4, 6,
      26, 33, 86–7, 90–2, 114–15,
      118, 217
Digital Chart of the World (DCW), 25,
      209, 225, 227
Digital Earth, 229
Digital elevation model (DEM), 72–3,
      101, 124
Digital landscape model (DLM),
      4, 6, 32–3, 87, 102,
      114–15, 217
Digital object identifyer (DOI), 53
Digital scale, 26
Digital terrain model(DTM), 124,
      126, 128
Digitizing, 15, 25–6, 30, 110, 112, 231
    on screen, 31–2
DIME, 34
Diverging maps, 81–3, 178
Documentation, 29, 53, 225–7
Dotmaps, 167, 179–81
Douglas-Peucker algorithm, 121–2
Draping functions, 128
Drones, 24
Dynamic variables, 196

E

Eckert IV projection, 106, 108, 110
Edge matching, 112, 138
Electronic atlases, 3, 18, 202–13
    interactive, 203
    on-line, 204
    stand-alone, 202–4
    view-only, 202
Electronic atlas functionality, 203–4
Electronic atlas information
      system, 18
Electronic chart and information system
      (ECDIS), 27
Ellipsoid, 103–4, 106
Endonym, 142–4
Environmental information system (EIS),
    16–17
Environmental maps, 66–7
Equator, 103–4, 106–7
Error propagation, 14–15

ESRI, 27, 89, 131, 208
EuroBoundary map, 27, 138, 141
EuroGeographics, 27, 131, 138
Exonym, 142–3
Exploitation right, 231
Exploratory data analysis (EDA), 216
Eye-tracking, mobile, 236–7

F

Federal Geographical Data Committee
        (FGDC), 229
Figurative symbols, 175–6
Fitness for use, 15, 19, 226
Flow diagram, 178
Flow line map, 181
Freedom of information act, 234–5
Frequency diagram, 158
Fuzziness index, 39

G

Galileo, 25
Gazetteer, 142, 144
Generalization, 26, 34, 46, 69, 87, 101,
        112–19
    algorithms, 119, 121–2, 126
    (carto)graphic, 115–17, 121
    conceptual, 116–18, 121
    knowledge-based, 123
    model, 115
    object, 114
Generic term, 84
Geographical data, 1
Geographical data infrastructure (GDI),
        2; see also Spatial data
        infrastructure
Geographical information science/
        GIScience, 1–2, 4, 21, 194,
        237–8
Geographical information system
        (GIS), 1–2, 7–12, 18, 61, 86–9,
        228–9, 236–7
    GIS database, 32, 38, 91, 96, 130
    GIS function(ality), 87, 89, 203
    GIS (software) packages/programs,
        9, 15, 86–8, 97, 124, 128,
        208
Geographical names, 18, 51, 84, 101,
        142–5
Geography Markup Language (GML),
        92, 137
GeoJSON, 137
Geometric series, 56
Geometric proportional symbols, 176
Geometrical resolution, see Resolution
Geonyms, see Geographical names
GLONASS, 25
Geoportal, 227
Georeferenced data, 4, 142
Georeferencing, 29, 101–2
Geoservice, 19, 92, 229
Geospatial analysis operation, 2, 7,
        12–13, 15, 67
Geospatial data, 1–7, 70–1, 199,
        215–16, 225

Geospatial relationships, 1, 6–7, 16, 34,
        43–4
Geovisual analytics, 220
Geovisualization, 215–16, 220
Global Geospatial Information
        Management (GGIM), 227, 229
Global Map, 27, 229
Global navigation satellite system
        (GNSS), 24
Global positioning system (GPS), 24–5,
        104, 188
Global spatial data infrastructure,
        see Spatial data
        infrastructure
Gnomonic projection, 106
Google Earth, 1, 20–1, 124, 128, 207
Google Maps, 1, 20, 27, 56, 107, 112,
        131, 207
Graded circles, 177
Grain, differences, 51, 78–80, 82, 165
Graphic approach to data classification,
        158–9
Graphical grammar, 16, 76
Graphical user interface (GUI), 88,
        97, 205
Graphical density, 118
Graphic(al) variables, 44, 77–80,
        95–6, 98, 130, 163–8, 181,
        192, 196

H

Harmonic series, 160
Headline, 86
Hidden surface removal, 131
Hierarchy, 52, 85
    graphical or visual, 80–1, 116, 153,
        164
    information, 81, 164, 199
Hill shading, 77, 123–4, 126–7
Histogram, 165, 178
Homonym, 144
Hydrographic chart, 27, 54

I

Imprint statement, 226
Information graphics/infographics, 44
Information visualization, 4, 44, 216
INSPIRE, 18, 139, 142, 229
International Cartographic Association
        (ICA), 20–1, 45, 89, 229
International Hydrographic Organization
        (IHO), 27
International Standard Bibliographic
        Description (ISBD), 226
International Standard Book Numbers
        (ISBN), 53
Internet, 1, 3–4, 19–20, 53–4, 226–7,
        234
Interval scale, 5, 150
Invariant, 86, 163
Isochrones, 5, 39, 53, 193
Isoline maps, 173–5
Isopleth maps, 174–5
Isotype, 176

J

Java Script Object Notation (GeoJSON),
        34, 92, 137

L

Landsat, 28
Large-scale maps, 46, 57, 61, 63, 112
Laser altimetry, 24, 58
Laser printer, 93
Latitude, 102–4
Layer tints, 58, 123–4
Layer zones, 82, 84
Length of data components, 164–5
Liability, 230–1, 235–6
Lidar, 24, 57
Line diagram, 178
Lineage, 19, 32, 39
Liquid crystal display (LCD), 82
Longitude, 102–4

M

Map (definition), 43, 45, 49
    analysis, 213
    categories, 59
    design, 20, 55, 88–90
    functions, 56, 59
    interpretation, 50, 213
    machine, 207–8
    production, 88–9, 91
    reading, 50, 213
    scale, 45, 61
    title, 86, 163, 222
    types, 48, 56, 59, 87–8, 162–4, 214
Map use, 3, 54–5, 171, 212–14, 232
    functions, 208
    strategies, 10–11
Map use cube, 3
Mapathon, 25
Mapping methods, 70, 165–6
Mapping organizations, 63, 104, 131
Marginal information, 55, 87, 136, 226
Mathematical approach to data
        classification, 159
Mathematical data, 226
Maximum likelihood, 39
Measurement levels, see
        Measurement scale
Measurement scale, 150, 156, 164–8
Mental map, 49–50
Menu, 30
Mercator projection, 106–9
Meridian, 103
Metadata/meta-information, 11–12, 19,
        41, 131, 212, 226–7
Missing Maps, 25, 32
Mobile maps, 56, 237
Modifiable areal unit problem, 47
Moral right, 231–2, 235
Mosaic map, see Chorochromatic map
Multimedia (information) system,
        18, 179
Multimedia maps, 96–8, 204
Multipurpose cadastre, 62

N

National (geo) spatial data
infrastructure, *see* Spatial
data infrastructure
National Geospatial-Intelligence Agency
(NGA), 26
National grid, 104–5
Natural Earth, 68, 89
Natural Earth projection, 108
Nearest neighbour index, 152
Nominal scale, 5, 150
Non area-related ratio maps, 171

O

Open Geospatial Consortium (OGC), 19,
88, 92, 137
Open Nautical Chart (ONC), 28
OpenStreetMap (OSM), 27, 32,
107, 112, 114, 138, 140,
207–8, 227
Operational Navigation Chart (ONC),
227–8
Ordinal scale, 5, 150
Ordnance Survey (OS), 138
OS Mastermap, 140
Overlay operation, 2, 13–15, 62

P

Panorama map, 124–5, 128–9
Parallel coordinate plots, 218
Parallel, 103–4, 109
Permanent map, 26, 49–50, 87
Photogrammetrical survey, 24
Pixel, 12, 26, 28, 39, 122–3
Pop-up menu, 87
Portable document file (PDF), 88, 91–3
Potential, 152–3
Printers, 88, 92–3
Probability, 11–12, 39, 41
Projections, map, 101, 104–10
    conformal, 106, 109
    equal-area, 106, 109
    equidistant, 106
Public access, 11
Public lending right, 234

Q

Quantiles, 160

R

Range of data components, 156, 164, 177
Raster data, 33, 92, 119, 122
Ratio scale, 5, 164
Reference atlas, 112–13, 116, 199–200,
207
Reliability diagram, 23, 39, 41
Relief, 95, 123–4, 126–7, 155, 228
Relief model, 45, 84, 230
Representativity, 11
Resolution, 26, 28, 36, 52, 93
    geometrical/(geo)spatial, 10, 28, 226

radiometric, 28
temporal, 190
Resource Description and Access
standard (RDA), 227
Right of integrity, 232
Right of possession, 234
Rubber-sheeting, 23, 110–12

S

Satellite data, 24–5, 39
Satellite imagery, 28, 32, 97, 112, 128,
138, 207, 216
Scalable vector graphics (SVG), 88
Scale, 26, 30, 45, 61
Scanner, 24, 27–8, 30–1, 34
Scanning maps, 26
Schematisation, 69
School atlases, 112–13, 199–201, 229
Scientific visualization, 44, 215–16
Search engine, 19, 53, 204
Sensor web enablement (SWE), 91
Sentinel, 28, 66–7, 131
Sequential scale, 82–3
Sliver polygon, 14–15
Small-scale maps, 46, 63, 65, 112
Socio-economic data, 65, 101, 147,
153, 173
Sound, 98
Space-time cube, 189, 217
Spatial data, *see* Geospatial data
Spatial data infrastructure (SDI), 2, 90,
202, 204, 227, 229
    global spatial data infrastructure
    (GSDI), 227, 229
    national (geo) spatial data
    infrastructure (NSDI/NGDI), 2,
    90, 204, 227, 229
Spatial information policy, 228
Spatialization, 216
SPOT, 28
Statistical mapping, 147–85
Statistical surface, 156–7, 162–3,
167–9, 182
Statistical surveys, 147
Story map, 208–9
StoryMaps, 208
Sustainable development goals, 18, 25,
230
Symbolization, 116–19

T

Temporal map, 49–50
Terrain visualization, 126
Terrestrial survey, 24
Texture filter, 95
Thematic maps, 47–8, 65, 75, 109, 206
Think aloud method, 236–7
Three-dimensional maps, 95, 125, 128
Three-dimensional symbols, 177
Time, mapping, 187–97
    database time, 189, 194
    display time, 194, 196
    world time, 188
Topographer, 51

Topographic maps, 47–8, 75, 101,
119, 123, 132–8, 147,
194, 206
Topography, 10
Topological information, 30
Topology, 4, 33–5, 70, 102–3, 123
Toponymic guidelines, 142
Toponyms, *see* Geographical names
Training of pixels, 39
Transcription, 144
Transformation, 111, 165
    affine, 110–11
    coordinate/projection, 26, 43,
    104, 130
    curvilinear, 110
    geometric, 109, 111, 128
    mapping methods, 168–70, 173
Transliteration, 144
Triangular irregular network (TIN),
126–7
Typography, 84

U

Ubiquitous maps, 53
Uncertainty, 39–42, 95, 237
United Nations Group of Experts
on Geographical Names
(UNGEGN), 142–4
Universal transverse Mercator projection
(UTM), 30, 106, 108
UTM (grid) coordinate system, 30, 108–9
Usability, 18, 55, 216, 236–7
User-centred design, 55
Utility maps, 62–4, 212

V

Vector data, 92
Vector map, 167, 181
Vedette, *see* Headline
Virtual map, 26, 49–50, 92, 202
Virtual reality environment (VR), 98,
130–1
Visibility map, 125, 128
Visualization, 2, 41–2, 44, 126,
128, 215
Vmap0 26
Volunteered geographical information
(VGI), 25, 91

W

Web, 19–20
    atlases, 202, 204
    coverage services (WCS), 92
    feature services (WFS), 92
    map design, 91, 94–6, 202
    map services (WMS), 92, 107
World Geodetic System, 1984
(WGS84), 104
World Wide Web Consortium (W3C), 92

Y

Yandex, 112

Printed and bound by CPI Group (UK) Ltd, Croydon, CR0 4YY

24/10/2024

01778290-0020